ServSafe®

COURSEBOOK | Fourth Edition

National Restaurant Association
EDUCATIONAL FOUNDATION

DISCLAIMER

The information presented in this book has been compiled from sources and documents believed to be reliable and represents the best professional judgment of the National Restaurant Association Educational Foundation (NRAEF). The accuracy of the information presented, however, is not guaranteed, nor is any responsibility assumed or implied, by the National Restaurant Association Educational Foundation for any damage or loss resulting from inaccuracies or omissions.

Laws may vary greatly by city, county, or state. This book is not intended to provide legal advice or establish standards of reasonable behavior. Operators who develop food safety-related policies and procedures as part of their commitment to employee and customer safety are urged to use the advice and guidance of legal counsel.

Requests to use or reproduce material from this book should be directed to:

Copyright Permissions
National Restaurant Association Educational Foundation
175 West Jackson Boulevard, Suite 1500
Chicago, IL 60604-2814
Email: permissions@nraef.org

Coursebook without Exam—CB4
ISBN 13: 978-1-58280-182-7
Coursebook with Exam—CBX4
ISBN 13: 978-1-58280-181-0
Coursebook with Online Exam Voucher—CBV4
ISBN 13: 978-1-58280-189-6
Wiley Coursebook without Exam—CB4-W
ISBN 13: 978-0-471-77572-0
Wiley Coursebook with Exam—CBX4-W
ISBN 13: 978-0-471-77569-0
Wiley Coursebook with Online Exam Voucher—CBV4-W
ISBN 13: 978-0-471-77570-6

Printed in the U.S.A.

10 9 8 7 6

TABLE OF CONTENTS

A MESSAGE FROM

The National Restaurant Association Educational Foundation

The NRAEF is pleased to bring you the fourth edition of *ServSafe® Coursebook.*

By opening this ServSafe book, you are taking the first step in your commitment to food safety. The information in this book will help you apply critical food safety practices to every meal you serve. You can feel good knowing that the ServSafe program was created by the foodservice industry, for the foodservice industry, and leads the way in setting high food safety standards.

In fact, ServSafe training and certification is recognized by more federal, state, and local jurisdictions than any other food safety certification.

Use this book to:

- **Learn Food Safety Principles.** *ServSafe Coursebook* provides critical food safety knowledge that will help you implement and maintain safe foodhandling practices in your establishment.
- **Teach Your Team Members.** The new **"Take It Back" reference guides** at the end of selected chapters will help you bring essential food safety knowledge back to your employees. Visit ***www.ServSafe.com/FoodSafety/resource*** for more information.
- **Sharpen Your Training Skills.** The information presented in the **Employee Food Safety Training chapter** will help you enhance your skills as a trainer. The training concepts in this chapter demonstrate different ways to convey essential food safety information to your employees.

For more information on the NRAEF and its programs, visit ***www.nraef.org*** or ***www.ServSafe.com.***

National Restaurant Association Educational Foundation

The NRAEF is a not-for-profit organization dedicated to fulfilling the educational mission of the National Restaurant Association. As the nation's largest private sector employer, the restaurant and foodservice industry is the cornerstone of the American economy, of career-and-employment opportunities, and of local communities. Focusing on three key strategies of risk management, recruitment, and retention, the NRAEF is the premier provider of educational resources, materials, and programs, which address attracting, developing and retaining the industry's workforce. Sales from all NRAEF products and services benefit the industry by directly supporting the NRAEF's educational initiatives.

The National Restaurant Association Educational Foundation's

INTERNATIONAL FOOD SAFETY COUNCIL®

The International Food Safety Council's mission is to heighten the awareness of the importance of food safety education throughout the restaurant and foodservice industry. The council envisions a future in which foodborne illness no longer exists.

Initiatives

The NRAEF encourages restaurant and foodservice professionals to become involved by participating in council activities such as:

- Food Safety Event—Washington, D.C.
- National Food Safety Education Month®—September, visit ***www.nraef.org/nfsem***

For more information about the NRAEF's International Food Safety Council, sponsorship opportunities, and initiatives, please call 312.261.5336, or visit the NRAEF's Web site at ***www.nraef.org/ifsc.***

Founding Sponsors

American Egg Board

The Beef Checkoff

Ecolab Inc.

FoodHandler Inc.

SYSCO Corporation

Tyson Foods, Inc.

Campaign Sponsors

Cintas Corporation

Rubbermaid Commercial Products

Past Founding Sponsors

Heinz North America

San Jamar/Katch All

UBF Foodsolutions North America

Acknowledgements

The development of the *ServSafe Coursebook* text would not have been possible without the expertise of our many advisors and manuscript reviewers. The NRAEF is pleased to thank the following organizations for their time, effort, and dedication to creating this fourth edition.

American Egg Board

Applebee's International, Inc.

The Beef Checkoff

Black Angus Steakhouse

Buffalo Wild Wings Grill & Bar

Burger King Corporation

Cargill, Inc.

Carlson Restaurants Worldwide, Inc.

Ecolab Inc.

Enrico's Italian Dining

FoodHandler Inc., Safety Management Services

Food Allergy and Anaphylaxis Network

Jack in the Box, Inc.

Jesse Brown Veterans Affairs Medical Center

National Restaurant Association

North American Meat Processors Association

Sodexho, Inc.

SYSCO Corporation

Taylor Farms, Inc.

United Fresh Fruit and Vegetable Association

Walt Disney World Co.

HOW TO USE *SERVSAFE COURSEBOOK*

The plan below will help you study and retain the food safety principles in this textbook that are vital to keeping your establishment safe.

Beginning Each Chapter

Before you begin reading each chapter, you can prepare by:

- **Reviewing the learning objectives.** Located on the front page of each chapter, the learning objectives identify tasks you should be able to do after finishing the chapter. They are linked to the essential practices for keeping your establishment safe.
- **Completing the Test Your Food Safety Knowledge questions.** Five True or False questions at the beginning of each chapter will test your prior food safety knowledge. The questions include page references for you to explore the topics further. Answers are located in the Answer Key.

Throughout Each Chapter

Use the following learning tools to help you identify and reinforce the key principles as you read each chapter:

- **Key Terms.** These terms are important for a thorough understanding of the chapter content. They are **highlighted** throughout the chapter, where either they are explicitly defined or their meanings are made clear within the paragraphs in which they appear. Each key term is also defined in the Glossary.
- **Exhibits.** These are placed throughout each chapter to visually reinforce the key principles presented in the text. They include charts, photographs, illustrations, and tables.
- **Icons.** Two types of icons appear in *ServSafe Coursebook.*
    - In Chapters 5 through 10, an icon representing the various points in the flow of food appears in the left margin. As you read through these chapters, you will notice the highlighted portion of the icon changes according to the point within the flow of food being discussed.

- Throughout the text you will see icons that reinforce critical food safety principles such as cross-contamination and proper cooling. While Key Point icons are the most common type, icons related to personal hygiene, cross-contamination, and time-temperature abuse are also included. Additionally, icons were developed by the International Association for Food Protection. (See page xi for an example.)

- **Something to Think About.** Several food safety stories appear throughout the text to provoke discussion about various food safety topics. Some of these real-world stories focus on foodborne illnesses that have occurred when food was not handled safely. They emphasize the importance of following food safety practices and allow you to apply what you have learned by asking how the incident could have been prevented. Other stories showcase real-world solutions to food safety problems. These solutions may help you address similar problems in your own establishment.

At the End of Each Chapter

Several activities have been provided at the end of each chapter to test your knowledge. These include:

- **Case in Point Activities.** These food safety case studies ask you to identify the errors made by the foodhandlers in each story and the proper practices that should have been followed.

- **Discussion Questions.** These questions are designed to make you think about some of the important food safety concepts presented in the chapter. Answers to the questions are provided in the Answer Key.

- **Multiple-Choice Study Questions.** These multiple-choice questions are directly based on the learning objectives. If you have difficulty answering them, you should review the content further. Answers are located at the back of each chapter.

- **Additional Resources.** In this section, you will find resources—books, articles, and Web sites—that will enable you to further explore the food safety concepts presented in each chapter.

Take It Back

The ServSafe program provides you with the essential knowledge to keep food safe in your establishment. It is your responsibility to take that knowledge back to your employees. "Take It Back," a new reference guide, has been added to the ServSafe program to help you do this. At the end of selected chapters, you will find a "Take It Back" guide that identifies food safety concepts critical for employees to understand. For each concept, the guide points to training tools that can be used to teach the concept in fifteen minutes or less. These tools use content and language appropriate for employees and include:

- ServSafe Steps to Food Safety Video Series
- *ServSafe Employee Guide*
- ServSafe Posters and Quiz Sheets
- ServSafe Fact Sheets and Optional Activities

Choose the tool or tools that work best for you. The ServSafe Posters, and the ServSafe Fact Sheets and Optional Activities are available on the ServSafe Instructor Deluxe CD-ROM. You can also download these tools from the ServSafe Food Safety Resource Center Web site at ***www.ServSafe.com/FoodSafety/resource.***

International Food Safety Icons

Hot Holding

Temperature Danger Zone

Refrigeration/ Cold Holding

Do Not Work if Ill

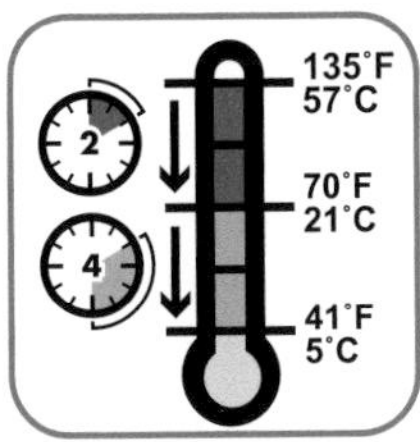

Cooling

Cross-Contamination

Handwashing

No Bare-Hand Contact

Wash, Rinse, and Sanitize

Potentially Hazardous Food

Cooking

Icons used with permission from the International Association for Food Protection

Unit 1

The Sanitation Challenge

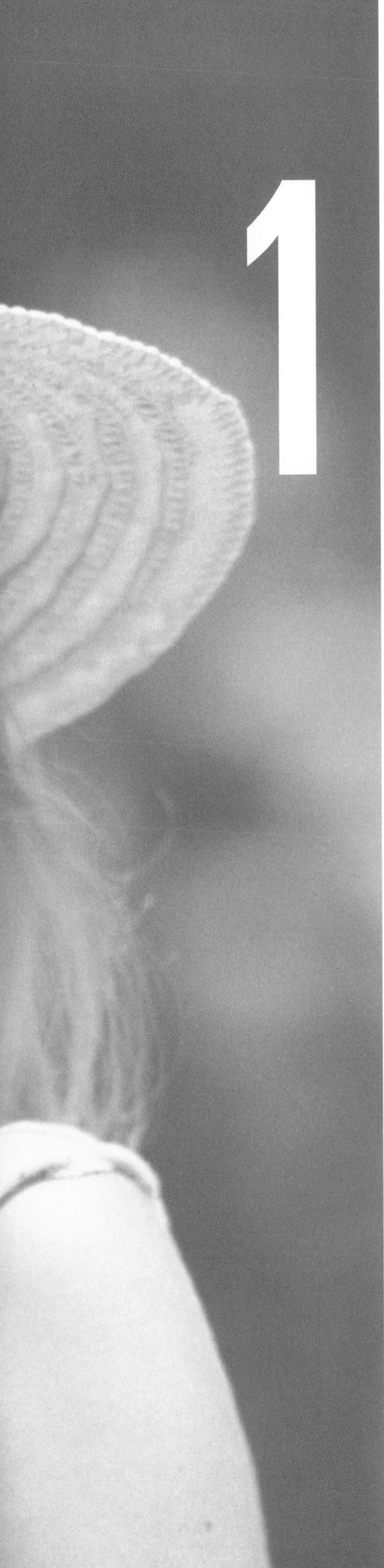

Providing Safe Food

Inside this chapter:

- The Dangers of Foodborne Illness
- Preventing Foodborne Illness
- How Food Becomes Unsafe
- Key Practices for Ensuring Food Safety
- The Food Safety Responsibilities of a Manager

After completing this chapter, you should be able to:

- Analyze evidence to determine the presence of foodborne-illness outbreaks.
- Recognize risks associated with high-risk populations.
- Identify the characteristics of potentially hazardous food.

Key Terms

- Foodborne illness
- Foodborne-illness outbreak
- Warranty of sale
- Reasonable care defense
- Immune system
- Potentially hazardous food
- *FDA Food Code*
- Ready-to-eat food
- Contamination
- Biological, chemical, and physical hazards
- Time-temperature abuse
- Cross-contamination
- Personal hygiene
- Food-contact surface

Apply Your Knowledge

Check to see how much you know about the concepts in this chapter. Use the page references provided with each question to explore the topic.

Test Your Food Safety Knowledge

1. **True or False:** A foodborne-illness outbreak has occurred when two or more people experience the same illness after eating the same food. *(See page 1-3.)*

2. **True or False:** Potentially hazardous food is usually moist. *(See page 1-6.)*

3. **True or False:** Adults are more likely than preschool-age children to become ill from contaminated food. *(See page 1-5.)*

4. **True or False:** People taking certain medications, such as antibiotics, are at high risk for foodborne illness. *(See page 1-5.)*

5. **True or False:** Cooked vegetables are *not* potentially hazardous. *(See page 1-7.)*

For answers, please turn to the Answer Key.

INTRODUCTION

When diners eat out, they expect safe food, clean surroundings, and well-groomed workers. Overall, the restaurant and foodservice industry does a good job of meeting these demands, but there is still room for improvement. Several factors account for this and likely include:

- Emergence of new foodborne pathogens (disease-causing microorganisms)
- Importation of food from countries lacking well-developed food safety practices
- Increases in the purchase of take-out food and home meal replacements
- Changing demographics, with an increased number of individuals at high risk for contracting foodborne illness, such as preschool-age children and elderly people
- Employee turnover rates that make it difficult to manage food safety programs

In the face of these challenges, establishments must take the necessary steps to help ensure that the food they serve is safe.

Health Alert

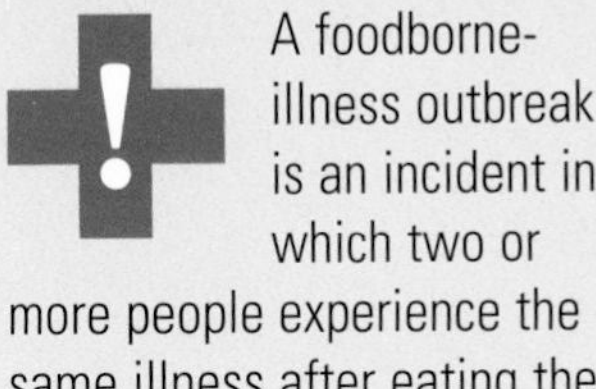

A foodborne-illness outbreak is an incident in which two or more people experience the same illness after eating the same food.

THE DANGERS OF FOODBORNE ILLNESS

A **foodborne illness** is a disease carried or transmitted to people by food. The Centers for Disease Control and Prevention (CDC) define a **foodborne-illness outbreak** as an incident in which two or more people experience the same illness after eating the same food. A foodborne illness is confirmed when laboratory analysis shows that a specific food is the source of the illness.

Each year, millions of people are affected by foodborne illness, although the majority of cases are not reported and do not occur at restaurants or foodservice establishments. However, the cases that are reported and investigated help the industry understand some of the causes of illness. They also heighten awareness of what can be done to control them. Fortunately, every establishment, no matter how large or small, can take steps to ensure the safety of the food it prepares and serves to its customers.

The Costs of Foodborne Illness

Foodborne illness costs the United States billions of dollars each year in lost productivity, hospitalization, long-term disability claims, and even death. National Restaurant Association figures show that a foodborne-illness outbreak can cost an establishment thousands of dollars. It can even result in closure.

If an establishment is implicated in a foodborne-illness outbreak, costs may include legal fees. The establishment may have to pay for testing food supplies and employees and may spend time and money cleaning and sanitizing the establishment. Food supplies that may or may not be contaminated will have to be discarded. Other costs are highlighted in *Exhibit 1a* on the next page.

Exhibit 1a

Costs of a Foodborne Illness to an Establishment

Loss of customers and sales	Lowered employee morale
Loss of prestige and reputation	Employee absenteeism
Lawsuits resulting in legal fees	Need for retraining employees
Increased insurance premiums	Embarrassment

Today, customers are very willing to sue to obtain compensation for injuries they feel they have suffered as a result of the food they were served. Under the federal *Uniform Commercial Code,* a plaintiff bringing about a lawsuit must prove all of the following:

- The food was unfit to be served.
- The food caused the plaintiff harm.
- In serving the food, the establishment violated the **warranty of sale,** that is, the rules stating how the food must be handled.

If the plaintiff wins the lawsuit, he or she can be awarded compensatory and punitive damages. Compensatory damages are awarded for lost work, lost wages, and medical bills. Punitive damages are awarded to punish the defendant for wanton and willful neglect. They are given in addition to normal compensation.

Exhibit 1b

People at High Risk for Foodborne Illness

Young children

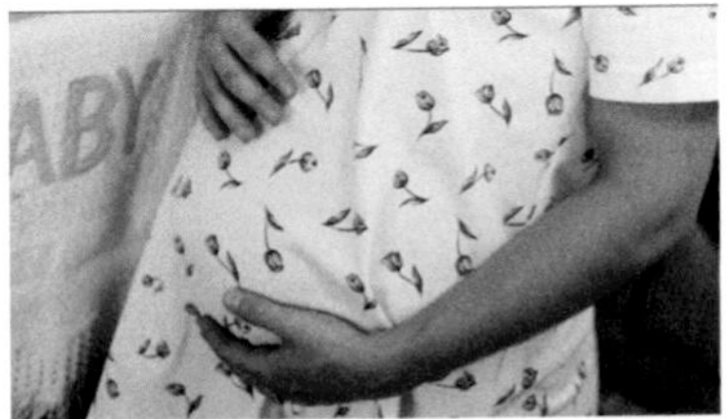

Pregnant women

Elderly people

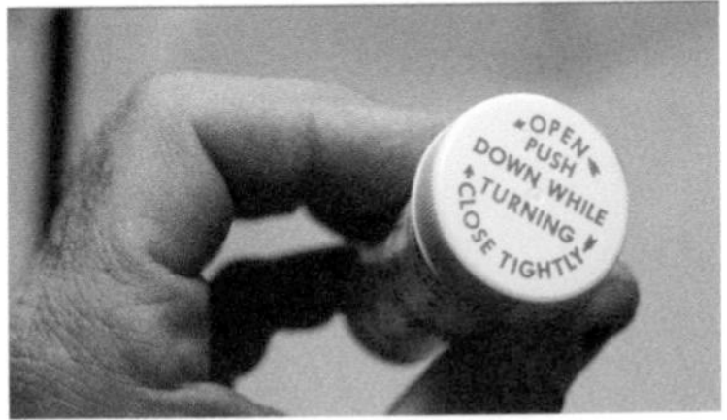

People taking certain medications

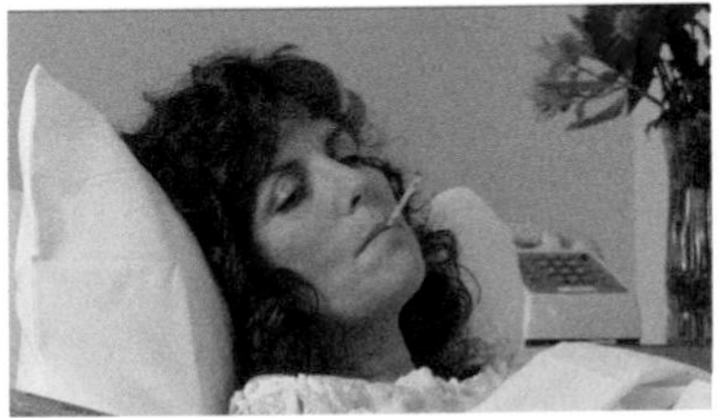

People who are seriously ill

If an establishment has a food safety management system in place, however, it can use a **reasonable care defense** against a food-related lawsuit. A reasonable care defense requires proof that the establishment did everything that could be reasonably expected to ensure that the food served was safe. Evidence of written standards, training practices, procedures and documentation, and positive inspection results are key to this defense. Since decisions in these suits follow the law in their respective states, check applicable state law as to the appropriate defense in any action.

PREVENTING FOODBORNE ILLNESS

There are many challenges to preventing foodborne illness. These include high employee-turnover rates, service to an increasing number of high-risk customers, and the service of potentially hazardous food. Preventing foodborne illness in your establishment requires a comprehensive approach. This includes setting up appropriate food safety programs and training employees to handle food safely, as well as identifying food that is most likely to become unsafe and the potential hazards that can contaminate it. Informing high-risk patrons of the risk of consuming raw or undercooked food in your establishment can also prevent foodborne illnesses.

Populations at High Risk for Foodborne Illness

Studies of U.S. demographics show that certain groups of people have a higher risk of contracting a foodborne illness than others, sometimes with serious results. (See *Exhibit 1b.*) They include:

- Infants and preschool-age children
- Pregnant women
- Elderly people
- People taking certain medications, such as antibiotics and immunosuppressants
- People who are seriously ill (especially those who have recently had major surgery, are organ-transplant recipients, or who have preexisting or chronic illnesses)

Young children are at a higher risk for contracting foodborne illnesses because they have not yet built up adequate **immune systems**—the body's defense system against illness.

Elderly people are at a higher risk because their immune systems have weakened with age. In addition, as people age their senses of smell and taste are diminished, so they may be less likely to detect "off" odors or tastes that indicate food may be spoiled.

Because these groups of people are more likely to become sick with a foodborne illness, it is of particular concern when they consume potentially hazardous food or ingredients that are raw or have not been fully cooked. In all cases, these high-risk guests should be informed of any potentially hazardous food or ingredients that are raw or not fully cooked. You should also tell them to consult a physician before regularly consuming this type of food. **Additionally, check with your regulatory agency for specific requirements.**

Key Point

Potentially hazardous food typically contains moisture and protein, has a neutral or slightly acidic pH, and requires time-temperature control to prevent the growth of microorganisms and toxin production.

Food Most Likely to Become Unsafe

Although any type of food can become contaminated, some types are better able to support the rapid growth of microorganisms than others. These items are called **potentially hazardous food.** Not only do they have a history of being involved in foodborne-illness outbreaks, but they also have a natural potential for contamination due to the way they are produced and processed. Potentially hazardous food typically:

- Contains moisture
- Contains protein
- Has a neutral or slightly acidic pH (see Chapter 2)
- Requires time-temperature control to prevent the growth of microorganisms and the production of toxins

The ***FDA Food Code,*** which is issued by the Food and Drug Administration (FDA) and serves as the basis for many state and local food regulations, identifies the food in *Exhibit 1c* as potentially hazardous food.

Care must be taken to prevent contamination when handling **ready-to-eat food,** which is food that is edible without any further washing or cooking.

Ready-to-eat food includes:

- Washed, whole, or cut fruit and vegetables
- Deli meat
- Bakery items
- Sugars, spices, and seasonings
- Properly cooked food

Exhibit 1c

Potentially Hazardous Food			
Milk and milk products		Meat: beef, pork, lamb	
Eggs (except those treated to eliminate *Salmonella* spp.)		Raw sprouts and sprout seeds	
Shellfish and crustaceans		Heat-treated plant food, such as cooked rice, beans, and vegetables	
Fish		Poultry	
Baked potatoes		Tofu or other soy-protein food	
Sliced melons		Untreated garlic-and-oil mixtures	Milk; Minced Garlic and Oil
Synthetic ingredients, such as textured soy protein in meat alternatives			

Potential Hazards to Food Safety

Unsafe food usually results from **contamination,** which is the presence of harmful substances in the food. Some food safety hazards are introduced by humans or by the environment, while others occur naturally.

These hazards are divided into three categories: **biological hazards, chemical hazards,** and **physical hazards.**

- **Biological hazards** include certain bacteria, viruses, parasites, and fungi, as well as certain plant, mushroom, and seafood toxins.
- **Chemical hazards** include pesticides, food additives and preservatives, cleaning supplies, and toxic metals leached from nonfood-grade cookware and equipment.
- **Physical hazards** consist of foreign objects that accidentally get into the food, such as hair, dirt, metal staples, and broken glass, as well as naturally occurring objects, such as bones in fillets.

Key Point

Illness-causing microorganisms are responsible for the majority of foodborne-illness outbreaks.

By far, biological hazards pose the greatest threat to food safety. Illness-causing microorganisms are responsible for the majority of foodborne-illness outbreaks.

HOW FOOD BECOMES UNSAFE

The CDC have identified some common factors that are responsible for foodborne illness. These include:

- Purchasing food from unsafe sources
- Failing to cook food adequately
- Holding food at improper temperatures
- Using contaminated equipment
- Poor personal hygiene

Each of these factors—with the exception of purchasing food from unsafe sources—is related to **time-temperature abuse, cross-contamination,** or poor **personal hygiene.** Often, cases of foodborne illness involve several factors.

Exhibit 1d

Time-Temperature Abuse

Food has been time-temperature abused any time it has been allowed to remain too long at temperatures favorable to the growth of microorganisms.

Exhibit 1e

Cross-Contamination

Cross-contamination occurs when microorganisms are transferred from one surface or food to another.

Time-Temperature Abuse

Food has been time-temperature abused any time it has been allowed to remain too long at temperatures favorable to the growth of foodborne microorganisms. (See *Exhibit 1d.*) A foodborne illness can result if food is time-temperature abused in the following manner:

- It is not held or stored at required temperatures.
- It is not cooked or reheated to temperatures that kill microorganisms.
- It is not cooled properly.

Cross-Contamination

Cross-contamination occurs when microorganisms are transferred from one surface or food to another. (See *Exhibit 1e.*) A foodborne illness can result if cross-contamination is allowed to occur in any of these ways:

- Contaminated ingredients are added to food that receives no further cooking.
- Cooked or ready-to-eat food is allowed to touch **food-contact surfaces**—surfaces that come in contact with food—that have not been cleaned and sanitized.
- Contaminated food is allowed to touch or drip fluids onto cooked or ready-to-eat food.
- A foodhandler touches contaminated food and then touches cooked or ready-to-eat food.
- Contaminated cleaning cloths are not cleaned and sanitized before being used on other food-contact surfaces.

Poor Personal Hygiene

Individuals with poor personal hygiene can offend customers, contaminate food or food-contact surfaces, and cause illness. A foodborne illness can result if employees:

- Fail to wash their hands properly after using the restroom or whenever their hands become contaminated
- Cough or sneeze on food

- Touch or scratch sores, cuts, or boils, and then touch food they are handling
- Come to work while sick

KEY PRACTICES FOR ENSURING FOOD SAFETY

The keys to food safety lie in controlling time and temperature throughout the flow of food, practicing good personal hygiene, and preventing cross-contamination. It is important to establish standard operating procedures that focus on these areas. The ServSafe program will provide you with the knowledge to properly design these procedures.

Something to Think About... **Turkey Trouble**

Many customers fell ill after eating at a buffet in a country club in New Mexico. Dozens required medical treatment. The culprit? Roast turkey, stuffing, and gravy contaminated with the bacteria *Staphylococcus aureus.*

More than one factor led to the outbreak: Several of the foodhandlers had the bacteria. Poor personal hygiene practices led them to contaminate the turkey. The problem was made worse when the cooked turkey was not cooled properly. Finally, the bacteria spread when the foodhandlers used the same utensils to handle the turkey and a variety of other food.

What should have been done to prevent this incident?

THE FOOD SAFETY RESPONSIBILITIES OF A MANAGER

Managers have a basic responsibility to serve safe food and train employees in safe foodhandling practices. They must be up to date on regulations affecting the establishment and, most important, have a positive and supportive attitude toward food safety.

Meeting Food Safety Regulations

To stay in operation, your establishment must comply with city, county, and state food regulations. The regulatory agency evaluating your establishment shares your commitment to serving safe food. However, they may have the authority to assess fines and close an establishment that serves unsafe food. Therefore, it is in your best interest to work with local authorities.

The FDA recommends that state and local health departments hold the person in charge of a restaurant or foodservice establishment responsible for knowing and demonstrating the following information:

- Illnesses carried or transmitted by food and their symptoms
- Types of toxic materials used in the operation and how to safely store, dispense, use, and dispose of them
- Major food allergens and their symptoms
- Relationship between personal hygiene and the spread of illness, especially concerning cross-contamination, bare-hand contact with ready-to-eat food, and handwashing
- Reporting system that ensures employees inform their manager of illnesses
- How to keep injured or ill employees from contaminating food or food-contact surfaces
- Clear guidelines as to when employees must be excluded from the establishment or restricted from handling food and equipment
- The need to control the length of time potentially hazardous food remains at temperatures that support the growth of microorganisms
- Hazards involved in the consumption of raw or undercooked meat, poultry, eggs, and fish
- Safe times and temperatures for cooking potentially hazardous food items, such as meat, poultry, eggs, and fish

- Safe times and temperatures for storing, holding, cooling, and reheating potentially hazardous food
- Correct procedures for cleaning and sanitizing utensils and food-contact surfaces of equipment
- The need for equipment that is sufficient in number and capacity and is properly designed, constructed, located, installed, operated, maintained, and cleaned
- Approved sources of potable water and the importance of keeping it safe
- The principles of a food safety management system
- How the establishment's food safety procedures meet regulatory requirements
- Rights, responsibilities, and authorities the local code assigns to employees, managers, and the local health department

Marketing Food Safety

Marketing your food safety efforts will help show employees and customers that you are committed to serving safe food. Make it clear that your operation takes food safety seriously.

Key Point

Show employees through actions that management is involved in and supports food safety policies.

Show employees through actions that management is involved in and supports food safety policies and that food safety training for managers and all employees is a high priority. Offer training courses, and evaluate and update them regularly. Discuss food safety expectations. Document foodhandling procedures, and update them as necessary. Show employees that safe foodhandling is appreciated—consider awarding certificates for training and giving out small rewards for good food safety records. Set a good example by following all food safety rules yourself.

Show customers that employees know and follow safety rules. Make sure employees' appearances reflect your concern for food safety. Consider using food safety place mats and posters to reinforce your message, and be sure your employees can answer simple food safety questions when asked by customers.

SUMMARY

Foodborne illness is a major concern to the restaurant and foodservice industry. A foodborne illness is a disease carried or transmitted to people by food. Infants and preschool-age children, pregnant women, the elderly, people who are taking certain medications, and people who are seriously ill are at a higher risk. This is especially true when they consume potentially hazardous food that is raw, undercooked, or contaminated.

An incident of foodborne illness can be very expensive for an establishment. Costs can include lawsuits, increased insurance premiums, and damage to the establishment's reputation.

Although any type of food can become contaminated, some types are better able to support the rapid growth of microorganisms than others. These food items are called potentially hazardous food. They typically contain moisture and protein and have a neutral or slightly acidic pH. Potentially hazardous food requires time-temperature control to prevent the growth of microorganisms and the production of toxins.

The keys to food safety lie in controlling time and temperature throughout the flow of food, practicing good personal hygiene, and preventing cross-contamination.

Apply Your Knowledge

Use these questions to review the concepts presented in this chapter.

Discussion Questions

1. What are some common characteristics of potentially hazardous food?
2. What are the potential costs associated with foodborne-illness outbreaks?
3. Why are the elderly at higher risk for contracting foodborne illnesses?
4. What are the three major types of hazards to food safety?

For answers, please turn to the Answer Key.

Apply Your Knowledge

Use these questions to test your knowledge of the concepts presented in this chapter.

Multiple-Choice Study Questions

1. Why are elderly people at a higher risk for foodborne illness?
   A. They are more likely to spend time in a hospital.
   B. Their immune systems have weakened with age.
   C. Their allergic reactions to chemicals used in food production might be greater than those of younger people.
   D. They are likely to have diminished appetites.
2. Which food would most likely cause a foodborne illness?
   A. Tomato juice
   B. Cooked rice
   C. Whole wheat flour
   D. Powdered milk
3. Which is *not* a common characteristic of potentially hazardous food?
   A. They are moist.
   B. They are dry.
   C. They have a pH that is neutral or slightly acidic.
   D. They contain protein.
4. For a foodborne illness to be considered an "outbreak," how many people must experience the same illness after eating the same food?
   A. One
   B. Two
   C. Ten
   D. Twenty
5. Which item is a potentially hazardous food?
   A. Raw carrots
   B. Dry rice
   C. Bread
   D. Raw bean sprouts

For answers, please turn to the Answer Key.

Take It Back*

The following food safety concepts from this chapter should be taught to your employees:

- Potentially hazardous food
- How food becomes unsafe

The tools below can be used to teach these concepts in fifteen minutes or less using the directions below. Each tool includes content and language appropriate for employees. Choose the tool or tools that work best.

<table>
<tr>
<td>Tool #1:
ServSafe Video 2:
Overview of Foodborne Microorganisms and Allergens</td>
<td>Tool #2:
ServSafe Employee Guide</td>
<td>Tool #3:
ServSafe Posters and Quiz Sheets</td>
<td>Tool #4:
ServSafe Fact Sheets and Optional Activities</td>
</tr>
<tr>
<td>
</td>
<td>
</td>
<td>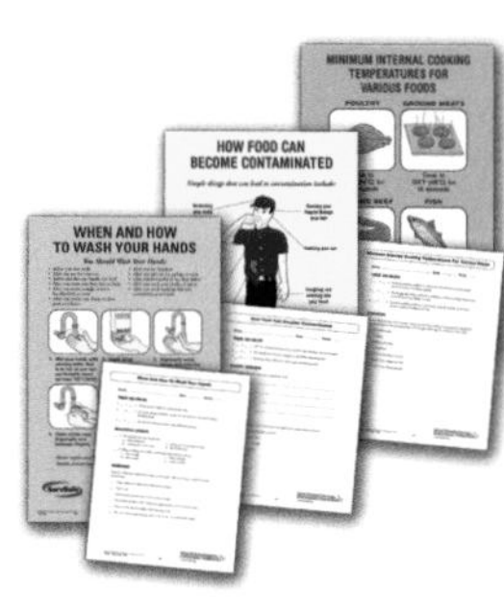
</td>
<td>
</td>
</tr>
<tr>
<td colspan="4">Potentially Hazardous Food</td>
</tr>
<tr>
<td>Video segment on food supporting growth of microorganisms
1 Show employees the segment.
2 Ask employees to identify food items that are potentially hazardous.</td>
<td>Section 1
What a Foodborne Illness Is
Discuss with employees food items that are potentially hazardous.</td>
<td></td>
<td></td>
</tr>
</table>

* **Visit the Food Safety Resource Center at *www.ServSafe.com/FoodSafety/resource* to download free posters, quiz sheets, fact sheets, and optional activities and to learn how to obtain *Employee Guides* and videos/DVDs.**

Take It Back*

Tool #1: ServSafe Video 1: *Introduction to Food Safety*

Tool #2: *ServSafe Employee Guide*

Tool #3: ServSafe Posters and Quiz Sheets

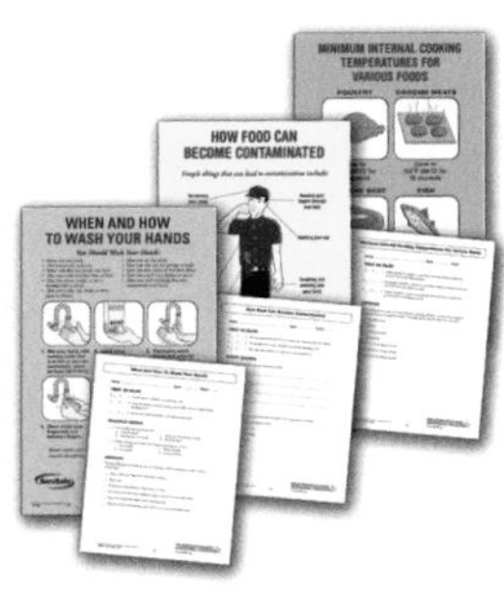

Tool #4: ServSafe Fact Sheets and Optional Activities

How Food Becomes Unsafe

Tool #1	Tool #2	Tool #3	Tool #4
Video segment on how food becomes unsafe 1 Show employees the segment. 2 Ask employees to identify the factors that can cause food to become unsafe.	**Section 1** **How Food Can Become Unsafe** 1 Discuss with employees the factors that can cause food to become unsafe. 2 Complete the Multiple Choice Questions.	**Poster: How Food Can Become Unsafe** 1 Discuss with employees the factors that can cause food to become unsafe as presented in the poster. 2 Have employees complete the Quiz Sheet: How Food Can Become Unsafe.	

* Visit the Food Safety Resource Center at *www.ServSafe.com/FoodSafety/resource* to download free posters, quiz sheets, fact sheets, and optional activities and to learn how to obtain *Employee Guides* and videos/DVDs.

ADDITIONAL RESOURCES

Articles and Texts

Gombas, David. E, Chen Yuhan, Rocelle S. Clavero, Virgina N. Scott. 2003. Survey of *Listeria monocytogenes* in Ready-to-Eat Foods. *Journal of Food Protection.* 66 (4): 559.

Web Sites

American Council on Science and Health

www.acsh.org

The American Council on Science and Health (ACSH) is a consumer-education consortium concerned with issues related to food, nutrition, chemicals, pharmaceuticals, lifestyle, the environment, and health. ACSH advocates sound science and the belief that Americans' continued physical and economic well-being depends on the advancement of scientific, medical, and technological research and innovation. Visit this Web site for publications, articles, editorials on food safety, and other timely health-related topics.

Centers for Disease Control and Prevention

www.cdc.gov

The Centers for Disease Control and Prevention (CDC) are focused on American health promotion, prevention, and preparedness. The CDC apply research and findings to improve people's daily lives and respond to health emergencies. This Web site provides information on disease and disease prevention, including foodborne illness, its causes, public-health impact, and methods for control.

Center for Infectious Disease Research & Policy

www.cidrap.umn.edu

The Center for Infectious Disease Research & Policy (CIDRAP) at the University of Minnesota functions to prevent illness and death from infectious diseases through epidemiologic research and the rapid translation of scientific information into real-world practical applications and solutions. The Center's Web site provides access to information on food safety and foodborne illness and news stories on food security and foodborne-illness surveillance.

Conference for Food Protection

www.foodprotect.org

The Conference for Food Protection is a nonprofit organization that provides a forum for regulators, industry professionals, academia, professional organizations, and consumers to identify problems, formulate recommendations, and develop and implement practices that ensure food safety. Though the conference has no formal regulatory authority, it is an organization that influences model laws and regulations among all government agencies. Visit this Web site for information regarding previously recommended changes to the *FDA Food Code,* standards for permanent outdoor cooking facilities, and other guidance documents, along with how to participate in the conference.

FDA Center for Food Safety and Applied Nutrition

www.cfsan.fda.gov/list.html

As the center within the Food and Drug Administration (FDA) responsible for food safety, the Center for Food Safety and Applied Nutrition (CFSAN) promotes and protects the public health by researching and implementing guidelines, policies, and standards to ensure that food is safe, nutritious, wholesome, and properly labeled. This Web site provides information relevant to all aspects of food safety and security, including corresponding guidelines, policies, and standards.

Food Marketing Institute

www.fmi.org

The Food Marketing Institute (FMI) represents food retailers and wholesalers in large multistore chains, regional firms, and independent supermarkets. FMI provides its members with a forum to work with the government, suppliers, employees, customers, and their communities. This Web site provides publications and resources that support FMI's efforts in research, education, public information, government relations, and industry relations.

Gateway to Government Food Safety Information

www.foodsafety.gov

This Web site provides links to selected government food safety-related information.

International Food Information Center

www.ific.org

Working with scientific experts who serve to translate research into understandable and useful information, the International Food Information Center (IFIC) collects and disseminates scientific information on food safety, nutrition, and health. Visit this Web site for information on these topics.

International Food Safety Council

www.nraef.org/ifsc/ifsc_about.asp

In 1993, the National Restaurant Association Educational Foundation recognized the need for food safety awareness and created the International Food Safety Council as its food safety awareness initiative. The council's mission is to heighten the awareness of the importance of food safety education throughout the restaurant and foodservice industry through its educational programs, publications, and awareness campaigns.

National Center for Infectious Diseases

www.cdc.gov/ncidod/index.htm

The National Center for Infectious Disease (NCID), one of the Centers for Disease Control and Prevention (CDC), works in partnership with local and state public-health officials, other federal agencies, medical and public-health professional associations, infectious-disease experts from academic and clinical practice, and international and public-service organizations to help prevent illness, disability, and death caused by infectious diseases in the United States. This Web site houses information on bacterial, viral, and parasitic diseases.

National Restaurant Association

www.restaurant.org

The National Restaurant Association (NRA) is the leading business association for the restaurant industry, representing more than 300,000 restaurant establishments of over 60,000 member companies. The association's mission is to represent, educate, and promote a rapidly growing industry that is comprised of 900,000 restaurant and foodservice outlets employing 12.2 million people. Visit this Web site for all issues and concerns related to your restaurant, including tips for running your establishment and vital data on your customers' spending habits.

National Restaurant Association Educational Foundation

www.nraef.org

The National Restaurant Association Educational Foundation is a nonprofit organization dedicated to fulfilling the educational mission of the National Restaurant Association. Focusing on three key strategies of risk management, recruitment, and retention, the NRAEF is the premier provider of educational resources, materials, and programs, all of which address attracting, developing and retaining the industry's workforce. The purchase of any NRAEF product helps fund outreach and scholarship activities on a national and state level.

Documents and Other Resources

Emerging Infectious Diseases

www.cdc.gov/ncidod/eid/index.htm

The online journal *Emerging Infectious Diseases* represents the scientific communications component of the Centers for Disease Control and Prevention's (CDC) efforts against the threat of emerging infections. The journal relies on a broad, international authorship base to publish reports of interest to researchers on infectious diseases and related sciences and reports on laboratory and epidemiologic findings within a broader public-health perspective. Visit this Web site for articles published about foodborne disease in the United States and around the world.

Evaluation and Definition of Potentially Hazardous Food

members.ift.org/IFT/Pubs/CRFSFS/archive/supplement2.htm

Visit this Web site to link to the evaluation of the definition of a potentially hazardous food by the Institute of Food Technologists (IFT). This document provides background on and reasons for the current *FDA Food Code* definition of a potentially hazardous food as a temperature-controlled-for-safety (TCS) food.

FDA Enforcement Report Index

www.fda.gov/opacom/Enforce.html

The Food and Drug Administration's Center for Food Safety and Applied Nutrition provides the *FDA Enforcement Report,* published weekly on this Web site. It contains information on actions such as recalls, injunctions, and seizures taken against food and drug products that have not met FDA regulatory requirements.

FDA Foodborne Illness

www.cfsan.fda.gov/~mow/foodborn.html

The Food and Drug Administration's Center for Food Safety and Applied Nutrition provides the FDA Foodborne Illness gateway page, which contains resources related to foodborne illness. Visit this Web site to access information about HACCP, foodborne pathogens, specific government food safety initiatives, and other government sources of foodborne-illness information.

2005 *FDA Food Code*

www.cfsan.fda.gov/~dms/fc05-toc.html

The Food and Drug Administration (FDA) publishes the *FDA Food Code,* a scientifically sound technical and legal document that serves as a model for regulating the retail and foodservice industry at the federal, state, and local level. Updates to the code are issued on the odd-numbered years. The *FDA Food Code* provides a system of safeguards designed to minimize foodborne illness and ensure employee health, food protection manager knowledge, safe food, nontoxic and cleanable equipment, and appropriate sanitation of the food establishment. It is used as the basis for information in this textbook.

Foodborne Diseases Active Surveillance Network (Food Net)

www.cdc.gov/foodnet

Together with the U.S. Department of Agriculture (USDA), the Food and Drug Administration (FDA), and ten data-collection sites throughout the United States, the Foodborne Diseases Active Surveillance Network (FoodNet) consists of active surveillance for foodborne diseases and related epidemiologic studies designed to help public-health officials better understand the causes, distribution, and control of foodborne diseases in the United States. Visit this Web site for data collected on the incidence of nine key foodborne pathogens, including their summary reports.

Foodborne Illness Cost Calculator

www.ers.usda.gov/data/foodborneillness

As a means to quantify the annual economic cost of foodborne illness, the U.S. Department of Agriculture Economic Research Service has developed a Foodborne Illness Cost Calculator. Visit this Web site to access this calculator to estimate medical costs due to illness, the cost of time lost from work due to nonfatal illness, and the cost of premature death for several foodborne pathogens. The calculator allows you to alter the assumptions to determine your own cost estimates.

Morbidity and Mortality Weekly Report

www.cdc.gov/mmwr

The *Morbidity and Mortality Weekly Report* (*MMWR*), prepared by the Centers for Disease Control and Prevention (CDC), is available free of charge on this Web site. The report contains data based on weekly accounts of reportable diseases from state health departments. Visit this Web site to search report databases for synopses of actual foodborne-illness outbreaks.

U.S. Department of Health and Human Services

www.hhs.gov

The Department of Health and Human Services (HHS) is the federal government's principal agency for protecting American public health. The department oversees more than three hundred programs executed by such agencies as the Food and Drug Administration (FDA) and the Centers for Disease Control and Prevention (CDC). Visit this Web site to access food safety information provided by the various agencies and departments overseen by HHS.

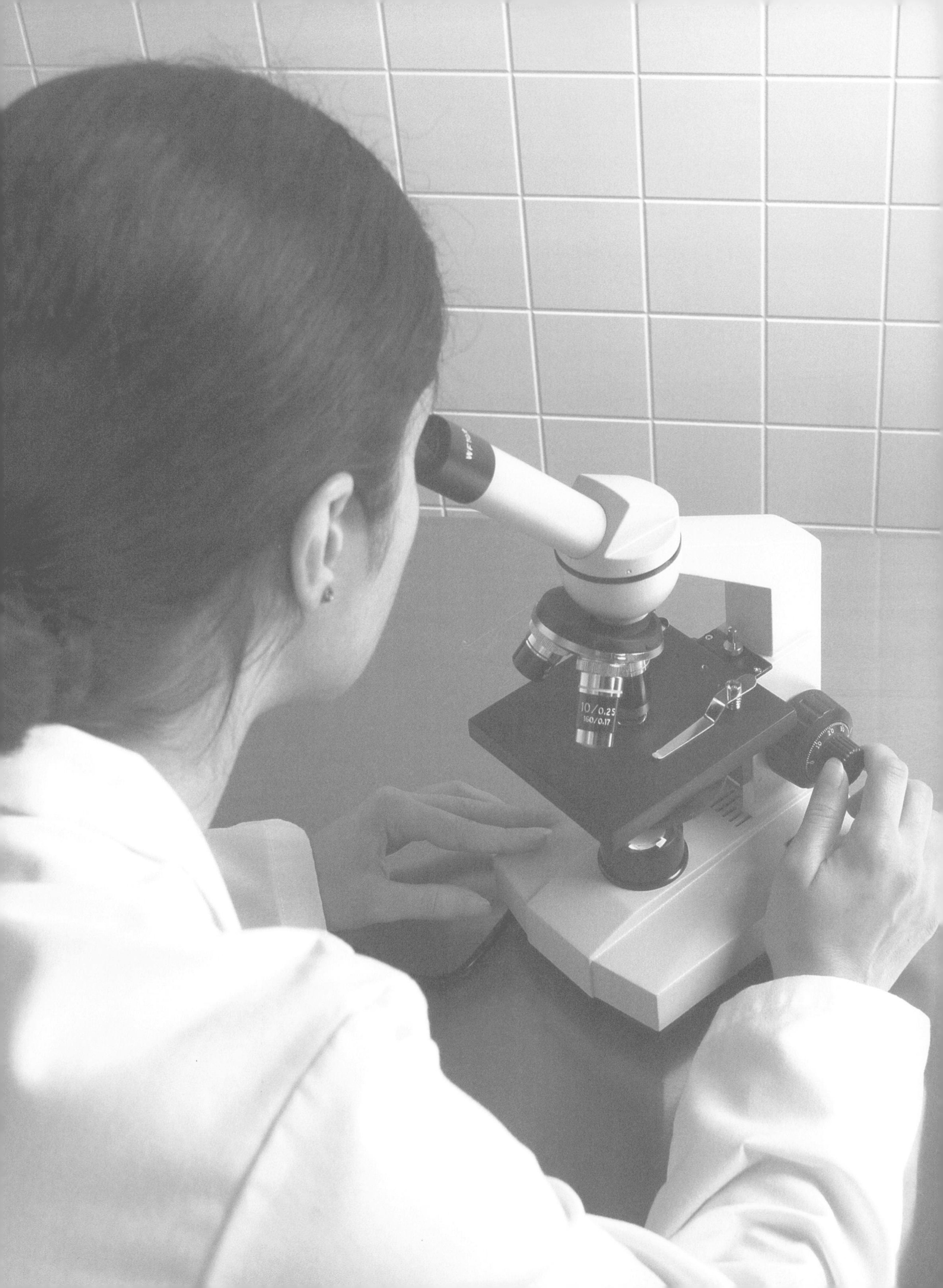
10/0.25
160/0.17

The Microworld

Inside this chapter:

- Microbial Contaminants
- Classifying Foodborne Illnesses
- Bacteria
- Viruses
- Parasites
- Fungi
- Emerging Pathogens and Issues
- Technological Advancements in Food Safety

After completing this chapter, you should be able to:

- Identify factors that affect the growth of foodborne pathogens (FAT TOM).
- Differentiate between foodborne infections, intoxications, and toxin-mediated infections.
- Identify major foodborne illnesses and their symptoms.
- Identify characteristics of major foodborne pathogens including sources, food involved in outbreaks, and methods of prevention.

Key Terms

- Microorganisms
- Pathogens
- Toxins
- Spoilage microorganism
- FAT TOM
- pH
- Temperature danger zone
- Water activity (a_w)
- Foodborne infection
- Foodborne intoxication
- Foodborne toxin-mediated infection
- Bacteria
- Spore
- Virus
- Parasite
- Fungi
- Mold
- Yeast

Apply Your Knowledge

Check to see how much you know about the concepts in this chapter. Use the page references provided with each question to explore the topic.

Test Your Food Safety Knowledge

1. **True or False:** *Bacillus cereus* is commonly associated with cereal crops, such as rice. *(See page 2-18.)*

2. **True or False:** A foodborne intoxication results when a person eats food containing pathogens, which then grow in the intestines and cause illness. *(See page 2-6.)*

3. **True or False:** Cooking food to the required minimum internal temperature can help prevent listeriosis. *(See page 2-15.)*

4. **True or False:** A person with shigellosis may experience bloody diarrhea. *(See page 2-14.)*

5. **True or False:** Highly acidic food typically does not support the growth of foodborne microorganisms. *(See page 2-4.)*

For answers, please turn to the Answer Key.

INTRODUCTION

In Chapter 1, you learned that foodborne microorganisms pose the greatest threat to food safety and that disease-causing microorganisms are responsible for the majority of foodborne-illness outbreaks. In this chapter you will learn about the microorganisms that cause foodborne illness and the conditions that allow them to grow. When you understand these conditions, you will begin to see how the growth of foodborne microorganisms can be controlled, a topic that will be covered in greater detail in later chapters.

Microorganisms are small, living organisms that can be seen only with a microscope. While not all microorganisms cause illness, some do. These are called **pathogens.** Eating food contaminated with foodborne pathogens or their **toxins** (poisons) is the leading cause of foodborne illness.

MICROBIAL CONTAMINANTS

There are four types of microorganisms that can contaminate food and cause foodborne illness: bacteria, viruses, parasites, and fungi.

These microorganisms can also be divided into two groups: **spoilage microorganisms** and pathogens. Mold is a spoilage microorganism. While its appearance, smell, and taste are not very appetizing, mold typically does not cause illness. Pathogens like *Salmonella* spp. and the hepatitis A virus can cause some form of illness when ingested. Unlike spoilage microorganisms, pathogens cannot be seen, smelled, or tasted in food.

What Microorganisms Need to Grow: FAT TOM

The six conditions that support the growth of foodborne microorganisms—with the exception of viruses—can be remembered by the acronym **FAT TOM**. (See *Exhibit 2a.*) A brief explanation of each condition follows.

Exhibit 2a

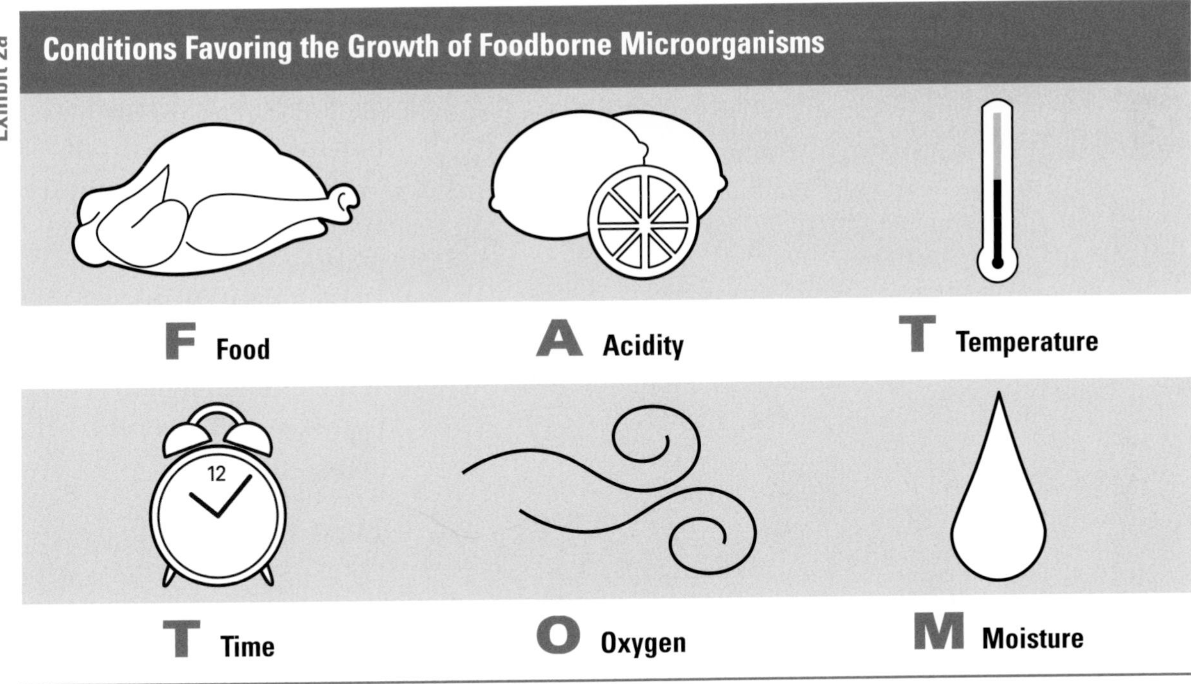

Food

Foodborne microorganisms need nutrients to grow—specifically proteins and carbohydrates. These are commonly found in potentially hazardous food such as meat, poultry, dairy products, and eggs.

Exhibit 2b

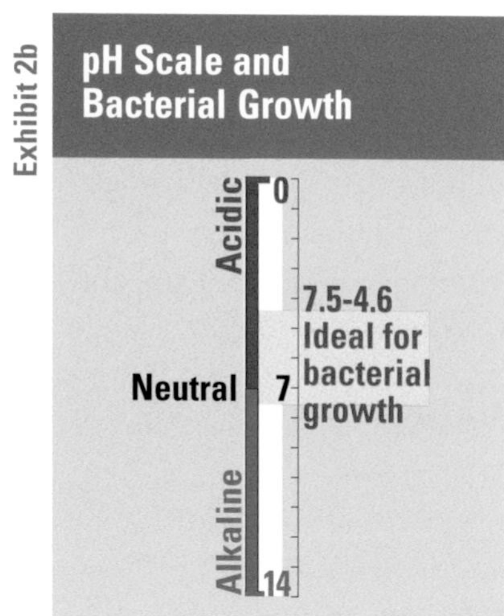

Acidity

pH is a measurement of how acidic or alkaline a food is. The pH scale ranges from 0.0 to 14.0. (See *Exhibit 2b.*) Food with a pH between 0.0 and 6.9 is acidic, while food with a pH between 7.1 and 14.0 is alkaline. A pH of 7.0 is neutral. Foodborne microorganisms typically do not grow in alkaline food, such as crackers, or highly acidic food, such as lemons. They grow best in food that has a neutral or slightly acidic pH (7.5 to 4.6). Unfortunately, the pH of most food falls into this range. (See *Exhibit 2c.*)

Temperature

Foodborne microorganisms grow well between the temperatures of 41°F and 135°F (5°C and 57°C). (See *Exhibit 2d.*) This range is known as the **temperature danger zone.** Exposing microorganisms to temperatures outside the danger zone does not necessarily kill them. Refrigeration temperatures, for example, may only slow their growth. Some bacteria actually grow at refrigeration temperatures. Food must be handled very carefully when it is thawed, cooked, cooled, and reheated as it is exposed to the temperature danger zone during these times.

Exhibit 2c

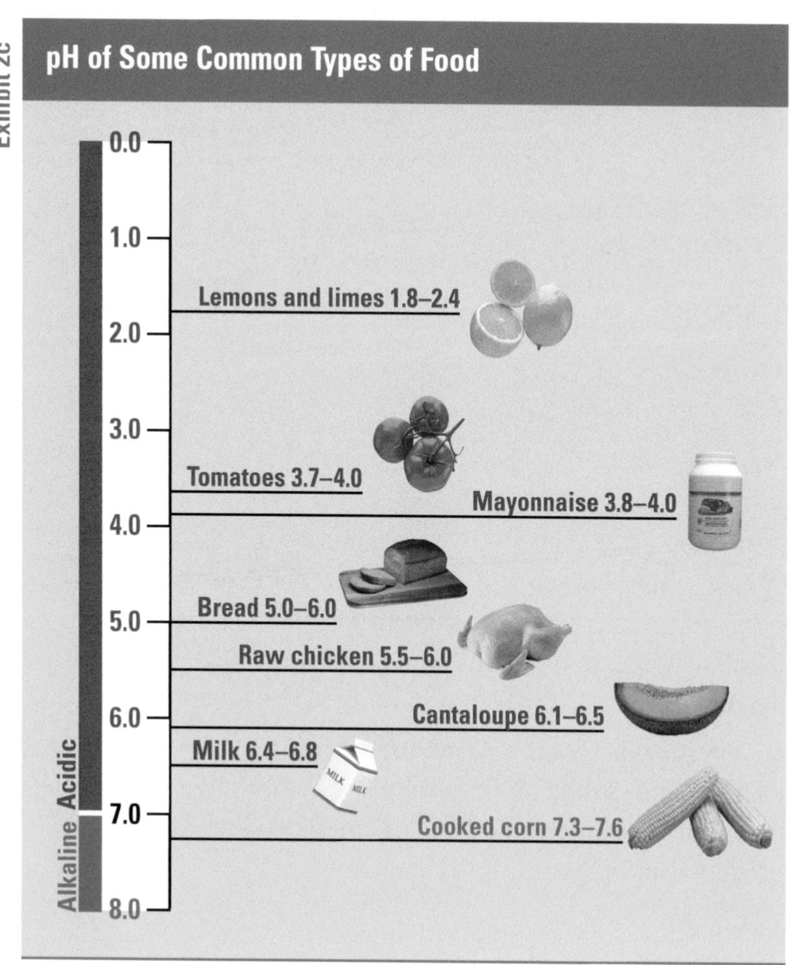

Time

Foodborne microorganisms need sufficient time to grow. This means that even under favorable conditions, microorganisms need enough time to move from the lag phase (slow growth) to the log phase (rapid growth). Keep in mind that some bacteria can double their population every twenty minutes.

Exhibit 2d

Temperature and Bacterial Growth

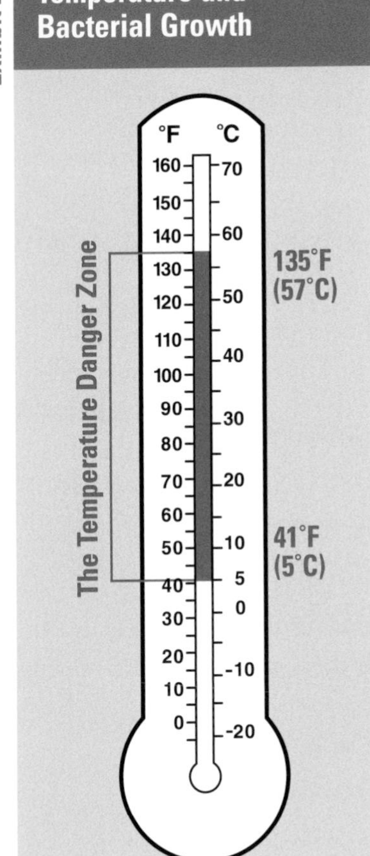

Most foodborne microorganisms grow well at temperatures between 41°F and 135°F (5°C and 57°C).

If potentially hazardous food remains in the temperature danger zone for four hours or more, pathogenic microorganisms can grow to levels high enough to make someone ill.

Oxygen

Some foodborne microorganisms require oxygen to grow, while others grow when oxygen is absent. Pathogens that grow without oxygen can occur in cooked rice, untreated garlic-and-oil mixtures, and baked potatoes that have been temperature abused.

Moisture

Most foodborne microorganisms require moisture to grow. The amount of moisture available in food for this growth is called its **water activity (a_w)**. It is measured on a scale of 0.0 to 1.0, with water having a water activity of 1.0. Potentially hazardous food typically has a water activity of .85 or higher.

Controlling the Growth of Microorganisms

FAT TOM is the key to controlling microorganisms. Denying any one of these conditions for growth can help keep food safe.

Food processors use several methods to keep microorganisms from growing, including:

- Adding lactic or citric acid to food to make it more acidic
- Adding sugar, salt, alcohol, or acid to a food to lower its water activity
- Using vacuum packaging to remove oxygen

While these control methods may not be practical for your establishment, there are two important conditions for growth that you can control—time and temperature. Remember that microorganisms grow well at temperatures between 41°F and 135°F (5°C and 57°C). To control growth, it is important to move food out of this temperature range by cooking it to the proper temperature, freezing it, or by refrigerating it at 41°F (5°C) or lower. In addition, foodborne microorganisms need sufficient time to grow. When preparing food, you must limit the amount of time it spends in the temperature danger zone.

CLASSIFYING FOODBORNE ILLNESSES

Foodborne illnesses are classified as infections, intoxications, or toxin-mediated infections. Each occurs in a different way.

- **Foodborne infections** result when a person eats food containing pathogens, which then grow in the intestines and cause illness. Typically, symptoms of a foodborne infection do not appear immediately.
- **Foodborne intoxications** result when a person eats food containing toxins that cause illness. The toxin may have been produced by pathogens found in the food or may be the result of a chemical contamination. The toxin might also be a natural part of a plant or animal consumed. Typically, symptoms of a foodborne intoxication appear quickly, within a few hours.
- **Foodborne toxin-mediated infections** result when a person eats food containing pathogens, which then produce illness-causing toxins in the intestines.

BACTERIA

Of all microorganisms, bacteria are of greatest concern to the manager. Knowing what bacteria are and understanding the environment in which they grow is the first step in controlling them.

Characteristics of Bacteria That Cause Foodborne Illness

Bacteria that cause foodborne illness have some basic characteristics:

- They are living, single-celled organisms.
- They may be carried by a variety of means: food, water, soil, animals, humans, or insects.
- Under favorable conditions, they can reproduce very rapidly.
- Some can survive freezing.

- Some change into a different form, called **spores,** to protect themselves.
- Some can cause food spoilage; others can cause illness.
- Some cause illness by producing toxins as they multiply, die, and break down. Typically, cooking does not destroy these toxins.

Bacterial Growth

Bacterial growth can be broken into four progressive stages (phases): lag, log, stationary, and death. (See *Exhibit 2e.*)

When bacteria are first introduced to food, they go through an adjustment period, called the lag phase. In this phase, their number is stable as they prepare for growth. To prevent food from becoming unsafe, it is important to prolong the lag phase as long as possible. You can accomplish this by controlling the conditions for growth: temperature, time, oxygen, moisture, and pH. For example, by refrigerating food, you can keep bacteria in the lag phase. If these conditions are not controlled, bacteria can enter the next phase, the log phase, in which they will grow remarkably fast.

Exhibit 2e

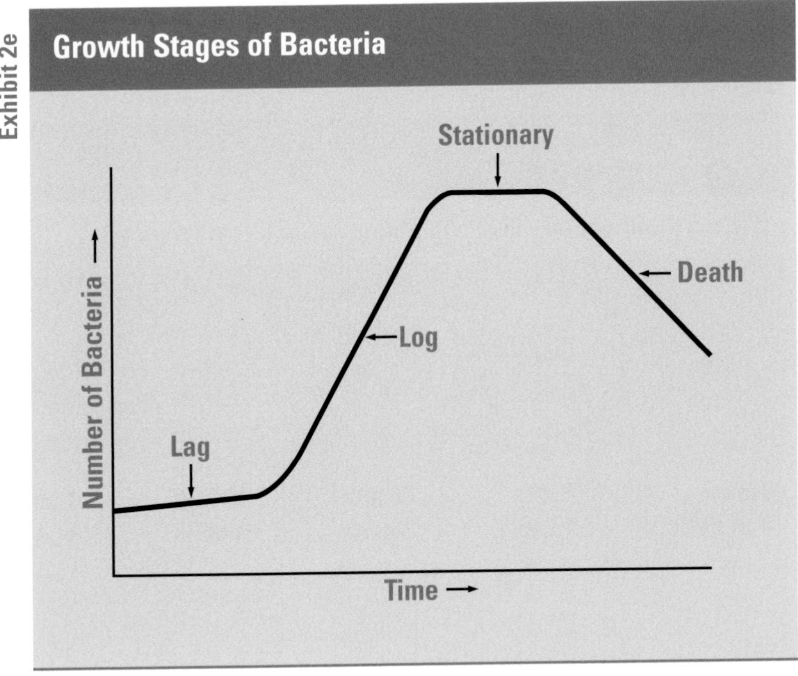

Bacteria reproduce by splitting in two. Those in the process of reproduction are called vegetative microorganisms. As long as conditions are favorable, bacteria can grow very rapidly, doubling their number as often as every twenty minutes. (See *Exhibit 2f.*) This is called exponential growth, and it occurs in the log phase. Food will rapidly become unsafe during the log phase.

Bacteria can continue to grow until nutrients and moisture become scarce or conditions become unfavorable. Eventually, the population reaches a stationary phase, in which just as many bacteria are growing as are dying. When the number of bacteria dying exceeds the number growing, the population declines. This is called the death phase.

The time required for bacteria to adapt to a new environment (lag phase) and to begin a rapid rate of growth (log phase) depends on several factors, such as temperature. *Exhibit 2g* shows how different temperatures affect the growth rate of *Salmonella* spp. As the graph shows, at warmer temperatures (95°F [35°C]), *Salmonella* spp. grows more quickly than at colder temperatures (44°F and 50°F [7°C and 10°C]). At even colder temperatures (42°F [6°C]), *Salmonella* spp. does not grow at all—but notice that it does not die either (prolonging the lag phase). This is why refrigerating food properly helps keep it safe.

Exhibit 2f

Rapid Bacterial Growth

Bacteria can double their population every twenty minutes.

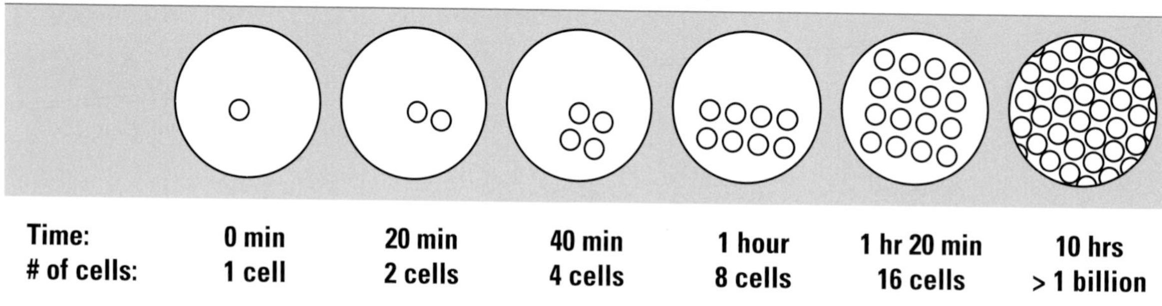

Exhibit 2g

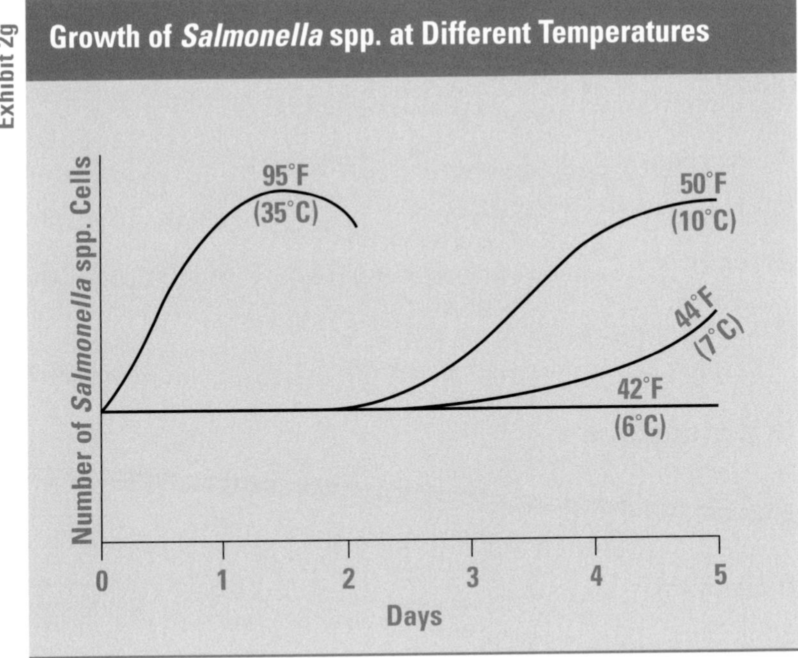

Spores

Certain bacteria can change into a different form, called spores, to protect themselves when nutrients are not available. Spores are commonly found in soil and can contaminate food grown there, such as vegetables, herbs, and cereal crops. They can also contaminate meat, poultry, fish, and other food exposed to soil or dust.

A spore can resist heat, allowing it to survive cooking temperatures. Spores can also revert back to a form capable of growth. This can occur when food is not stored at the proper temperature or when it is not held or cooled properly. You must prevent this to keep food safe.

Major Foodborne Illnesses Caused by Bacteria

The major foodborne illnesses caused by bacteria follow on the next several pages. (Biological toxins are discussed in Chapter 3.) For each illness, it is important to understand its common source, the food commonly associated with it, its most common symptoms, and the most important measures that can be taken to prevent the illness from occurring. The table on the next page provides a high-level overview of these characteristics. While not all-inclusive, this information will help you see the similarities and differences that will make it easier to remember each illness.

In addition, the bacterial illnesses on pages 2-12 through 2-22 are grouped by the way they are classified: infection, intoxication, or toxin-mediated infection.

Infections

Intoxications

Toxin-Mediated Infections

Major Foodborne Illnesses Caused by Bacteria

Illness Characteristics	Type of Illness										
	Infection						Intoxication			Toxin-Mediated Infection	
	Campylobacteriosis	Salmonellosis	Shigellosis	Listeriosis	*Vibrio parahaemolyticus* Gastroenteritis	*Vibrio vulnificus* Septicemia/Gastroenteritis	*Bacillus cereus* Gastroenteritis	Staphylococcal Gastroenteritis	Botulism	*Clostridium perfringens* Gastroenteritis	Hemorrhagic Colitis
Bacteria Characteristics											
Forms spores							x		x	x	
Produces toxins							x	x	x	x	x
Commonly Associated Food											
Poultry	x	x								x	
Eggs		x									
Meat		x		x			x			x	x
Fish											
Shellfish					x	x					
Ready-to-eat food		x	x	x				x			
Produce			x				x		x		x
Cereal crops							x				
Dairy		x		x							
Contaminated water	x		x	x							
Most Common Symptoms											
Diarrhea	x	x	x		x	x	x			x	x
Abdominal pain/cramps	x	x	x		x	x	x	x		x	x
Nausea					x	x	x	x	x		
Vomiting		x			x	x	x	x	x		
Fever	x	x	x		x	x					
Headache	x										
Most Important Prevention Measures											
Proper handwashing			x					x			
Proper cooking	x	x		x	x	x	x	x			x
Proper holding							x	x	x	x	
Proper cooling							x	x	x	x	
Proper reheating									x	x	
Approved suppliers				x	x	x					
Exclude foodhandlers		x	x								x
Prevent cross-contamination	x	x		x							x

Infection

Illness: Campylobacteriosis *(CAMP-ee-lo-BAK-teer-ee-O-sis)*

Bacteria: *Campylobacter jejuni*
(CAMP-ee-lo-BAK-ter jay-JUNE-ee)

Though *Campylobacter jejuni* is commonly associated with poultry, it has been known to contaminate water. Illness often occurs when poultry is improperly cooked and when raw poultry has been allowed to cross-contaminate other food and food-contact surfaces. Campylobacteriosis is best controlled through proper cooking and the prevention of cross-contamination.

Food Commonly Associated with the Bacteria

- Poultry
- Water contaminated with the bacteria

Most Common Symptoms

- Diarrhea (may be watery or bloody)
- Abdominal cramps
- Fever
- Headache

Most Important Prevention Measures

To reduce the bacteria in food:

- Cook food, particularly poultry, to required minimum internal temperatures.

To prevent the transfer of the bacteria to food:

- Prevent cross-contamination between raw poultry and ready-to-eat food.

Infection

Illness: Salmonellosis *(SAL-men-uh-LO-sis)*

Bacteria: *Salmonella* spp.
(SAL-ma-NEL-uh)

Many farm animals naturally carry *Salmonella* spp. It is often associated with poultry and eggs, dairy products, and beef. It has also been found in ready-to-eat food, such as produce that has come in contact with these animals or their waste. Since illness can occur after consuming only a small amount of this type of bacteria, it is critical to cook food properly and to prevent cross-contamination.

Food Commonly Associated with the Bacteria

- Poultry and eggs
- Dairy products
- Beef

Most Common Symptoms

- Diarrhea
- Abdominal cramps
- Vomiting
- Fever

The severity of these symptoms depends upon the health of the person and the amount of bacteria consumed.

Salmonella spp. is often present in a person's feces for several weeks after symptoms have ended.

Most Important Prevention Measures

To reduce the bacteria in food:

- Cook raw beef, poultry, and eggs to required minimum internal temperatures.

To prevent the transfer of the bacteria to food:

- Minimize cross-contamination between raw meat and poultry and ready-to-eat food.
- Exclude foodhandlers from working in the establishment if they have been diagnosed with salmonellosis.

Infection

Illness: Shigellosis *(SHIG-uh-LO-sis)*

Bacteria: ***Shigella*** spp. *(shi-GEL-uh)*

Shigella spp. is found in the feces of humans with shigellosis. Most illnesses occur when people consume food or water contaminated with this type of bacteria. It can be transferred to food when foodhandlers fail to wash their hands after using the restroom. Flies can also transfer the bacteria from feces to food. It is only necessary to consume a small amount of *Shigella* spp. to become ill. Preventing illness is dependent upon proper hygiene and excluding foodhandlers with diarrhea from the establishment.

Food Commonly Associated with the Bacteria

- Food that is easily contaminated by hands, such as salads containing potentially hazardous food
- Food that has made contact with contaminated water, such as produce

Most Common Symptoms

- Bloody diarrhea
- Abdominal pain and cramps
- Fever (occasionally)

Severe cases of illness can result in hemolytic uremic syndrome (HUS).

High levels of the bacteria can be found in the feces of people with shigellosis for weeks after symptoms have ended.

Most Important Prevention Measures

To prevent the transfer of the bacteria to food:

- Exclude foodhandlers from working in the establishment if:
  - They have diarrhea.
  - They have been diagnosed with shigellosis.
- Wash hands when necessary.
- Control flies inside and outside the establishment.

Infection

Illness: Listeriosis *(liss-TEER-ee-O-sis)*

Bacteria: *Listeria monocytogenes*

(liss-TEER-ee-uh MON-o-SI-TAHJ-uh-neez)

Listeria monocytogenes is naturally found in soil, water, and plants. Unlike other types of bacteria, it grows in cool, moist environments. It is commonly associated with ready-to-eat products, such as deli meats, hot dogs, and soft cheese. While illness is uncommon in healthy people, high-risk populations are especially vulnerable—particularly pregnant women. To prevent illness, it is critical to discard food that has passed its use-by or expiration date. It is also important to cook food properly and to avoid cross-contamination.

Food Commonly Associated with the Bacteria

- Raw meat
- Unpasteurized milk and milk products
- Ready-to-eat food including:
  - Deli meat
  - Hot dogs
  - Soft cheese

Most Common Symptoms

Pregnant women (third trimester):

- Spontaneous abortion of the fetus

Newborns:

- Sepsis
- Pneumonia
- Meningitis

Most Important Prevention Measures

- Discard any product that has passed its use-by or expiration date.
- Avoid using unpasteurized dairy products.

To reduce the bacteria in food:

- Cook raw meat to required minimum internal temperatures.

To prevent the transfer of the bacteria to food:

- Prevent cross-contamination between raw or undercooked and ready-to-eat food.

Infection

Illness: *Vibrio parahaemolyticus* Gastroenteritis

(VIB-ree-o PAIR-uh-hee-mo-LIT-ih-kus GAS-tro-EN-ter-I-tiss)

Bacteria: *Vibrio parahaemolyticus*

(VIB-ree-o PAIR-uh-hee-mo-LIT-ih-kus)

Vibrio parahaemolyticus is naturally found in the waters of the Gulf of Mexico, as well as the Atlantic and Pacific Coasts. It is commonly associated with raw or partially cooked oysters harvested from warm waters during the months of April to October. This type of bacteria can grow *very rapidly* at temperatures in the middle of the temperature danger zone. Preventing illness depends upon purchasing oysters from approved, reputable suppliers and cooking them to the required minimum internal temperature.

Food Commonly Associated with the Bacteria

- Raw or partially cooked oysters

Most Common Symptoms

- Diarrhea and abdominal cramps
- Nausea and vomiting
- Low grade fever and chills

Most Important Prevention Measures

- Purchase oysters from approved, reputable suppliers.
- Cook oysters to the required minimum internal temperature.

Infection

Illnesses:
Vibrio vulnificus Primary Septicemia
(VIB-ree-o vul-NIF-ih-kus SEP-ti-CEE-mee-uh)

Vibrio vulnificus Gastroenteritis
(VIB-ree-o vul-NIF-ih-kus GAS-tro-EN-ter-I-tiss)

Bacteria: *Vibrio vulnificus*
(VIB-ree-o vul-NIF-ih-kus)

Vibrio vulnificus is naturally found in the waters of the Gulf of Mexico, and the Atlantic and Pacific Coasts. It is commonly associated with oysters harvested from warm waters during the months of April to October. This type of bacteria causes two different illnesses: primary septicemia and gastroenteritis. The illnesses typically occur when raw or partially cooked oysters with the bacteria are eaten. Preventing these illnesses depends upon purchasing oysters from approved, reputable suppliers and cooking them to the required minimum internal temperature.

Food Commonly Associated with the Bacteria

- Raw or partially cooked oysters

Most Common Symptoms

Primary Septicemia: *Most common*
(Most frequently occurs in people with liver disease and diabetes; usually deadly for 70 to 80 percent of this population)

- Fever and chills
- Nausea
- Skin lesions
- Diarrhea and vomiting (possible)

Gastroenteritis: *Less common*
(Occurs in otherwise healthy people)

- Diarrhea
- Abdominal cramps

Most Important Prevention Measures

- Purchase oysters from approved, reputable suppliers.
- Cook oysters to the required minimum internal temperature.
- Inform people at risk to consult a physician before regularly consuming raw or partially cooked oysters.

Intoxication

Illness: ***Bacillus cereus*** **Gastroenteritis**
(ba-CIL-us SEER-ee-us GAS-tro-EN-ter-I-tiss)

Bacteria: ***Bacillus cereus***
(ba-CIL-us SEER-ee-us)

Bacillus cereus is a spore forming bacteria found in soil. It is commonly associated with plants and cereal crops, such as rice. This type of bacteria can produce two different toxins when allowed to grow to high levels. Each causes a different type of illness. Inside food, the bacteria produce an *emetic toxin;* while inside the human intestine, the bacteria create a *diarrheal toxin.* Preventing these illnesses is dependent upon preventing bacterial growth and toxin production. This can be accomplished by cooking, holding, and cooling food properly.

Food Commonly Associated with the Bacteria

Diarrheal toxin:
- Cooked corn
- Cooked potatoes
- Cooked vegetables
- Meat products

Emetic toxin:
- Cooked rice dishes including:
  - Fried rice
  - Rice pudding

Most Common Symptoms

Diarrheal toxin:
- Watery diarrhea
- Abdominal cramps and pain
- Vomiting is absent

Emetic toxin:
- Nausea
- Vomiting

Most Important Prevention Measures

To reduce the bacteria in food:
- Cook food to required minimum internal temperatures.

To prevent the growth of the bacteria in food:
- Hold food at the proper temperature.
- Cool food properly.

Intoxication

Illness: Staphylococcal Gastroenteritis

(STAF-ul-lo-KOK-al GAS-tro-EN-ter-I-tiss)

Bacteria: *Staphylococcus aureus*

(STAF-uh-lo-KOK-us OR-ee-us)

Staphylococcus aureus is primarily found in humans—particularly in the hair, nose, throat, and sores. It is often transferred to food when people carrying this type of bacteria touch these areas and handle food without washing their hands. If allowed to grow to large numbers in food, the bacteria can produce toxins that cause the illness when eaten. Since cooking cannot destroy these toxins, it is critical to prevent bacterial growth. Practicing good personal hygiene can prevent the transfer of the bacteria to food.

Food Commonly Associated with the Bacteria

Food that requires handling during preparation, including:

- Salads containing potentially hazardous food (egg, tuna, chicken, macaroni)
- Deli meat

Most Common Symptoms

- Nausea
- Vomiting and retching
- Abdominal cramps

Most Important Prevention Measures

To prevent the transfer of the bacteria to food:

- Wash hands when necessary, especially after touching the hair, face, or body.
- Properly cover cuts on hands and arms.
- Restrict foodhandlers with infected cuts on hands or arms from working with or around food and food equipment.

To prevent the growth of the bacteria in food:

- Minimize the time food spends in the temperature danger zone.
  - Cook, hold, and cool food properly.

Intoxication

Illness: Botulism *(BOT-chew-liz-um)*

Bacteria: *Clostridium botulinum*

(klos-TRID-ee-um BOT-chew-line-um)

While *Clostridium botulinum* forms spores that can be found in almost any food, it is commonly associated with produce grown in the soil, such as onions, potatoes, and carrots. This type of bacteria does not grow well in refrigerated or highly acidic food, nor in food with low water activity. However, it does grow without oxygen and can produce a deadly toxin when food is temperature abused. Without treatment, death is likely. Holding, cooling, and reheating food properly inhibits the growth of the bacteria and reduces the potential for illness.

Food Commonly Associated with the Bacteria

- Improperly canned food
- Reduced-oxygen-packaged (ROP) food
- Temperature-abused vegetables such as:
  - Baked potatoes
  - Untreated garlic-and-oil mixtures

Most Common Symptoms

Initially:

- Nausea and vomiting

Later:

- Weakness
- Double vision
- Difficulty speaking and swallowing

Most Important Prevention Measures

- Hold, cool, and reheat food properly.
- Inspect canned food for damage.

Toxin-Mediated Infection

Illness: *Clostridium perfringens* Gastroenteritis

(klos-TRID-ee-um per-FRIN-jins GAS-tro-EN-ter-I-tiss)

Bacteria: *Clostridium perfringens*

(klos-TRID-ee-um per-FRIN-jins)

Clostridium perfringens is found naturally in soil where it forms spores that allow it to survive. It is also carried in the intestines of both animals and humans. People become ill after eating this type of bacteria, which produces toxins once inside their intestines. While *Clostridium perfringens* does not grow at refrigeration temperatures, it grows *very rapidly* in food in the temperature danger zone. To prevent illness, it is critical to keep the bacteria from growing, especially when holding, cooling, and reheating food.

Food Commonly Associated with the Bacteria

- Meat
- Poultry
- Dishes made with meat and poultry, such as:
  - Stews
  - Gravies

Commercially prepared food is not often involved in outbreaks.

Most Common Symptoms

- Diarrhea
- Severe abdominal pain

Nausea is uncommon and fever and vomiting are absent.

Most Important Prevention Measures

To prevent the growth of the bacteria (especially in meat dishes):

- Cool and reheat food properly.
- Hold food at the proper temperature.

Toxin-Mediated Infection

Illness: Hemorrhagic Colitis *(hem-or-AJ-ik ko-LI-tiss)*

Bacteria: Shiga toxin-producing *Escherichia coli*

(ess-chur-EE-kee-UH KO-LI)
(including: O157:H7, O26:H11, O111:H8, O158:NM)

Shiga toxin-producing *E. coli* is naturally found in the intestines of cattle, which can contaminate the meat during the slaughtering process. Although it has been associated with contaminated produce, it is more commonly associated with undercooked ground beef. A person needs to consume only a small amount of this type of bacteria to become ill. Once eaten, it produces toxins in the intestines, which cause the illness. It is critical to cook food properly and avoid cross-contamination to prevent illness.

Food Commonly Associated with the Bacteria

- Ground beef (raw and undercooked)
- Contaminated produce

Most Common Symptoms

- Diarrhea (eventually becomes bloody)
- Abdominal cramps
- Severe cases can result in hemolytic uremic syndrome (HUS)

The bacteria can be present in the feces of infected individuals for several weeks after symptoms appear.

Most Important Prevention Measures

To reduce the bacteria in food:

- Cook food, particularly ground beef, to required minimum internal temperatures.

To prevent the transfer of the bacteria to food:

- Prevent cross-contamination between raw meat and ready-to-eat food.
- Exclude employees from the establishment if:
  - They have diarrhea.
  - They have been diagnosed with hemorrhagic colitis.

VIRUSES

Viruses are the smallest of the microbial contaminants. While a virus cannot reproduce in food, once inside a human cell, it will produce more viruses.

Viruses share some basic characteristics:

- Some may survive freezing.
- They can be transmitted from person to person, from people to food, and from people to food-contact surfaces.
- They usually contaminate food through a foodhandler's improper personal hygiene.
- They can contaminate both food and water supplies.
- They are classified as infections.

Practicing good personal hygiene is an important way to prevent foodborne viruses from contaminating food. It is especially important to minimize bare-hand contact with ready-to-eat food.

Key Point

Viruses can be transmitted from person to person, from people to food, and from people to food-contact surfaces.

Key Point

Practicing good personal hygiene and minimizing bare-hand contact with ready-to-eat food are important ways to prevent the contamination of food by foodborne viruses.

Major Foodborne Illnesses Caused by Viruses

There are two major foodborne illnesses caused by viruses. For each illness, it is important to understand its common source, the food commonly associated with it, its most common symptoms, and the most important measures that can be taken to prevent the illness from occurring. The table on the following page provides a high-level overview of these characteristics. While not all-inclusive, this information will help you see the similarities and differences that will make it easier to remember each illness.

Viral Foodborne Illnesses

Major Foodborne Illnesses Caused by Viruses					
Illness Characteristics	**Type of Illness**				
	Infection			**Intoxication**	**Toxin-Mediated Infection**
	Hepatitis A	Norovirus			
Commonly Associated Food					
Poultry					
Eggs					
Meat					
Fish					
Shellfish	x	x			
Ready-to-eat food	x	x			
Produce					
Cereal crops					
Dairy					
Contaminated water	x	x			
Most Common Symptoms					
Diarrhea		x			
Abdominal pain/cramps	x	x			
Nausea	x	x			
Vomiting		x			
Fever	x				
Headache					
Most Important Prevention Measures					
Proper handwashing	x	x			
Proper cooking					
Proper holding					
Proper cooling					
Proper reheating					
Approved suppliers	x	x			
Exclude foodhandlers	x	x			
Prevent cross-contamination	x				

Infection

Illness: Hepatitis A *(HEP-a-TI-tiss)*

Virus: Hepatitis A *(HEP-a-TI-tiss)*

Hepatitis A is primarily found in the feces of people infected with the virus. While water and many types of food can become contaminated, the virus is more commonly associated with ready-to-eat food items. It has also been found in shellfish contaminated by sewage. Hepatitis A is often transferred to food when infected foodhandlers touch food or equipment with fingers containing feces. It is only necessary to consume a small amount of the virus to become ill. Proper handwashing is critical to preventing the illness, since cooking does not destroy the hepatitis A virus.

Food Commonly Associated with the Virus

- Ready-to-eat food, including:
  - Deli meat
  - Produce
  - Salads
- Raw and partially cooked shellfish

Most Common Symptoms

Initially:

- Fever (mild)
- General weakness
- Nausea
- Abdominal pain

Later:

- Jaundice

An infected person may not show symptoms for weeks, but can be very infectious.

Most Important Prevention Measures

To prevent the transfer of the virus to food:

- Wash hands properly.
- Exclude employees from the establishment who have jaundice or have been diagnosed with hepatitis A.
- Minimize bare-hand contact with ready-to-eat food.

Other prevention measures:

- Purchase shellfish from approved, reputable suppliers.
- Inform high-risk populations to consult a physician before regularly consuming raw or partially cooked shellfish.

Infection

Illness: Norovirus Gastroenteritis *(NOR-o-VI-rus GAS-tro-EN-ter-I-tiss)*

Virus: Norovirus *(NOR-o-VI-rus)*

Norovirus is primarily found in the feces of people infected with the virus. It has also been found in contaminated water. Like hepatitis A, it is commonly associated with ready-to-eat food. The virus is very contagious and is often transferred to food when infected foodhandlers touch the food with fingers containing feces. Consuming even a small amount of the virus can lead to infection. People become contagious within a few hours of eating contaminated food. Proper handwashing is essential to prevent the illness. It is also critical to prevent foodhandlers from working with food if they have symptoms related to the illness.

Food Commonly Associated with the Virus

- Ready-to-eat food
- Shellfish contaminated by sewage

Most Common Symptoms

- Vomiting
- Diarrhea
- Nausea
- Abdominal cramps

Norovirus is often present in a person's feces for days after symptoms have ended.

Most Important Prevention Measures

To prevent the transfer of the virus to food:

- Exclude foodhandlers with diarrhea and vomiting from the establishment.
- Exclude employees who have been diagnosed with norovirus from the establishment.
- Wash hands properly.

Other prevention measures:

- Purchase shellfish from approved, reputable suppliers.

Key Point

Parasites are typically passed to humans through an animal host.

PARASITES

Parasites are living organisms that need a host to survive. They infect many animals, such as cows, chicken, pigs, and fish, and can be transmitted to humans. Parasites are larger than bacteria, but are still very small—often microscopic. They are a hazard to both food and water.

Major Foodborne Illnesses Caused by Parasites

There are four major foodborne illnesses caused by parasites. They are all classified as infections. For each illness, it is important to understand its common source, the food commonly associated with it, its most common symptoms, and the most important measures that can be taken to prevent the illness from occurring. The table on the following page provides a high-level overview of these characteristics. While not all-inclusive, this information will help you see the similarities and differences that will make it easier to remember each illness.

Parasitic Foodborne Illnesses

Major Foodborne Illnesses Caused by Parasites

Illness Characteristics	Type of Illness					
	Infection				Intoxication	Toxin-Mediated Infection
	Anisakiasis	Cyclosporiasis	Cryptosporidiosis	Giardiasis		
Commonly Associated Food						
Poultry						
Eggs						
Meat						
Fish	x					
Shellfish	x					
Ready-to-eat food						
Produce		x	x			
Cereal crops						
Dairy						
Contaminated water		x	x	x		
Most Common Symptoms						
Diarrhea	x	x	x			
Abdominal pain/cramps	x	x	x	x		
Nausea	x	x	x	x		
Vomiting	x					
Fever		x		x		
Headache						
Most Important Prevention Measures						
Proper handwashing		x	x	x		
Proper cooking	x					
Proper holding						
Proper cooling						
Proper reheating						
Approved suppliers	x	x	x			
Exclude foodhandlers		x	x	x		
Prevent cross-contamination	x					

Infection

Illness: Anisakiasis *(ANN-ih-SAHK-ee-AH-sis)*

Parasite: *Anisakis simplex*
(ANN-ih-SAHK-iss SIM-plex)

Anisakis simplex is a worm-like parasite found in certain fish and shellfish. An illness can develop when raw or undercooked seafood containing the parasite is eaten. It can be either invasive or noninvasive. In its noninvasive form, the person coughs the parasite from the body. In the invasive form, the parasite penetrates the lining of the stomach or small intestine and must be surgically removed. To prevent illness, it is critical to cook seafood properly. It is important to purchase seafood from an approved, reputable supplier.

Food Commonly Associated with the Parasite

Raw and undercooked:

- Herring
- Cod
- Halibut
- Mackerel
- Pacific salmon

Most Common Symptoms

Non-invasive:

- Tingling in throat
- Coughing up worms

Invasive:

- Stomach pain
- Nausea
- Vomiting
- Diarrhea

Most Important Prevention Measures

- Cook fish to required minimum internal temperatures.
- Purchase fish from approved, reputable suppliers.

If fish will be served raw or undercooked:

- Purchase sushi-grade fish.
- Ensure sushi-grade fish has been frozen by the supplier to the proper time-temperature requirements.

Infection

Illness: Cyclosporiasis *(SI-klo-spor-I-uh-sis)*

Parasite: *Cyclospora cayetanensis*

(SI-klo-SPOR-uh KI-uh-te-NEN-sis)

Cyclospora cayetanensis is a parasite that has been found in contaminated water and has been associated with produce irrigated or washed with contaminated water. It can also be found in the feces of people infected with the parasite. Foodhandlers can transfer the parasite to food when they touch it with fingers containing feces. For this reason, foodhandlers with diarrhea must be excluded from the establishment. It is also critical to purchase produce from an approved, reputable supplier.

Food Commonly Associated with the Parasite

- Produce irrigated or washed with water containing the parasite

Most Common Symptoms

- Nausea (mild to severe)
- Abdominal cramping
- Mild fever
- Diarrhea alternating with constipation

Symptoms are more severe in people with weakened immune systems.

Most Important Prevention Measures

- Purchase produce from approved, reputable suppliers.

To prevent the transfer of the parasite to food:

- Exclude foodhandlers with diarrhea from the establishment.
- Wash hands properly to minimize risk of cross-contamination.

Infection

Illness: Cryptosporidiosis *(KRIP-TOH-spor-id-ee-O-sis)*

Parasite: *Cryptosporidium parvum*

(KRIP-TOH-spor-ID-ee-um PAR-vum)

Cryptosporidium parvum is a parasite that has been found in contaminated water, produce that has been irrigated with contaminated water, and cows and other herd animals. It can also be found in the feces of people infected with the parasite. Foodhandlers can transfer the parasite to food when they touch it with fingers containing feces. It is common for the parasite to be spread from person to person in day-care and medical communities. For this reason, proper handwashing is essential to prevent the illness. It is also critical to purchase produce from an approved, reputable supplier.

Food Commonly Associated with the Parasite

- Untreated or improperly treated water
- Contaminated produce

Most Common Symptoms

- Watery diarrhea
- Stomach cramps
- Nausea
- Weight loss

Symptoms will be more severe in people with weakened immune systems.

Most Important Prevention Measures

- Purchase produce from approved, reputable suppliers.
- Use properly treated water.

To prevent the transfer of the parasite to food:

- Exclude foodhandlers with diarrhea from the establishment.
- Wash hands properly to minimize the risk of cross-contamination.

Infection

Illness: Giardiasis *(jee-are-dee-AH-sis)*

Parasite: *Giardia duodenalis*

(jee-ARE-dee-uh do-WAH-den-AL-is)
(also known as *G. lamblia,* or *G. intestinalsis*)

Giardia duodenalis is a parasite that has been found in improperly treated water. It can be found in the feces of infected people. Foodhandlers can transfer the parasite to food when they touch it with fingers contaminated with feces. It is common for the parasite to be spread from person to person in day-care centers. For this reason, proper handwashing is essential to prevent the illness. It is also critical to use water that has been properly treated.

Food Commonly Associated with the Parasite

- Improperly treated water

Most Common Symptoms

Initially:

- Fever

Later:

- Loose stool
- Abdominal cramps
- Nausea

Most Important Prevention Measures

- Use properly treated water.

To prevent the transfer of the parasite to food:

- Exclude foodhandlers with diarrhea from the establishment.
- Wash hands properly to minimize risk of cross-contamination.

FUNGI

Fungi range in size from microscopic single-celled organisms to very large multicellular organisms. They are found naturally in air, soil, plants, water, and some food. **Molds, yeasts,** and mushrooms are examples of fungi. (Mushroom toxins are discussed in Chapter 3.)

Molds

Individual mold cells can usually be seen only with a microscope. However, fuzzy or slimy mold colonies, consisting of a large number of cells, are often visible to the naked eye. Bread mold is an example. The spores produced by molds are not the same as the spores produced by bacteria. Molds use spores for reproduction.

Molds are responsible for the spoilage of food. This spoilage results in discoloration and the formation of odors and off-flavors. Molds are able to grow on almost any food at almost any storage temperature. They can also grow in environments that are moist or dry, have a high or low pH, and are salty or sweet. They typically prefer to grow in and on acidic food with low water activity. Molds often spoil fruit, vegetables, meat, cheese, and bread because of their water activity and pH.

Some molds produce toxins that can cause allergic reactions, nervous system disorders, and kidney and liver damage. For example, aflatoxin, produced by the molds *Aspergillus flavus* and *Aspergillus parasticus,* can cause liver disease.

Food such as corn and corn products, peanuts and peanut products, cottonseed, milk, and tree nuts (such as Brazil nuts, pecans, pistachio nuts, and walnuts) have been associated with aflatoxins.

Basic Characteristics of Foodborne Molds

Molds share some basic characteristics:

- They spoil food and sometimes cause illness.
- They grow under almost any condition but grow well in acidic food with low water activity.

Exhibit 2h

Mold on Cheese

Food with mold that is not a natural part of the product should always be discarded.

Exhibit 2i

Yeast on Jelly

Food that has been spoiled by yeast should always be discarded.

- Freezing temperatures prevent or reduce the growth of molds, but do not destroy them.
- Some molds produce toxins such as aflatoxins.

The Food and Drug Administration (FDA) recommends cutting away any moldy areas in hard cheese—at least one inch (2.5 centimeters) around them. However, to avoid illnesses caused by mold toxins, throw out all moldy food unless the mold is a natural part of the food (e.g., cheese such as Gorgonzola, Brie, and Camembert).

While cooking can destroy mold cells and spores, some toxins can remain. For this reason, food with molds that are not a natural part of the product should always be discarded. (See *Exhibit 2h.*)

Yeasts

Some yeasts have the ability to spoil food rapidly. Carbon dioxide and alcohol are produced as yeast consumes food. Yeast spoilage may, therefore, produce a smell or taste of alcohol. Yeast may appear as a pink discoloration or slime and may bubble.

Yeasts are similar to molds in that they grow well in acidic food with low water activity, such as jellies, jams, syrup, honey, and fruit juice. Food that has been spoiled by yeast should be discarded. (See *Exhibit 2i.*)

EMERGING PATHOGENS AND ISSUES

Progress has been made in preventing foodborne illnesses. For example, typhoid fever (caused by *Salmonella* Typhi) was very common in the early twentieth century. Typhoid is now almost forgotten in the United States due to improved methods of treating drinking water and sewage, milk sanitation and pasteurization, and shellfish sanitation. Similarly, cholera and trichinosis, once very common, are now also rare in the United States.

In the past, food items involved in outbreaks were often raw or inadequately cooked meat, poultry, seafood, or unpasteurized milk. Today, some food items previously considered safe have become vehicles for foodborne pathogens. For example, it was assumed for centuries that the internal contents of an egg were sterile and safe to eat raw. Now, however, there are clear indications that, in rare cases, eggs can be contaminated internally with *Salmonella* Enteritidis. Outbreaks of salmonellosis have been traced to undercooked, contaminated eggs included in traditional recipes such as eggnog, hollandaise sauce, Caesar salad, and French toast.

As detection techniques and foodborne-illness investigations improve, new pathogens causing foodborne illness are being recognized. In addition, as the population continues to grow, more food is needed. Food is produced in greater quantities and at centralized sources. More food is shipped from greater distances, across state and national borders. It remains important for establishments to have sound practices in place. This includes purchasing from approved, reputable suppliers and using proper storage, preparation, and serving methods.

Issues Affecting Produce

Consumers are looking for healthy alternatives when they go out to eat. Year-round demand for fresh fruit and vegetables continues to increase. To meet this demand, many produce items are now imported from the global food supply. Produce and highly acidic food items that were once considered safe are now considered potential vehicles for foodborne pathogens. Most produce is typically consumed raw, so there is virtually no step during preparation to eliminate, reduce, or control pathogens. To help ensure that the produce items you serve are safe, work only with an approved, reputable supplier, and ensure safe foodhandling practices in your establishment.

Avian Flu

With a growing global food supply, the foodservice and regulatory communities continue to watch for emerging pathogens and food safety threats. Recently, outbreaks of avian influenza, or "bird flu," have occurred in certain parts of the world. The H5N1 strain of avian influenza (AI) is very contagious to birds and can cause severe illness and death in them, particularly in poultry flocks. As a result of recent outbreaks, millions of birds were destroyed in Cambodia, China, Indonesia, Japan, Laos, South Korea, Thailand, and Vietnam. Poultry and eggs from any country affected by the H5N1 outbreak are banned in the United States.

Although the risk is low, public-health officials are concerned that the H5N1 strain has the potential to infect humans. Contact with the feces, saliva, and tissue of infected birds could potentially cause infection in humans. The population most at risk for illness has been poultry workers in those areas where an outbreak has occurred. It is highly unlikely that the virus will infect persons working in non-agricultural jobs.

Although the H5N1 strain of AI has been found in the eggs and meat of infected poultry, the virus is destroyed by proper cooking and controlled through proper foodhandling procedures.

Bovine Spongiform Encephalopathy

Some food-related issues affect the food industry and the general public even though they are not considered food safety issues. One of these is bovine spongiform encephalopathy (BSE), which attacks the brain and nervous system of cattle. BSE has been linked to the human illness known as variant Creutzfeldt-Jakob disease (vCJD).

BSE was first diagnosed in cattle in the United Kingdom. It has been found in cattle in other countries as well as in the United States. To protect the public, the United States has banned imported meat products and animal protein products from countries with confirmed BSE unless the animal is less than thirty months old. The importation from these countries of live cattle over thirty months of age has also been banned. The United States also has a surveillance program to detect BSE in domestic cattle. Due to extensive preventive measures throughout the world, the risk of contracting this disease through the food supply is low.

TECHNOLOGICAL ADVANCEMENTS IN FOOD SAFETY

Food irradiation, often called cold pasteurization, is a common method for controlling foodborne microorganisms. It involves exposing food to an electron beam or gamma rays, similar to the way microwaves are used to heat food. The FDA has approved the use of irradiation on certain types of food in the United States. To date, irradiation is allowed for:

- Raw meat and meat products (to eliminate shiga toxin-producing *E.coli* and to reduce levels of *Listeria* spp., *Salmonella* spp., and *Campylobacter* spp.)
- Pork (to control the parasite *Trichinella spiralis*)
- Poultry (to control bacterial contaminants)
- Fruit and vegetables (to prevent premature maturation and to control insects)
- Herbs, spices, teas, and other dried vegetable substances (to control various microbial contaminants)

Benefits of food irradiation include:

- Reduction or elimination of pathogens and spoilage microorganisms
- Replacement of chemical treatment of food
- Extended shelf life of food

Irradiated food has only recently become widely available in the United States, despite regulatory approval and repeated endorsements by professional health and nutrition organizations. No evidence exists of harmful effects resulting from the amount or types of radiation used. Nutritional value, appearance, and taste are virtually not affected. However, some consumers remain apprehensive about food exposed to radiation, due to their limited knowledge of the irradiation process. Studies show that once the process and benefits are explained to them, consumers are more receptive to the idea of irradiation and would purchase irradiated products.

It is important to note that food irradiation does not replace proper food storage and handling practices, since food could still become unsafe if cross-contaminated. Therefore, when handling irradiated food, you should follow the same food safety practices you would follow with other food.

SUMMARY

Microorganisms are responsible for the majority of foodborne illness. Understanding how they grow, contaminate food, and affect humans is critical to understanding how to prevent the foodborne illnesses they cause.

The acronym FAT TOM—which stands for food, acidity, temperature, time, oxygen, and moisture—is the key to controlling the growth of microorganisms.

Foodborne illnesses are classified as infections, intoxications, or toxin-mediated infections. Foodborne infections result when a person eats food containing pathogens, which then grow in the intestines and cause illness. Typically, symptoms do not appear immediately. Foodborne intoxications result when a person eats food containing toxins produced by pathogens found on the food or which are the result of a chemical contamination. The toxin might also be a natural part of a plant or animal consumed. Typically, symptoms of foodborne intoxication appear quickly, within a few hours. Foodborne toxin-mediated infections result when a person eats food that contains pathogens which then produce illness-causing toxins in the intestines.

Of all foodborne microorganisms, bacteria are of greatest concern to the restaurant and foodservice manager. Under favorable conditions, bacteria can reproduce very rapidly. Although they may be resistant to low—even freezing—temperatures, they can be killed by high temperatures, such as those reached during cooking. Certain bacteria, however, can change into a different form, called spores, to protect themselves. Since spores are commonly found on food grown in or exposed to soil, it is important to hold, cool, and reheat food properly. This will prevent spores that might be present from reverting back to a form capable of growth and causing illness.

Viruses are the smallest of the microbial contaminants. While a virus cannot reproduce in food once consumed, it will cause illness. Practicing good personal hygiene and minimizing bare-hand contact with ready-to-eat food is an important defense against viral foodborne illnesses.

Parasites are organisms that need to live in a host organism to survive. This includes animals such as cows, chickens, pigs, and fish. Proper cooking and freezing can kill parasites.

Fungi, such as molds and yeasts, are generally responsible for spoiling food. Some molds, however, can produce harmful toxins. For this reason, food containing mold should always be discarded unless the mold is a natural part of the product. Yeasts can spoil food rapidly. Food spoiled by yeast should also be discarded.

Apply Your Knowledge

1. Based on the information given, what type of microorganism caused the illness—bacteria, virus, or parasite?
2. What is the name of the microorganism most likely responsible for the outbreak?
3. Is this illness an infection, intoxication, or toxin-mediated infection?

A Case in Point

A day-care center decided to prepare stir-fried rice to serve for lunch the next day. The rice was cooked to the proper temperature at 1:00 p.m. It was then covered and placed on a countertop where it was allowed to cool at room temperature. At 6:00 p.m., the cook placed the rice in the refrigerator. At 9:00 a.m. the following day, the rice was combined with the other ingredients for stir-fried rice and cooked to 165°F (74°C) for at least fifteen seconds. The cook covered the rice and left it on the range until noon when she reheated it. Within an hour of eating the rice, several of the children complained that they were nauseous and began to vomit.

For answers, please turn to the Answer Key.

Apply Your Knowledge

Use these questions to review the concepts presented in this chapter.

Discussion Questions

1. Why should potentially hazardous food *not* be left in the temperature danger zone?
2. What is the difference between a foodborne infection, a foodborne intoxication, and a toxin-mediated infection?
3. How can an outbreak of salmonellosis be prevented?
4. What are some basic characteristics of foodborne viruses?
5. What two FAT TOM conditions for the growth of microorganisms are easiest for an establishment to control? Give examples of how each can be controlled.

For answers, please turn to the Answer Key.

Apply Your Knowledge

Use these questions to test your knowledge of the concepts presented in this chapter.

Multiple-Choice Study Questions

1. Foodborne microorganisms grow well at temperatures between
   A. 32°F and 70°F (0°C and 21°C).
   B. 38°F and 155°F (3°C and 68°C).
   C. 41°F and 135°F (5°C and 57°C).
   D. 70°F and 165°F (21°C and 74°C).
2. Which condition does *not* typically support the growth of microorganisms?
   A. Moisture
   B. Protein
   C. Time
   D. High acidity
3. Which microorganism is primarily found in the hair, nose, and throat of humans?
   A. Hepatitis A virus
   B. *Giardia duodenalis*
   C. *Staphylococcus aureus*
   D. *Clostridium botulinum*
4. While commonly associated with ground beef, which microorganism has also been associated with contaminated lettuce?
   A. *Salmonella* spp.
   B. *Campylobacter jejuni*
   C. Shiga toxin-producing *E. coli*
   D. Norovirus
5. A person who has campylobacteriosis may experience
   A. chills and skin lesions.
   B. weakness and double vision.
   C. headache and bloody diarrhea.
   D. diarrhea alternating with constipation.

Apply Your Knowledge

Multiple-Choice Study Questions

6. Which practice can help prevent salmonellosis?
   A. Purchasing sushi-grade fish
   B. Inspecting canned food for damage
   C. Cooking poultry and eggs to the proper temperature
   D. Purchasing oysters from reputable, approved suppliers

7. Which practice can help prevent staphylococcal gastroenteritis?
   A. Prohibiting the use of unpasteurized dairy products
   B. Controlling flies inside and outside the establishment
   C. Purchasing shellfish from reputable, approved sources
   D. Restricting foodhandlers with infected cuts from working around food

8. Which microorganism has been associated with produce irrigated with contaminated water?
   A. *Anisakis simplex*
   B. *Vibrio parahaemolyticus*
   C. *Cyclospora cayetanensis*
   D. *Clostridium perfringens*

9. Which statement about foodborne mold is *not* true?
   A. Some types produce toxins.
   B. It grows well in acidic food.
   C. Freezing temperatures destroy it.
   D. It grows well in food with little moisture.

10. The type of illness that results when a person eats food containing pathogens which then grow in the intestines and cause illness is called a
    A. foodborne infection.
    B. foodborne intoxication.
    C. foodborne toxin-mediated infection.
    D. foodborne gastroenteritis.

For answers, please turn to the Answer Key.

Take It Back*

The following food safety concept from this chapter should be taught to your employees:

- How microorganisms make food unsafe

The tools below can be used to teach these concepts in fifteen minutes or less using the directions below. Each tool includes content and language appropriate for employees. Choose the tool or tools that work best.

Tool #1: ServSafe Video 2: *Overview of Foodborne Microorganisms and Allergens*	**Tool #2:** *ServSafe Employee Guide*	**Tool #3:** ServSafe Posters and Quiz Sheets	**Tool #4:** ServSafe Fact Sheets and Optional Activities
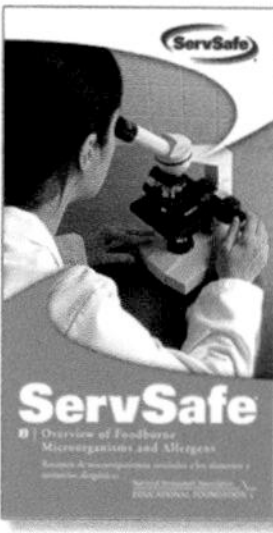		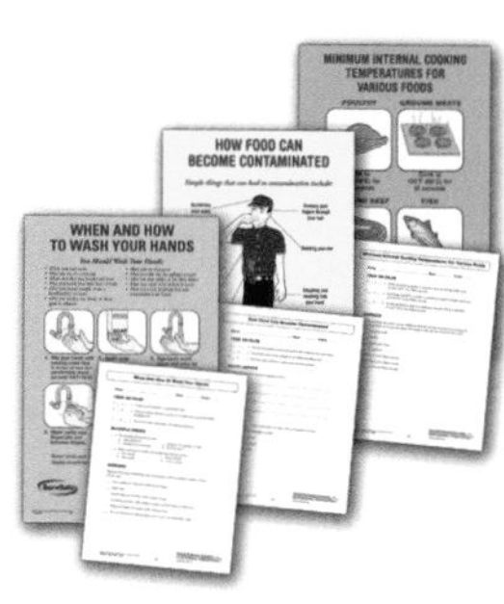	

How Microorganisms Make Food Unsafe

Tool #1	Tool #2	Tool #3	Tool #4
Video segment on microbial contaminants 1 Show employees the segment. 2 Ask employees to identify the four types of microorganisms and explain how bacteria can make food unsafe.	**Section 1** **What a Foodborne Illness Is** Discuss with employees the four types of microorganisms. Explain how bacteria can make food unsafe.		

* Visit the Food Safety Resource Center at *www.ServSafe.com/FoodSafety/resource* to download free posters, quiz sheets, fact sheets, and optional activities and to learn how to obtain *Employee Guides* and videos/DVDs.

Notes

ADDITIONAL RESOURCES

Articles and Texts

Belay, Ermias D., James J. Seyvar, Wun-Ju Shieh, et. al. 2005. Variant Creutzfeldt-Jakob Disease Death, United States. *Emerging Infectious Disease.* 11 (9): 1351.

Doyle, Michael, Larry R. Buechat, and Thomas J. Montville, eds. *Food Microbiology: Fundamentals and Frontiers, 2nd edition.* Washington: ASM Press, 2001.

Jay, James M., David A. Golden, and Martin J. Loessner. *Modern Food Microbiology, 7th edition.* New York: Springer, 2005.

Heymann, David, editor. *Control of Communicable Disease Manual, 18th edition.* Washington: American Public Health Association Publication, 2004.

Miller, R.B. *Electronic Irradiation of Foods: An Introduction to the Technology.* New York: Springer, 2005.

Pitt, J.I., Alisa D. Hocking, and Ulf Thrane. *Advances in Food Mycology.* Proceedings of the International Commission on Food Mycology. New York: Kluwer Academic/Plenum Publishers, 2005.

Ryser, Elliot T. and Elmer H. Marth. *Listeria, Listeriosis and Food Safety, 3rd edition.* New York: Marcel Dekker, Incorporated, 2005.

Web Sites

American Council on Science and Health

www.acsh.org

The American Council on Science and Health (ACSH) is a consumer-education consortium concerned with issues related to food, nutrition, chemicals, pharmaceuticals, lifestyle, the environment, and health. ACSH advocates sound science and the belief that Americans' continued physical and economic well-being depends on the advancement of scientific, medical, and technological research and innovation. Visit this Web site for publications, articles, editorials on food safety, and other timely health-related topics.

American Egg Board

www.aeb.org

The American Egg Board (AEB) is the U.S. egg producers' link to the consumer. This Web site provides resources for the restaurant and foodservice industry, including ways to ensure safe receiving, storage, and preparation of eggs, and general information about eggs and egg products.

Bad Bug Book

www.cfsan.fda.gov/~mow/intro.html

Produced by the Food and Drug Administration's Center for Food Safety and Applied Nutrition (FDA CFSAN), this online handbook provides basic facts about foodborne pathogenic microorganisms and biological toxins. The information brings together information from the U.S. Food and Drug Administration, the Centers for Disease Control and Prevention, the U.S. Department of Agriculture, Food Safety Inspection Service, and the National Institutes of Health.

Centers for Disease Control and Prevention

www.cdc.gov

The Centers for Disease Control and Prevention (CDC) are focused on American health promotion, prevention, and preparedness. The CDC apply research and findings to improve people's daily lives and respond to health emergencies. This Web site provides information on disease and disease prevention, including foodborne illness, its causes, public-health impact, and methods for control.

Center for Infectious Disease Research & Policy

www.cidrap.umn.edu

The Center for Infectious Disease Research & Policy (CIDRAP) at the University of Minnesota functions to prevent illness and death from infectious diseases through epidemiologic research and the rapid translation of scientific information into real-world practical applications and solutions. The Center's Web site provides access to information on food safety and foodborne illness and news stories on food security and foodborne-illness surveillance.

Food and Drug Administration

www.fda.gov

The U.S. Food and Drug Administration (FDA) is responsible for promoting and protecting public health by helping safe and effective products reach consumers, monitoring products for continued safety, and helping the public obtain accurate, science-based information needed to improve health. Visit this Web site for information about all programs regulated by the FDA, including food-product labeling, bottled water, and all food products except meat and poultry.

FDA Center for Food Safety and Applied Nutrition

www.cfsan.fda.gov/list.html

As the center within the Food and Drug Administration (FDA) responsible for food safety, the Center for Food Safety and Applied Nutrition (CFSAN) promotes and protects the public health by researching and implementing guidelines, policies, and standards to ensure that food is safe, nutritious, wholesome, and properly labeled. This Web site provides information relevant to all aspects of food safety and security, including corresponding guidelines, policies, and standards.

Gateway to Government Food Safety Information

www.foodsafety.gov

This Web site provides links to selected government food safety-related information.

Institute of Food Technologists

www.ift.org/cms

The Institute of Food Technologists (IFT) is a nonprofit international scientific society with 22,000 members working in food science, technology and related professions in the industry, academia, and government. As the society for food science and technology, IFT brings sound science to the public discussion of food issues. Informational scientific reports and summaries can be found on this Web site, as well as other useful publications.

International Food Information Center

www.ific.org

Working with a roster of scientific experts who serve to translate research into understandable and useful information, the International Food Information Center (IFIC) collects and disseminates scientific information on food safety, nutrition, and health. This Web site provides information on topics in food safety and health.

International Fresh-Cut Produce Association

www.fresh-cuts.org

The International Fresh-Cut Produce Association (IFPA) represents processors, distributors, retail, and foodservice buyers of fresh-cut produce and companies that supply goods and services in the fresh-cut industry. IFPA provides its members with the knowledge and technical information necessary to deliver convenient, safe, and wholesome food. This Web site offers a detailed list of frequently asked questions about the fresh-cut produce industry and has a contact feature in which to ask a question. Free publications on food safety and best practices for fresh-cut produce are also available.

National Center for Infectious Diseases

www.cdc.gov/ncidod/index.htm

The National Center for Infectious Disease (NCID), one of the centers of the CDC, works in partnership with local and state public-health officials, other federal agencies, medical and public-health professional associations, infectious-disease experts from academic and clinical practice, and international and public-service organizations to fulfill its mission to prevent illness, disability, and death caused by infectious diseases in the United States. This Web site houses information on bacterial, viral, and parasitic diseases.

National Chicken Council

www.nationalchickencouncil.com

The National Chicken Council (NCC) is the national, nonprofit trade association for the U.S. chicken industry, representing integrated chicken producer/processors, poultry distributors, and allied firms. Visit this Web site for consumer consumption behaviors, product information, and research related to poultry-associated foodborne illnesses.

The Produce Marketing Association

www.pma.com

The Produce Marketing Association (PMA) is a nonprofit association serving members who market fresh fruit, vegetables, and related products worldwide. Their members represent the production, distribution, retail, and foodservice sectors of the industry. Visit this Web site for articles and publications on produce food safety information within the foodservice and retail industry.

United Fresh Fruit & Vegetable Association

www.uffva.org

The United Fresh Fruit & Vegetable Association is a national trade organization that represents the interests of growers, shippers, brokers, wholesalers, and distributors of produce. The association promotes increased produce consumption and provides scientific and technical expertise to all segments of the industry. This Web site contains news updates and free publications and lists upcoming events and initiatives on produce safety and quality.

Documents and Other Resources

Avian Influenza

www.cdc.gov/flu/avian

The Centers for Disease Control and Prevention (CDC) provide an expert resource page on avian influenza (AI). Visit this Web site to find key facts about the virus and disease, frequently asked questions, background on known outbreaks, information for specific groups, and links to additional references and resources.

Avian Influenza

www.who.int/csr/disease/avian_influenza/en

Visit the World Health Organization's Web site for information about avian influenza (AI) around the world. Available information includes a list of frequently asked questions, status of the global pandemic threat, outbreak summaries, guidelines and recommendations, and many other references and resources.

Avian Influenza and Food Safety

www.cfsan.fda.gov/~dms/avfluqa.html

Visit this Web site from the Food and Drug Administration's Center for Food Safety and Applied Nutrition (FDA CFSAN) for questions and answers on avian influenza (AI) and its impact on food safety.

Bovine Spongiform Encephalopathy

www.cdc.gov/ncidod/dvrd/bse/index.htm

The Centers for Disease Control and Prevention (CDC) provide background information on bovine spongiform encephalopathy (BSE), or mad cow disease, on this Web site, along with references and links to other Web resources.

Bovine Spongiform Encephalopathy

www.fda.gov/oc/opacom/hottopics/bse.html

The Food and Drug Administration's (FDA) resource Web site provides comprehensive information about bovine spongiform encephalopathy (BSE), or mad cow disease, including a fact sheet, a list of frequently asked questions, links to government actions to strengthen safeguards, and information for specific groups.

Emerging Infectious Diseases

www.cdc.gov/ncidod/eid/index.htm

An online journal, *Emerging Infectious Diseases* represents the scientific communications component of Centers for Disease Control and Prevention's (CDC) efforts against the threat of emerging infections. The journal relies on a broad, international authorship base to publish reports of interest to researchers in infectious diseases and related sciences and reports on laboratory and epidemiologic findings within a broader public-health perspective. Visit this Web site for articles published about foodborne disease in the Untied States and around the world.

Food Irradiation

www.cdc.gov/ncidod/dbmd/diseaseinfo/foodirradiation.htm

Visit this Web site to find a listing of frequently asked questions regarding food irradiation from the Centers for Disease Control and Prevention (CDC).

Morbidity and Mortality Weekly Report

www.cdc.gov/mmwr

The *Morbidity and Mortality Weekly Report* (*MMWR*) prepared by the Centers for Disease Control and Prevention, is available, free of charge, on this Web site. The report contains data based on weekly accounts of reportable diseases from state health departments. Visit this Web site to search report databases for synopses of actual foodborne-illness outbreaks.

National Shellfish Sanitation Program Guide for the Control of Molluscan Shellfish

www.cfsan.fda.gov/~ear/nss2-toc.html

The National Shellfish Sanitation Program (NSSP) Guide for the Control of Molluscan Shellfish, available at this Web site, consists of a Model Ordinance, supporting guidance documents, recommended forms, and other related materials associated with the program. The Model Ordinance includes guidelines to ensure that the shellfish produced in the United States in compliance with the guidelines are safe and sanitary. In addition, the Model Ordinance provides readily adoptable standards and administrative practices necessary for the sanitary control of molluscan shellfish.

Retail Food Stores and Food Service Establishments: Food Security Preventive Measures Guidance

www.cfsan.fda.gov/~dms/secguid5.html

Visit this Web site for guidance that represents the Food and Drug Administration's (FDA) current approach on the different measures that restaurant and foodservice establishments may use to minimize the risk of tampering or other criminal activities to the food under their control.

Variant Creutzfeldt-Jakob Disease

www.cdc.gov/ncidod/dvrd/vcjd/index.htm

Visit this Centers for Disease Control and Prevention (CDC) Web site to find information about the variant Creutzfeldt-Jakob disease; its possible connection to bovine spongiform encephalopathy (BSE), or mad cow disease; prevention measures; and additional references and resources.

Notes

10
HOTEL SIZE
SOAP
PADS
Manual Pot and Pan Detergent
Détergent pour le nettoyage à la main des casseroles
Detergente para el Lavado a Mano de Ollas y Sartenes
3.78 L / 1 GALLON (US)

3 Contamination, Food Allergens, and Foodborne Illness

Inside this chapter:

- Biological Contamination
- Chemical Contamination
- Physical Contamination
- The Deliberate Contamination of Food
- Food Allergens

After completing this chapter, you should be able to:

- Identify biological, chemical, and physical contaminants.
- Identify methods to prevent biological, chemical, and physical contamination.
- Identify the eight most common allergens, associated symptoms, and methods of prevention.

Key Terms

- Biological contaminants
- Biological toxins
- Chemical contaminants
- Toxic-metal poisoning
- Physical contaminants
- Food security
- Food allergy

Apply Your Knowledge

Check to see how much you know about the concepts in this chapter. Use the page references provided with each question to explore the topic.

Test Your Food Safety Knowledge

1. **True or False:** A person with ciguatera fish poisoning often sweats and experiences a burning sensation in the mouth. *(See page 3-6.)*
2. **True or False:** Cooking can destroy the toxins in toxic, wild mushrooms. *(See page 3-12.)*
3. **True or False:** Copper utensils and equipment can cause an illness when used to prepare acidic food. *(See page 3-13.)*
4. **True or False:** When transferring a cleaning chemical to a spray bottle, it is unnecessary to label the bottle if the chemical is clearly visible. *(See page 3-13.)*
5. **True or False:** A person with a shellfish allergy who unknowingly eats soup made with clam juice may experience a tightening in the throat. *(See page 3-17.)*

For answers, please turn to the Answer Key.

INTRODUCTION

Food is considered contaminated when it contains hazardous substances. These substances may be biological, chemical, or physical. The most common food contaminants are **biological contaminants** that belong to the microworld—bacteria, parasites, viruses, and fungi. Most foodborne illnesses result from these contaminants, but biological and chemical toxins are also responsible for many foodborne illnesses. While these hazards pose a significant threat to food, the danger from physical hazards should also be recognized.

A thorough understanding of the causes of biological, chemical, and physical contamination will aid in the prevention of foodborne illnesses and help you keep food safe.

BIOLOGICAL CONTAMINATION

As you learned in Chapter 2, foodborne intoxication occurs when a person eats food that contains toxins. The toxin may have been produced by pathogens found in the food or may be the result of a chemical contamination. The toxin could also come from a plant or animal that was eaten.

Toxins in seafood, plants, and mushrooms are responsible for many cases of foodborne illness in the United States each year. Most of these **biological toxins** are a natural part of the plant or animal. Some occur in animals as a result of their diet.

Fish Toxins

Some fish toxins are systemic—produced by the fish itself. Pufferfish, moray eels, and freshwater minnows all produce systemic toxins. Cooking does not destroy them. Due to the extreme risk it poses, pufferfish should not be served unless the chef has been licensed to prepare it.

While some fish toxins are systemic, microorganisms on fish produce others. Some occur when predatory fish consume smaller fish that have eaten the toxin.

Major Foodborne Illnesses Caused by Fish Toxins

There are two major foodborne illnesses caused by fish toxins. It is important to understand the most common types of fish associated with each toxin, the most common symptoms, and the most important measures that can be taken to prevent an illness from occurring. The table on the next page provides a high-level overview of these characteristics. While not all-inclusive, this information will help you see the similarities and differences that will make it easier to remember each illness.

In general, purchasing fish from an approved, reputable supplier is the best way to guard against an illness associated with fish toxins. Additionally, check the temperature of fish upon delivery, making sure it has been received at 41°F (5°C) or lower. Refuse product that has been thawed and refrozen, as this is a sign that the fish has been time-temperature abused.

Fish Toxin Illnesses

Major Foodborne Illnesses Caused by Fish Toxins

Illness Characteristics	Type of Illness: Infection	Intoxication: Scombroid poisoning	Intoxication: Ciguatera fish poisoning	Intoxication	Toxin-Mediated Infection
Commonly Associated Food					
Fish		x	x		
Shellfish					
Most Common Symptoms					
Diarrhea		x			
Abdominal pain/cramps					
Nausea			x		
Vomiting		x	x		
Fever					
Headache		x			
Neurological symptoms		x	x		
Most Important Prevention Measures					
Proper handwashing					
Proper cooking					
Proper holding		x			
Proper cooling					
Proper reheating					
Approved suppliers		x	x		
Exclude foodhandlers					
Prevent cross-contamination					

Intoxication

Illness: Scombroid poisoning *(SCOM-broyd)*

Toxin: Histamine *(HISS-ta-meen)*

Scombroid poisoning—also known as histamine poisoning—is an illness caused by consuming high levels of histamine in scombroid and other species of fish. When the fish are time-temperature abused, bacteria on the fish produce histamine. It cannot be destroyed by freezing, cooking, smoking, or curing. For this reason, it is critical to purchase scombroid fish and other affected species from approved, reputable suppliers.

Food Commonly Associated with the Toxin

- Tuna
- Bonito
- Mackerel
- Mahi mahi

Most Common Symptoms

Initially:

- Reddening of the face and neck
- Sweating
- Headache
- Burning or tingling sensation in the mouth or throat

Later: (possibly)

- Diarrhea
- Vomiting

Most Important Prevention Measures

- Purchase fish from approved, reputable suppliers.
- Prevent time-temperature abuse during storage and preparation.

Intoxication

Illness: Ciguatera fish poisoning *(SIG-wa-TAIR-ah)*

Toxin: Ciguatoxin *(SIG-wa-TOX-in)*

Ciguatoxin is found in certain marine algae. It is commonly associated with predatory reef fish from the Pacific Ocean, western Indian Ocean, and the Caribbean Sea. The toxin accumulates in these fish when they consume smaller fish that have eaten the toxic algae. Ciguatoxin cannot be detected by smell or taste. Since cooking or freezing the fish will not eliminate the toxin, it is critical to purchase reef fish from approved, reputable suppliers. These suppliers take steps to ensure that the fish has been harvested from waters free of the toxic algae.

Food Commonly Associated with the Toxin

Predatory reef fish from the Pacific Ocean, western Indian Ocean, and the Caribbean Sea, including:

- Barracuda
- Grouper
- Jacks
- Snapper

Most Common Symptoms

- Reversal of hot and cold sensations
- Nausea
- Vomiting
- Tingling in fingers, lips, or toes
- Joint and muscle pain

Symptoms may last months or years depending upon the extent of the illness.

Most Important Prevention Measures

- Purchase reef fish from approved, reputable suppliers.

Exhibit 3a

Shellfish Toxins

Shellfish toxins cannot be smelled or tasted, and they are not destroyed by freezing or cooking.

Shellfish Toxins

Many toxins associated with shellfish are found in toxic marine algae. The shellfish become contaminated as they filter this toxic algae from the water. These toxins cannot be smelled or tasted, and they are not destroyed by freezing or cooking. (See *Exhibit 3a.*)

Major Foodborne Illnesses Caused by Shellfish Toxins

There are three major foodborne illnesses caused by shellfish toxins. It is important to understand the shellfish associated with each toxin, the most common symptoms, and the most important measures that can be taken to prevent the illness from occurring. The table on the next page provides a high-level overview of these characteristics. While not all-inclusive, this information will help you see the similarities and differences that will make it easier to remember each illness.

As with fish, purchasing shellfish from approved, reputable suppliers is the most important safeguard against foodborne illness.

Shellfish Toxin Illnesses

Major Foodborne Illnesses Caused by Shellfish Toxins

Illness Characteristics	Type of Illness				
	Infection	Intoxication			Toxin-Mediated Infection
		Paralytic Shellfish Poisoning (PSP)	Neurotoxic Shellfish Poisoning (NSP)	Amnesic Shellfish Poisoning (ASP)	
Commonly Associated Food					
Fish					
Shellfish		x	x	x	
Most Common Symptoms					
Diarrhea		x	x	x	
Abdominal pain/cramps				x	
Nausea		x			
Vomiting		x	x	x	
Fever					
Headache					
Neurological symptoms		x	x	x	
Most Important Prevention Measures					
Proper handwashing					
Proper cooking					
Proper holding					
Proper cooling					
Proper reheating					
Approved suppliers		x	x	x	
Exclude foodhandlers					
Prevent cross-contamination					

Intoxication

Illness: Paralytic Shellfish Poisoning (PSP) *(PAIR-ah-LIT-ik)*

Toxin: Saxitoxin *(SAX-ih-TOX-in)*

Saxitoxin is found in certain toxic marine algae in colder waters, such as those of the Pacific and New England Coasts. Shellfish, including clams, mussels, oysters, and scallops become contaminated as they filter these algae from the water. Paralytic Shellfish Poisoning (PSP) is the illness that results when these shellfish are eaten. Since the toxin cannot be smelled or tasted and it is not destroyed by cooking or freezing, it is critical to purchase shellfish from approved, reputable suppliers.

Food Commonly Associated with the Toxin

Shellfish found in colder waters such as those of the Pacific and New England Coasts. They include:

- Clams
- Mussels
- Oysters
- Scallops

Most Common Symptoms

- Numbness
- Tingling of the mouth, face, arms, and legs
- Dizziness
- Nausea
- Vomiting
- Diarrhea

Death due to paralysis may result if high levels of the toxin are consumed.

Most Important Prevention Measures

- Purchase shellfish from approved, reputable suppliers.

Intoxication

Illness: Neurotoxic Shellfish Poisoning (NSP) *(NUR-o-TOX-ik)*

Toxin: Brevetoxin *(BREV-ih-TOX-in)*

Brevetoxin is found in certain toxic marine algae in warmer waters, such as those of the west coast of Florida, the Gulf of Mexico and the Caribbean. Shellfish, including oysters, clams, and mussels, become contaminated as they filter these algae from the water. Neurotoxic Shellfish Poisoning (NSP) is the illness that results when these shellfish are eaten. Since the toxin cannot be smelled or tasted and it is not destroyed by cooking or freezing, it is critical to purchase shellfish from approved, reputable suppliers.

Food Commonly Associated with the Toxin

Shellfish found in warmer waters, such as those found in the west coast of Florida, the Gulf of Mexico, and the Caribbean. They include:

- Clams
- Mussels
- Oysters

Most Common Symptoms

- Tingling and numbness of the lips, tongue, and throat
- Dizziness
- Reversal of the sensations of hot and cold
- Vomiting
- Diarrhea

Most Important Prevention Measures

- Purchase shellfish from approved, reputable suppliers.

Intoxication

Illness: Amnesic Shellfish Poisoning (ASP) *(am-NEE-zik)*

Toxin: Domoic acid *(do-MO-ik)*

Domoic acid is a toxin found in certain toxic marine algae in cooler coastal waters, such as those of the Pacific Northwest and the east coast of Canada. Shellfish, including oysters, scallops, clams, and mussels become contaminated as they filter these algae from the water. Amnesic Shellfish Poisoning (ASP) is the illness that results when these shellfish are eaten. Since the toxin cannot be smelled or tasted and it is not destroyed by cooking or freezing, it is critical to purchase shellfish from approved, reputable suppliers.

Food Commonly Associated with the Toxin

Shellfish found in the coastal waters of the Pacific Northwest and the east coast of Canada. They include:

- Clams
- Mussels
- Oysters
- Scallops

Most Common Symptoms

Initially:

- Vomiting
- Diarrhea
- Abdominal pain

Later: (possibly)

- Confusion
- Memory loss
- Disorientation
- Seizure
- Coma

The severity of symptoms depends on the amount of toxin consumed and the health of the individual.

Most Important Prevention Measures

- Purchase shellfish from approved, reputable suppliers.

Exhibit 3b

Mushroom Toxins

Foodborne illnesses associated with mushrooms are almost always caused by the consumption of toxic, wild mushrooms collected by amateur mushroom hunters.

Mushroom Toxins

Foodborne illnesses associated with mushrooms are almost always caused by the consumption of toxic, wild mushrooms collected by amateur mushroom hunters. (See *Exhibit 3b.*) Most cases occur when toxic mushroom species are confused with edible species. The symptoms of intoxication vary depending on the species consumed. Cooking or freezing will not destroy toxins found in toxic, wild mushrooms. Restaurant and foodservice establishments should not use mushrooms picked in the wild or products made with them unless the mushrooms have been purchased from approved suppliers.

Establishments that serve mushrooms picked in the wild should have written buyer specifications that:

- Identify the mushroom's common name, its Latin name, and its author.
- Ensure that the mushroom was identified in its fresh state.
- Indicate the name of the person who identified the mushroom, including a statement regarding his or her qualifications.

Plant Toxins

Plant toxins are another form of biological contamination. Foodborne illnesses from plant toxins have occurred from consumption of the following:

- Fava beans
- Rhubarb leaves, jimsonweed, and water hemlock
- Apricot kernels
- Honey from bees that have gathered nectar from mountain laurel or rhododendrons
- Milk from cows that have eaten snakeroot

Some plants are toxic when raw but safe when properly cooked. Fava beans and red kidney beans are examples.

In general, toxic plant species and products prepared with them should be avoided. Only commercially processed honey and properly cooked beans should be used.

Exhibit 3c

Toxic Metals

Acidic food prepared in equipment made from toxic metals, such as copper, can cause illness.

CHEMICAL CONTAMINATION

Chemical contaminants are responsible for many cases of foodborne illness. Contamination can come from a variety of substances normally found in restaurants and foodservice establishments. These include toxic metals, pesticides, cleaning products, sanitizers, and lubricants.

Toxic Metals

Utensils and equipment that contain toxic metals—such as lead in a pewter pitcher, copper in a saucepan (see *Exhibit 3c*), or zinc in a galvanized bucket—can cause **toxic metal poisoning.** If acidic food is stored in or prepared with this equipment, the toxic metals can be transferred to the food. Only food-grade utensils and equipment should be used to prepare and store food.

Carbonated-beverage dispensers that are improperly installed can also create a hazard. If carbonated water is allowed to flow back into the copper supply lines, it could leach copper from the line and contaminate the beverage. Beverage-dispensing systems should be installed and maintained by professionals who will ensure that a proper backflow-prevention device is installed.

Exhibit 3d

Chemicals and Pesticides

Store chemicals and pesticides away from food, utensils, and equipment used for food.

Chemicals and Pesticides

Chemicals such as cleaning products, polishes, lubricants, and sanitizers can contaminate food if they are improperly used or stored. To prevent contamination and keep food safe:

- Follow the directions supplied by the manufacturer when using chemicals.
- Exercise caution when using chemicals during operating hours to prevent contamination of food and food-preparation areas.
- Store chemicals away from food, utensils, and equipment used for food. (See *Exhibit 3d.*) Keep them in a separate storage area in their original container.
- If chemicals must be transferred to smaller containers or spray bottles, label each container appropriately.

Pesticides are often used to control pests in food-preparation and food-storage areas. They should only be applied by a licensed pest control operator (PCO). All food should be wrapped or stored before pesticides are applied. If pesticides are stored in the establishment, exercise the same care as with other chemicals used there.

See *Exhibit 3e* for a summary of common chemical contaminants.

Exhibit 3e

Chemical Contaminants

Source of Contamination	Associated Food	Prevention Measures
Toxic Metals		
Utensils and equipment containing toxic metals, such as lead, copper, and zinc	■ Any food, but especially high-acid food, such as sauerkraut, tomatoes, and citrus products ■ Carbonated beverages	■ Use only food-grade containers. ■ Use only food-grade brushes on food; do not use paintbrushes or wire brushes. ■ Do not use enamelware, which may chip and expose the underlying metal. ■ Do not use equipment or utensils made of toxic metals. ■ Use a backflow-prevention device on carbonated beverage dispensers.
Chemicals		
Cleaning products, polishes, lubricants, and sanitizers	■ Any food	■ Follow manufacturers' directions for storage and use; use only recommended amounts. ■ Store away from food, utensils, and equipment used for food. ■ Store in original, labeled containers. ■ Utensils used for dispensing chemicals should never be used to handle food. ■ If chemicals must be transferred to smaller containers or spray bottles, label each container appropriately. ■ Use only food-grade lubricants or oils on kitchen equipment or utensils.
Pesticides		
Used in food-preparation and storage areas	■ Any food	■ Pesticides should only be applied by a licensed professional. ■ Wrap or store all food before pesticides are applied.

Exhibit 3f

Physical Contamination

Metal shavings from the lid of a can might contaminate the food inside.

PHYSICAL CONTAMINATION

Physical contamination results when foreign objects are accidentally introduced into food, or when naturally occurring objects, such as bones in fillets, pose a physical hazard. Common **physical contaminants** may include:

- Metal shavings from cans (see *Exhibit 3f*)
- Staples from cartons
- Glass from broken light bulbs
- Blades from plastic or rubber scrapers
- Fingernails, hair, and bandages
- Dirt
- Bones

Closely inspect the food you receive, and take steps to ensure it will not become contaminated during the flow of food in your operation.

THE DELIBERATE CONTAMINATION OF FOOD

While the food safety principles discussed in the ServSafe program help an establishment address the accidental contamination of food, managers must also be aware of how to prevent or eliminate deliberate contamination. In addition to biological, chemical, and physical contaminants, nuclear and radioactive contaminants are also a concern.

Threats to **food security** in the restaurant and foodservice industry might occur at any level in the food-supply chain. These attacks are usually focused on a specific food item, process, company, or business.

Those who would knowingly contaminate a food product include, but are not limited to, organized terrorist or activist groups, individuals posing as customers, current or former employees, vendors, and competitors.

The key to protecting food is to make it as difficult as possible for tampering to occur. An effective food security program will consider all of the points in which food is vulnerable. Potential threats can come from these three areas:

- Human elements
- Interior elements
- Exterior elements

Managers must ensure that employees are aware of their role in keeping food secure in the operation. This can be accomplished by developing procedures and training to address each potential threat. *Exhibit 3g* lists things to consider when determining how to handle food security in your establishment.

Exhibit 3g

Addressing Food Security Threats in Your Operation*

Human Elements

- Verify the identity of applicants—ask for references, verify them, and check identification.
- Train employees in food security and establish food security awareness in your establishment.
- Train employees to report suspicious activity.
- Establish a system to ensure that only on-duty employees are allowed in work areas.
- Establish rules for opening the back doors of the facility—determine who is authorized to open these doors and under what circumstances.
- Control access to food-production and food-storage areas by nonemployees.
- Allow employees to bring only essential items to work.
- Consider a two-employee rule during food preparation—employees should not be alone in food-preparation areas.
- Monitor preparation areas regularly via video cameras, windows, other employees, or management.

Interior Elements

- Limit access to doors, windows, roofs, and food-storage areas.
- Control entrances and exits to food displays, storage areas, and kitchens.
- Eliminate hiding places in all areas of the operation.
- Inspect all incoming food items; never accept suspect food.
- Restrict traffic in food-preparation and storage areas.
- Monitor self-service areas, and food items and equipment on display, such as salad bars, condiments, and exposed tableware.

Continued on the next page...

Exhibit 3g

Addressing Food Security Threats in Your Operation* *continued*

Exterior Elements

- Ensure that the building's exterior is well lit.
- Control access to the ventilation system.
- Identify all food suppliers and consider using tamper-evident packages. Check the identification of the delivery person and the scheduled times of delivery, and document those deliveries.
- Tell suppliers that food security is a priority and ask what steps they are taking to ensure their products are secure.
- Verify and preapprove all service personnel and providers.
- Prevent access to the facility by nonemployees after normal business hours.

* For more information on food security, visit *www.nraef.org/foodsecurity.*

FOOD ALLERGENS

Nearly seven million Americans have food allergies. A **food allergy** is the body's negative reaction to a particular food protein. Depending on the person, allergic reactions may occur immediately after the food is eaten or several hours later. The reaction could include some or all of the following symptoms:

- Itching in and around the mouth, face, or scalp
- Tightening in the throat
- Wheezing or shortness of breath
- Hives
- Swelling of the face, eyes, hands, or feet
- Gastrointestinal symptoms, including abdominal cramps, vomiting, or diarrhea
- Loss of consciousness
- Death

You and your employees should be aware of the most common food allergens, including milk and dairy products, eggs and egg products, fish, shellfish, wheat, soy and soy products, peanuts, and tree nuts.

You and your employees should be able to inform customers of menu items that contain these potential allergens. Designate one person per shift to answer customers' questions regarding menu items. To help customers with allergies enjoy a safe meal at your establishment, keep the following points in mind:

- Be able to fully describe each of your menu items when asked. Tell customers how the item is prepared and identify any "secret" ingredients if necessary.
- If you or your employees do not know if an item is allergen free, say so. Urge the customer to order something else.
- When preparing food for a customer with allergies, ensure that the food makes no contact with the ingredient to which the customer is allergic. Make sure all cookware, utensils, and tableware are allergen free to prevent food contamination.

Something to Think About...

Now That's a Bright Idea!

Ten years ago, a Midwestern restaurant group decided to review its food-allergy prevention policies. After looking at the entire operation, the group developed a system that uses brightly colored tickets to call out meals for guests with food allergies.

When a guest informs his or her server of a food allergy, the server records the individual's order on a separate, brightly colored ticket. The ticket includes the guest's table number, position at the table, the food allergy, and the order. The server then delivers the ticket to the kitchen and confirms it with the chef. The ticket stays with the meal and is signed by the manager before it is served.

SUMMARY

A foodborne intoxication occurs when a person eats food that contains toxins. The toxin may have been produced by pathogens found on the food or may be the result of a chemical contamination. The toxin could also come from a plant or animal that was eaten.

While some fish toxins are produced by the fish itself, microorganisms on fish are responsible for others. Some fish toxins occur as a result of the fish's diet. Ciguatoxin, for example, is found in certain predatory reef fish that have eaten smaller fish that have consumed the toxin. It is critical to purchase fish from approved, reputable suppliers since these toxins cannot be destroyed by cooking or freezing.

Many toxins associated with shellfish are found in certain types of toxic marine algae. The shellfish become contaminated as they filter the toxic algae from the water. As with fish, purchasing shellfish from approved, reputable suppliers is the most important safeguard against foodborne illness.

Foodborne illnesses associated with mushrooms are almost always caused by the consumption of toxic, wild mushrooms collected by amateur mushroom hunters. Do not use mushrooms picked in the wild or products made with them unless they have been purchased from approved, reputable suppliers.

Chemical contaminants can come from a variety of substances normally found in restaurants and foodservice establishments. These include toxic metals, pesticides, cleaning products, sanitizers, and lubricants. To prevent contamination, only use food-grade utensils and equipment to prepare and store food. Cleaning products, polishes, lubricants, and sanitizers should be used as directed and stored properly. If used, pesticides should be applied by a licensed professional.

Physical contamination can occur when physical objects are accidentally introduced into food or when naturally occurring objects, such as the bones in fish, pose a physical hazard. Closely inspect the food you receive and take steps to ensure food will not become physically contaminated during its flow through your operation.

Food security addresses the prevention or elimination of the deliberate contamination of food. Contamination can occur in biological, chemical, physical, nuclear, or radioactive form. The key to protecting food is to make it as difficult as possible for tampering to occur.

Many people have food allergies. Managers and employees should be aware of the most common food allergens, which include milk and dairy products, eggs and egg products, fish, shellfish, wheat, soy and soy products, peanuts, and tree nuts. You and your employees should be able to inform customers of these and other potential food allergens that may be included in food served at your establishment.

Apply Your Knowledge

Explain why the mahi-mahi steaks caused an outbreak of scombroid poisoning.

A Case in Point

Roberto received a shipment of frozen mahi-mahi steaks. The steaks were frozen solid at the time of delivery and the packages were sealed and contained a large amount of ice crystals, indicating they had been time-temperature abused. Roberto accepted the mahi-mahi steaks and thawed them in the refrigerator at a temperature of 38°F (3°C). The thawed fish steaks were then held at this temperature during the evening shift and were cooked to order. The chefs followed the appropriate guidelines for preparing, cooking, and serving the fish, monitoring time and temperature throughout the process. Unfortunately, the fish steaks caused several scombroid fish poisonings.

For answers, please turn to the Answer Key.

Apply Your Knowledge

Use these questions to review the concepts presented in this chapter.

Discussion Questions

1. What measures should be taken to prevent a seafood-specific foodborne illness?
2. How can toxic-metal poisoning occur and what are some ways to prevent it?
3. What are some ways to keep chemicals from contaminating food?
4. What measures can be taken to help ensure the safety of customers with food allergies?

For answers, please turn to the Answer Key.

Apply Your Knowledge

Use these questions to test your knowledge of the concepts presented in this chapter.

Multiple-Choice Study Questions

1. Fresh tuna steaks have been delivered to your establishment at an internal temperature of 50°F (5°C). You should reject the tuna since serving it could lead to which foodborne illness?
   A. Anisakiasis
   B. Scombroid poisoning
   C. Ciguatera fish poisoning
   D. Vibrio vulnificus gastroenteritis

2. Which is *not* a common food allergen?
   A. Eggs
   B. Dairy products
   C. Peanuts
   D. Pork

3. All of these practices can lead to toxic-metal poisoning *except*
   A. cooking tomato sauce in a copper pot.
   B. storing orange juice in a pewter pitcher.
   C. using a backflow-prevention device on a carbonated beverage dispenser.
   D. serving fruit punch in a galvanized tub.

4. What is the *best* method for preventing a foodborne illness from seafood toxins?
   A. Purchasing smoked or cured seafood
   B. Freezing seafood prior to cooking it
   C. Purchasing seafood from approved, reputable suppliers
   D. Cooking seafood to the required minimum internal temperature

Apply Your Knowledge

Multiple-Choice Study Questions

5. A man who ate oysters later became disoriented and suffered memory loss. What illness was most likely the cause?
   A. Amnesic shellfish poisoning
   B. Paralytic shellfish poisoning
   C. Neurotoxic shellfish poisoning
   D. *Vibrio vulnificus* gastroenteritis

6. Which practice will *not* prevent food from becoming contaminated?
   A. Labeling chemical spray bottles
   B. Inspecting food during receiving
   C. Storing products in food-grade containers
   D. Storing high-acid food away from other food

7. Which is not a symptom of an allergic reaction to peanuts?
   A. Diarrhea
   B. Shortness of breath
   C. Swelling of the feet
   D. Reversal of hot and cold sensations

8. Which statement about mushrooms is true?
   A. Freezing will destroy toxins found in toxic, wild mushrooms.
   B. Cooking will not destroy toxins found in toxic, wild mushrooms.
   C. Most cases of foodborne illness occur when edible species are temperature abused.
   D. Foodborne illnesses almost always occur when mushrooms are purchased from approved suppliers.

For answers, please turn to the Answer Key.

ADDITIONAL RESOURCES

Web Sites

American Council on Science and Health

www.acsh.org

The American Council on Science and Health (ACSH) is a consumer-education consortium concerned with issues related to food, nutrition, chemicals, pharmaceuticals, lifestyle, the environment, and health. ACSH advocates sound science and the belief that Americans' continued physical and economic well-being depends on the advancement of scientific, medical, and technological research and innovation. Visit this Web site for publications, articles, editorials on food safety, and other timely health-related topics.

Bad Bug Book

www.cfsan.fda.gov/~mow/intro.html

Produced by the Food and Drug Administration's Center for Food Safety and Applied Nutrition (FDA CFSAN), this online handbook provides basic facts about foodborne pathogenic microorganisms and biological toxins.

Centers for Disease Control and Prevention

www.cdc.gov

The Centers for Disease Control and Prevention (CDC) are collectively the nation's premiere health promotion, prevention, and preparedness agency and a global leader in public health. The CDC apply research and findings to improve people's daily lives and respond to health emergencies. This Web site provides information on disease and disease prevention, including foodborne illness, its causes, public-health impact, and methods for control.

Center for Infectious Disease Research & Policy

www.cidrap.umn.edu

The Center for Infectious Disease Research & Policy (CIDRAP) at the University of Minnesota functions to prevent illness and death from infectious diseases through epidemiologic research and the rapid translation of scientific information into real-world practical applications and solutions. The center's Web site provides access to information on food safety and foodborne illness and news stories on food security and foodborne-illness surveillance.

Chlorine Chemistry Council®

www.c3.org

The Chlorine Chemistry Council® is a national trade association representing the manufacturers and users of chlorine and chlorine-related products. Visit this Web site for facts on how chlorine contributes to the safety of food served in restaurants.

Environmental Protection Agency

www.epa.gov

Since 1970, the U.S. Environmental Protection Agency (EPA) has been focused on creating a cleaner environment and protecting American public health. EPA staff research and set national standards for a variety of environmental programs. Visit this Web site to find out the latest news and information relating to food safety and the environment.

Food Allergy & Anaphylaxis Network

www.foodallergy.org

The Food Allergy & Anaphylaxis Network (FAAN) is a non-profit organization representing Americans with food allergies. FAAN's mission is to increase awareness, provide education and advocacy, and advance research on behalf of all those affected by food allergies and anaphylaxis. Visit this Web site for information on food allergies and their causes and other useful allergy-related information.

FDA Center for Food Safety and Applied Nutrition

www.cfsan.fda.gov/list.html

As the center within the Food and Drug Administration (FDA) responsible for food safety, the Center for Food Safety and Applied Nutrition (CFSAN) promotes and protects the public health by researching and implementing guidelines, policies, and standards to ensure that food is safe, nutritious, wholesome, and properly labeled. This Web site provides information relevant to all aspects of food safety and security, including corresponding guidelines, policies, and standards.

Food Safety and Inspection Service

www.fsis.usda.gov

The Food Safety and Inspection Service (FSIS) is the public health agency within the U.S. Department of Agriculture responsible for ensuring the nation's commercial supply of meat, poultry, and egg products is safe, wholesome, and correctly labeled and packaged. Visit this Web site for food safety information and regulations related to meat, poultry, and eggs.

Gateway to Government Food Safety Information

www.foodsafety.gov

This Web site provides links to selected government food safety-related information.

Institute of Food Technologists

www.ift.org/cms

The Institute of Food Technologists (IFT) is a nonprofit international scientific society with 22,000 members working in food science, technology and related professions in industry, academia, and government. Informational scientific reports and summaries can be found on this Web site, as well as other useful publications.

Mushroom Council

www.mushroomcouncil.org

The Mushroom Council, composed of fresh-market producers or importers of mushrooms, administers a national promotion, research, and consumer information program to maintain and expand markets for fresh mushrooms. Information about shipping, handling, and preparation of mushrooms as well as risks associated with the use of mushrooms harvested by inexperienced mushroom hunters is available on this Web site.

National Institute for Occupational Safety and Health

www.cdc.gov/niosh/homepage.html

The National Institute for Occupational Safety and Health (NIOSH) is part of the Centers for Disease Control and Prevention (CDC) and the federal agency responsible for helping assure safe and healthful working conditions for working men and women. This Web site provides access to information about NIOSH programs and training sessions; guidance documents on illness; accident prevention and safety, including proper handling of chemicals; and preventing latex allergies.

Occupational Safety and Health Administration

www.osha.gov

The Occupational Safety and Health Administration's (OSHA) role is to assure the safety and health of American workers by setting and enforcing standards; providing training, outreach, and education; and encouraging continual improvement in workplace safety and health. This Web site provides links to regulation and compliance documents and valuable information on the control of occupational hazards.

Other Documents and Resources

2005 *FDA Food Code*

www.cfsan.fda.gov/~dms/fc05-toc.html

The Food and Drug Administration (FDA) publishes the *FDA Food Code,* a scientifically sound technical and legal document that serves as a model for regulating the retail and food service industry at the federal, state, and local level. Updates to the code are issued every other year on the odd-numbered years. The *FDA Food Code* provides a system of safeguards designed to minimize foodborne illness and ensure employee health, food protection manager knowledge, safe food, nontoxic and cleanable equipment, and appropriate sanitation of the food establishment. It is used as the basis for information in this textbook.

Food Allergens

www.cfsan.fda.gov/~dms/wh-alrgy.html

The Food and Drug Administration's Center for Food Safety and Applied Nutrition (FDA CFSAN) produces information on food allergens, their management, and relevant legislation at this Web site.

Food Safety and Terrorism

www.cfsan.fda.gov/~dms/fsterr.html

The Food and Drug Administration's Center for Food Safety and Applied Nutrition (FDA CFSAN) provides this Web site. It contains resources related to food security and food terrorism and information about facility registration, prior notice of imported food shipments, and other guidance documents for the food industry.

Guide for the Control of Molluscan Shellfish

www.cfsan.fda.gov/~ear/nss2-toc.html

The Food and Drug Administration's Center for Food Safety and Applied Nutrition (FDA CFSAN) provides the National Shellfish Sanitation Program's *Guide for the Control of Molluscan Shellfish.* This Web site consists of a Model Ordinance, supporting guidance documents, recommended forms, and other related materials associated with the program. The Model Ordinance includes guidelines to ensure that the shellfish produced in the United States in compliance with the guidelines are safe and sanitary. In addition, the Model Ordinance provides readily adoptable standards and administrative practices necessary for the sanitary control of molluscan shellfish.

Interstate Certified Shellfish Shippers List

www.info1.cfsan.fda.gov/shellfish/sh/shellfis.cfm

The Food and Drug Administration's Center for Food Safety and Applied Nutrition (FDA CFSAN) provides the Interstate Certified Shellfish Shippers list. Visit this Web site, which is updated monthly, for those shippers that have been certified by regulatory authorities in the United States, Canada, Chile, Korea, Mexico, and New Zealand under the uniform sanitation requirements of the national shellfish program.

Morbidity and Mortality Weekly Report

www.cdc.gov/mmwr

The *Morbidity and Mortality Weekly Report* (*MMWR*) is available free of charge on this Web site. The report contains data based on weekly accounts of reportable diseases from state health departments. Visit this Web site to search report databases for synopses of actual foodborne-illness outbreaks.

Seafood Information and Resources

www.cfsan.fda.gov/seafood1.html

The Food and Drug Administration's Center for Food Safety and Applied Nutrition (FDA CFSAN) provides a wealth of seafood information and resources on this Web site. Visit this site for an overview of the FDA's seafood regulatory program, information on seafood-related foodborne pathogens and contaminants, and access to seafood guidance and regulation documents.

The Safe Foodhandler

Inside this chapter:

- How Foodhandlers Can Contaminate Food
- Diseases Not Transmitted Through Food
- Components of a Good Personal Hygiene Program
- Management's Role in a Personal Hygiene Program

After completing this chapter, you should be able to:

- Identify personal behaviors that can contaminate food.
- Identify proper handwashing procedures.
- Identify when hands should be washed.
- Identify appropriate hand antiseptics and when to use them.
- Identify hand maintenance requirements.
- Identify the proper procedure for covering cuts, wounds, and sores.
- Identify procedures that must be followed when using gloves.
- Identify jewelry that poses a hazard to food safety.
- Identify requirements for employee work attire.
- Identify the regulatory exceptions for allowing bare-hand contact with ready-to-eat and cooked food.
- Identify criteria for excluding employees from the establishment or restricting them from working with or around food.
- Identify criteria for excluding or restricting employees from working within establishments that serve high-risk populations.
- Identify illnesses that must be reported to the health agency.
- Identify policies that should be implemented regarding eating, drinking, and smoking while working with food.

Key Terms

- Gastrointestinal illness
- Infected lesion
- Carriers
- Finger cot
- Hair restraint

Apply Your Knowledge

Check to see how much you know about the concepts in this chapter. Use the page references provided with each question to explore the topic.

Test Your Food Safety Knowledge

1. **True or False:** During handwashing, foodhandlers must vigorously scrub their hands and arms for five seconds. *(See page 4-5.)*
2. **True or False:** Gloves should be changed before beginning a different task. *(See page 4-9.)*
3. **True or False:** Foodhandlers must wash their hands after smoking. *(See page 4-7.)*
4. **True or False:** A foodhandler diagnosed with shigellosis cannot continue to work at an establishment while he or she has the illness. *(See page 4-12.)*
5. **True or False:** Hand antiseptics should only be used before handwashing. *(See page 4-7.)*

For answers, please turn to the Answer Key.

INTRODUCTION

At every step in the flow of food through the operation—from receiving through service—foodhandlers can contaminate food and cause customers to become ill. Good personal hygiene is a critical protective measure against foodborne illness, and customers expect it.

You can minimize the risk of foodborne illness by establishing a personal hygiene program for your operation that spells out specific hygiene policies. You must also train your employees on these policies and enforce them. When employees have the proper knowledge, skills, and attitudes toward personal hygiene, you are one step closer to keeping food safe.

HOW FOODHANDLERS CAN CONTAMINATE FOOD

In previous chapters, you learned that foodhandlers can cause illness by transferring microorganisms to food they touch. Many times these microorganisms come from the foodhandlers themselves. Foodhandlers can contaminate food when they:

- Have a foodborne illness
- Show symptoms of **gastrointestinal illness** (an illness relating to the stomach or intestine)
- Have **infected lesions** (infected wounds or cuts)
- Live with or are exposed to a person who is ill
- Touch anything that may contaminate their hands

Key Point

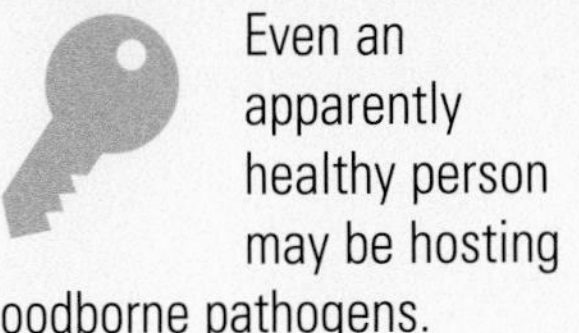

Even an apparently healthy person may be hosting foodborne pathogens.

Even an apparently healthy person may be hosting foodborne pathogens. With some illnesses, such as hepatitis A, a person is at the most infectious stage of the illness for several weeks before symptoms appear. With other illnesses, the pathogens can remain in a person's system for months after all signs of infection have ceased. Some people are called **carriers** because they might carry pathogens and infect others, yet never become ill themselves.

The next three paragraphs will help illustrate some of the routes by which employees can contaminate food.

1. A deli foodhandler who was diagnosed with salmonellosis failed to inform his manager that he was ill for fear of losing wages. It was later determined that he was the cause of an outbreak that involved more than two hundred customers through twelve different products.

2. A foodhandler suffering from diarrhea, a symptom of gastrointestinal illness, did not wash his hands after using the restroom. He made approximately five thousand people ill when he mixed a vat of buttercream frosting with his bare hands and arms. Another large foodborne-illness outbreak was caused by a foodhandler who scratched an infected facial lesion and then handled a large amount of sliced pepperoni.

3 A foodborne-illness outbreak was traced to a woman who prepared food for a dinner party. The investigation revealed that the woman was caring for her infant son, who had diarrhea. The woman could not recall washing her hands after changing the infant's diaper. As a result, twelve of her dinner guests became violently ill with symptoms that included diarrhea and vomiting.

Key Point

Simple acts such as rubbing an ear, scratching the scalp, or touching a pimple or sore can contaminate food.

Simple acts such as nose picking, rubbing an ear, scratching the scalp, touching a pimple or an open sore, or running fingers through the hair can contaminate food. Thirty to 50 percent of healthy adults carry *Staphylococcus aureus* in their noses, and about 20 to 35 percent carry it on their skin. If these microorganisms contaminate a foodhandler's hands that then touch food, the consequences can be severe. For this reason, foodhandlers must pay close attention to what they do with their hands and maintain good personal hygiene.

Key Point

Diseases such as AIDS, hepatitis B and C, and tuberculosis are not spread through food.

DISEASES NOT TRANSMITTED THROUGH FOOD

In recent years, the public has expressed growing concern over communicable diseases spread through intimate contact or by direct exchange of bodily fluids. Diseases such as Acquired Immune Deficiency Syndrome (AIDS), hepatitis B and C, and tuberculosis are not spread through food.

As a manager, you should be aware of the following laws concerning employees who have tested positive for the Human Immunodeficiency Virus (HIV) or have tuberculosis or hepatitis B or C.

- The Americans with Disabilities Act (ADA) provides civil-rights protection to individuals who are HIV positive or have hepatitis B, and thus prohibits employers from firing people or transferring them out of foodhandling duties simply because they have these diseases.
- Employers must maintain the confidentiality of employees who have any nonfoodborne illness.

Exhibit 4a

Good Personal Hygiene

Good personal hygiene is the key to the prevention of foodborne illness.

COMPONENTS OF A GOOD PERSONAL HYGIENE PROGRAM

Good personal hygiene is key to the prevention of foodborne illness. (See *Exhibit 4a.*) Good personal hygiene includes:

- Following hygienic hand practices
- Maintaining personal cleanliness
- Wearing clean and appropriate uniforms and following dress codes
- Avoiding unsanitary habits and actions
- Maintaining good health
- Reporting illnesses

Hygienic Hand Practices

Handwashing

Handwashing is the most critical aspect of personal hygiene. While it may appear fundamental, many foodhandlers fail to wash their hands properly and as often as needed. As a manager, it is your responsibility to train your foodhandlers and then monitor them. Never take this simple action for granted.

Thorough handwashing only takes about twenty seconds. To ensure proper handwashing in your establishment, train your foodhandlers to follow these five steps (see *Exhibit 4b* on the next page):

1. **Wet your hands with running water as hot as you can comfortably stand (at least 100°F [38°C]).**
2. **Apply soap.** Apply enough soap to build up a good lather.
3. **Scrub hands and arms vigorously for ten to fifteen seconds.** Lather well beyond the wrists, including the exposed portions of the arms. Clean under fingernails and between fingers. A nailbrush might be helpful.
4. **Rinse thoroughly under running water.**
5. **Dry hands and arms with a single-use paper towel or warm-air hand dryer.** Use a paper towel to turn off the faucet. When in a restroom, use a paper towel to open the door.

Personal Hygiene

Managers must train foodhandlers when and how to properly wash their hands, and then they must monitor them.

Exhibit 4b

Proper Handwashing Procedure

The whole process should take approximately twenty seconds.

1 **Wet your hands with running water as hot as you can comfortably stand (at least 100°F [38°C]).**

2 **Apply soap.**

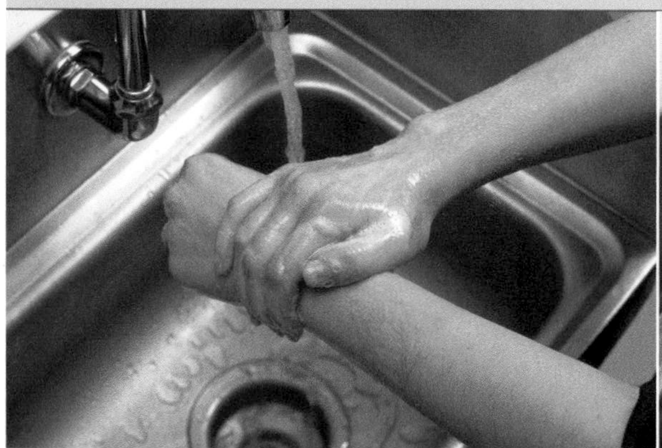

3 **Vigorously scrub hands and arms for ten to fifteen seconds.** Clean under fingernails and between fingers.

4 **Rinse thoroughly under running water.**

5 **Dry hands and arms with a single-use paper towel or warm-air hand dryer.** Use a paper towel to turn off the faucet. When in a restroom, use a paper towel to open the door.

Exhibit 4c

Hand Care for Foodhandlers

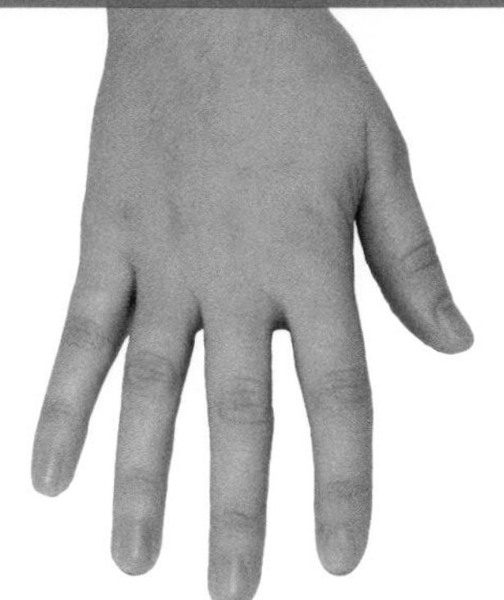

Keep fingernails short and clean.

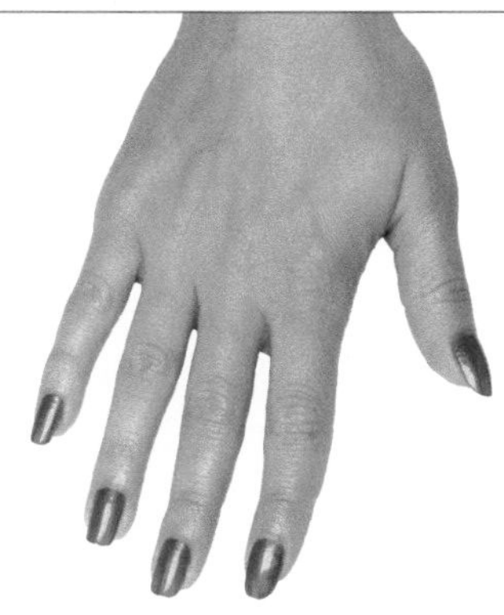

Do not wear false nails or nail polish.

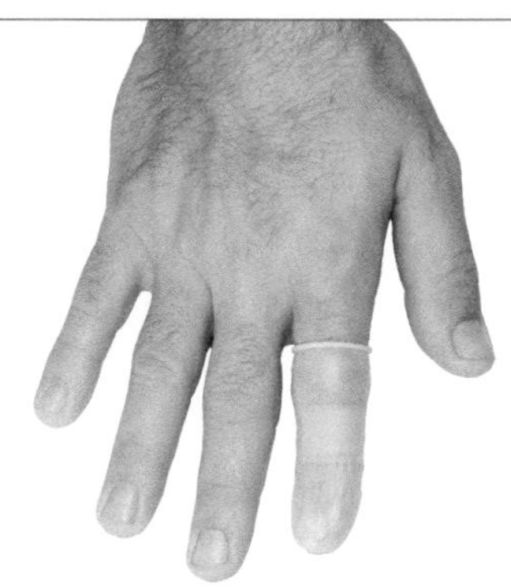

Bandage cuts and cover bandages.

Some establishments require foodhandlers to use liquid or gel antiseptics on their hands to reduce the microorganisms on their skin. These substances must comply with Food and Drug Administration (FDA) standards and should only be used after proper handwashing—never in place of it. Once an antiseptic is applied, foodhandlers should not touch food or equipment until the substance has dried.

Foodhandlers must wash their hands before they start work and after:

- Using the restroom
- Handling raw meat, poultry and fish (before *and* after)
- Touching the hair, face, or body
- Sneezing, coughing, or using a tissue
- Smoking, eating, drinking, or chewing gum or tobacco
- Handling chemicals that might affect the safety of food
- Taking out the garbage
- Clearing tables or bussing dirty dishes
- Touching clothing or aprons
- Touching anything else that may contaminate hands, such as unsanitized equipment, work surfaces, or washcloths

Bare-Hand Contact with Ready-to-Eat Food

Proper handwashing minimizes the risk of contamination associated with bare-hand contact with ready-to-eat food. Establishments that allow bare-hand contact with this food should have written policies and procedures on employee health, handwashing, and other hygiene practices. Check with your regulatory agency for requirements in your jurisdiction.

Hand Maintenance

In addition to proper washing, hands need other regular care to ensure they will not transfer microorganisms to food. To keep food safe, make sure foodhandlers follow these guidelines (see *Exhibit 4c*):

- **Keep fingernails short and clean.** Long fingernails may be difficult to keep clean.

Key Point

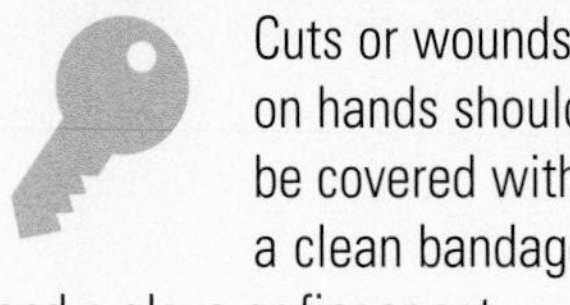

Cuts or wounds on hands should be covered with a clean bandage and a glove or finger cot.

- **Do not wear false fingernails.** False and acrylic nails should not be worn while handling food because they can be difficult to keep clean and can break off into food. Some jurisdictions allow false nails if single-use gloves are worn. Check your local requirements.
- **Do not wear nail polish.** It can disguise dirt under nails and may flake off into food. Some jurisdictions allow nail polish if single-use gloves are worn. Check your local requirements.
- **Cover all hand cuts and wounds with clean bandages.** If there is a bandage on the hand, clean gloves or **finger cots** (a protective covering for individual fingers) should be worn at all times to protect the bandage and prevent it from falling off into food. You may need to move a foodhandler with an infected wound to a nonfoodhandling position until it heals. The new position should not involve contact with food or food-contact surfaces.

Something to Think About...

More than They Bargained for

At a restaurant on the East Coast, the salad bar was a popular attraction. One afternoon while preparing the lettuce, an employee cut her finger. She immediately bandaged it and returned to work. Unfortunately, while she was tossing the salad, the bandage fell off the employee's finger and into the lettuce. A short time later, a customer notified the restaurant manager that she had found a used bandage in her salad. The manager made the necessary apologies and quickly comped her meal. Fortunately, the customer was very understanding, and the rest of the evening proceeded without incident.

What should have been done to prevent this situation?

Glove Use

Gloves can help keep food safe by creating a barrier between hands and food. (See *Exhibit 4d.*) When purchasing gloves for handling food, managers should:

- **Buy disposable gloves.** Gloves used to handle food are for single use only. They should never be washed and reused.

Exhibit 4d

Gloves

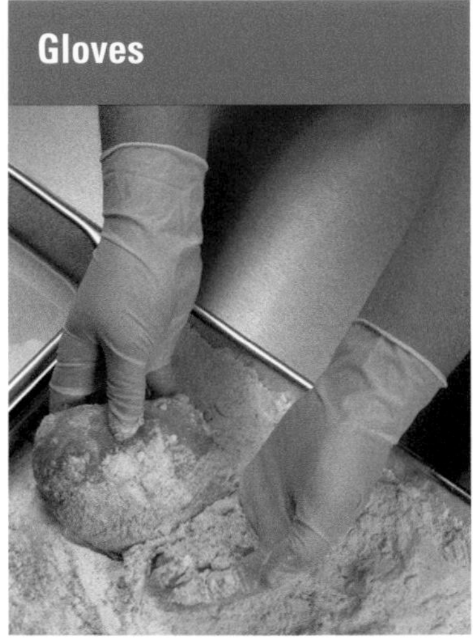

Gloves can help keep food safe by creating a barrier between hands and food.

Key Point

Gloves must never be used in place of handwashing.

- **Buy gloves for different tasks.** Long gloves, for example, should be used for hand-mixing salads. Colored gloves can also be used to help prevent cross-contamination.
- **Provide a variety of glove sizes.** Gloves that are too big will not stay on the hand, while those that are too small will tear or rip easily.
- **Consider latex alternatives for employees who are sensitive to the material.**
- **Focus on safety, durability, and cleanliness.** Make sure you purchase gloves specifically designed for food contact, which include gloves bearing the NSF International mark. (This mark is discussed in Chapter 11.)

Gloves must never be used in place of handwashing. Hands must be washed before putting gloves on and when changing to a new pair. Gloves should be removed by grasping them at the cuff and peeling them off inside out over the fingers while avoiding contact with the palm and fingers.

Foodhandlers should change their gloves:

- As soon as they become soiled or torn
- Before beginning a different task
- At least every four hours during continual use, and more often when necessary
- After handling raw meat and before handling cooked or ready-to-eat food

Often foodhandlers consider gloves more sanitary than bare hands. Because of this false sense of security, they might not change gloves as often as necessary. For this reason, it is critical to reinforce the importance of proper glove use with foodhandlers.

Other Good Personal Hygiene Practices

Personal hygiene can be a sensitive subject for some people, but because it is vital to food safety, managers must address the subject with every foodhandler.

Exhibit 4e

Hair Restraint

Foodhandlers should wear a clean hat or other hair restraint to keep hair away from food and to keep them from touching it.

General Personal Cleanliness

Foodhandlers must maintain personal cleanliness. This includes bathing or showering before work and keeping hair clean, since oily, dirty hair can harbor pathogens.

Proper Work Attire

A foodhandler's attire plays an important role in the prevention of foodborne illness. Dirty clothes may harbor pathogens, as well as give customers a bad impression of your establishment. Therefore, managers should make sure foodhandlers observe strict dress standards.

Foodhandlers should:

- **Wear a clean hat or other hair restraint.** A hair restraint will keep hair away from food and keep the foodhandler from touching it. (See *Exhibit 4e.*) Foodhandlers with facial hair should also wear beard restraints.
- **Wear clean clothing daily.** The type of clothing chosen should minimize contact with food and equipment, and should reduce the need for adjustments. If possible, foodhandlers should change into work clothes at the establishment.
- **Remove aprons when leaving food-preparation areas.** For example, aprons should be removed and properly stored prior to taking out the garbage or using the restroom.
- **Remove jewelry from hands and arms prior to preparing or serving food and when working around food-preparation areas.** (See *Exhibit 4f.*) Jewelry may contain microorganisms, and foodhandlers may be tempted to touch it. Wearing jewelry may also pose a hazard when working around equipment. Remove rings (except for a plain band), bracelets (including medical information jewelry), and watches. Your company may also require you to remove other types of jewelry, including earrings, necklaces, and facial jewelry such as nose rings.
- **Wear appropriate shoes.** Foodhandlers should wear clean, closed-toe shoes with a sensible, nonskid sole.

Exhibit 4f

Jewelry

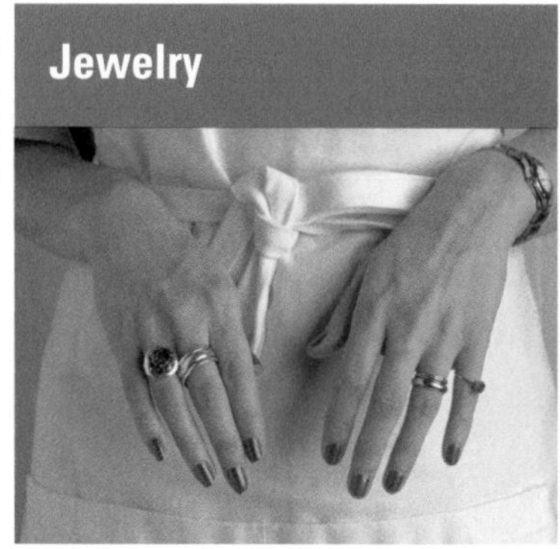

Remove jewelry from hands and arms prior to preparing or serving food and when working around food-preparation areas.

Exhibit 4g

Safe and Unsafe Food-Tasting Practices

Safe

Unsafe

Check with your local regulatory agency regarding requirements. These requirements should be reflected in written policies that are consistently monitored and enforced. All potential employees should be made aware of these policies prior to employment.

Policies Regarding Eating, Drinking, Smoking, and Chewing Gum and Tobacco

Small droplets of saliva can contain thousands of disease-causing microorganisms. In the process of eating, drinking, chewing gum, or smoking, saliva can be transferred to a foodhandler's hands or directly to the food the person is handling. For this reason, foodhandlers must not smoke, chew gum or tobacco, or eat or drink while preparing or serving food, while working in food-preparation areas, or while working in areas used to clean utensils and equipment.

Some jurisdictions allow employees to drink from a covered container with a straw while in these areas. Check with your local regulatory agency. Foodhandlers should eat, drink, chew gum, or use tobacco products only in designated areas, such as an employee break room. Employees should never be allowed to spit in the establishment.

If food must be tasted during preparation, it must be placed in a separate dish and tasted with a clean utensil. (See *Exhibit 4g.*) The dish and utensil should then be removed from the food-preparation area for cleaning and sanitizing.

Policies for Reporting Illness and Injury

Foodhandlers must be encouraged to report health problems to the manager of the establishment before working. If they become ill while working, they must immediately report their condition. Additionally, if food or equipment could become contaminated, the foodhandler must stop working and see a doctor. There are several instances when a foodhandler must either be *restricted* from working with or around food or *excluded* from working within the establishment. (See *Exhibit 4h* on the next page.)

If a foodhandler must refrigerate medication while working and it will be stored with food, he or she must store it inside a covered, leak-proof container that is clearly labeled.

Exhibit 4h

Handling Employee Illnesses

If	Then
The foodhandler has a sore throat with fever.	**Restrict the employee from working with or around food.** **Exclude the employee from the establishment if you primarily serve a high-risk population.**
The foodhandler has one or more of the following symptoms: ■ Vomiting ■ Diarrhea ■ Jaundice	**Exclude the employee from the establishment.** If the person is vomiting or has diarrhea, do not allow the individual to return to work unless he or she: ■ Has been symptom free for twenty-four hours, **or** ■ Has a written release from a medical practitioner Do not allow employees with jaundice to return to work unless they have a written release from a medical practitioner.
The foodhandler has been diagnosed with a foodborne illness caused by one of the following pathogens: ■ *Salmonella* Typhi ■ *Shigella* spp. ■ Shiga toxin-producing *E. coli* ■ Hepatitis A virus ■ Norovirus	**Exclude the employee from the establishment and notify the local regulatory agency.** Work with the employee's medical practitioner and/or the local regulatory agency to determine when the person can safely return to work.

Any cuts, burns, boils, sores, skin infections, or infected wounds should be reported to the manager. They should be covered with a bandage when the foodhandler is working with or around food or food-contact surfaces. Bandages should be clean and dry and must prevent leakage from the wound. As previously mentioned, disposable gloves or finger cots should be worn over bandages on hands. Foodhandlers wearing bandages may need to be temporarily reassigned to duties not involving contact with food or food-contact surfaces.

Vaccination for Hepatitis A

Hepatitis A is a disease that causes inflammation of the liver. It is often transmitted to food through poor personal hygiene. It infects many people each year, resulting in community-wide outbreaks. Of all foodborne illnesses facing the foodservice industry, hepatitis A is the only one that can be prevented by vaccine. While effective handwashing is a critical practice to prevent contamination, vaccinating your foodhandlers for hepatitis A can provide an additional barrier. This may be especially recommended in areas where hepatitis A outbreaks are highly prevalent.

Exhibit 4i

Modeling Proper Personal Hygiene

Managers must model proper behavior for foodhandlers at all times.

MANAGEMENT'S ROLE IN A PERSONAL HYGIENE PROGRAM

Management plays a critical role in the effectiveness of a personal hygiene program. Responsibilities include:

- Establishing proper personal hygiene policies
- Training foodhandlers on personal hygiene policies and retraining them when necessary
- Modeling proper behavior for foodhandlers at all times (see *Exhibit 4i*)
- Supervising sanitary practices continuously
- Revising policies when laws and regulations change and when changes are recognized in the science of food safety

Job Assignments

When job descriptions are developed and responsibilities are assigned, consider the risk of cross-contamination, and plan tasks to prevent it. The risk may be higher if foodhandlers are required to perform several different duties than if specific foodhandlers are assigned to a single duty. For example, an employee expected to prepare and wrap food, clear off tables, and then return to food-preparation duties could more easily contaminate food than an employee who is assigned to just one of these tasks. By planning tasks to prevent cross-contamination, you will minimize the amount of time needed for supervision and enable employees to follow food safety rules more easily.

Something to Think About... **Who's Game?**

A restaurant chain on the East Coast was looking for ways to increase its employees' knowledge of general food safety principles and the importance of handwashing. Management decided that, in addition to traditional training, they would hold a voluntary competition. Spurred by a chance to win a cash prize, employees formed teams and competed against each other in a test of handwashing basics and food safety knowledge. The program was a success, with most employees scoring better than 90 percent on the food safety quiz. The competition is now an annual event.

SUMMARY

Foodhandlers can contaminate food at every step in its flow through the establishment. Good personal hygiene is a critical protective measure against contamination and foodborne illness. A successful personal hygiene program depends on trained foodhandlers who possess the knowledge, skills, and attitude necessary to keep food safe.

Foodhandlers have the potential to contaminate food when they have been diagnosed with a foodborne illness, show symptoms of a gastrointestinal illness, have infected lesions, or touch anything that might contaminate their hands. Foodhandlers must pay close attention to what they do with their hands since simple acts such as nose picking or running fingers through the hair can contaminate food. Proper handwashing must always be practiced. This is especially important before starting work; after using the restroom; after sneezing, coughing, smoking, eating, or drinking; and before and after handling raw food. It is up to the manager to monitor handwashing to make sure it is thorough and frequent. In addition, hands need other care to ensure they will not transfer contaminants to food. Fingernails should be kept short and clean. Cuts and wounds should be covered with clean bandages. Hand cuts should also be covered with gloves or finger cots.

Gloves can create a barrier between hands and food; however, they should never be used in place of handwashing. Hands must be washed before putting on gloves and when changing to a new pair. Gloves used to handle food are for single use and should never be washed and reused. They must be changed when they become soiled or torn, when beginning a new task, and whenever contamination occurs.

Personal hygiene can be a sensitive subject for some people, but, because it is vital to food safety, it must be addressed with every employee. All employees must maintain personal cleanliness. They should bathe or shower before work and keep their hair clean.

Prior to handling food, employees must put on clean clothing, appropriate shoes, and a clean hair restraint. They must also remove jewelry from hands and arms. Aprons should always be removed and properly stored when the employee leaves food-preparation areas.

Establishments should implement strict policies regarding eating, drinking, smoking, and chewing gum and tobacco. These activities should not be allowed when the foodhandler is preparing or serving food or working in food-preparation areas.

Employees must be encouraged to report health problems to management before working with food. If their condition could contaminate food or equipment, they must stop working and see a doctor. Managers must not allow foodhandlers to work if they have been diagnosed with a foodborne illness caused by *Salmonella* Typhi, *Shigella* spp., shiga toxin-producing *E. coli,* the hepatitis A virus, or Norovirus. Foodhandlers must also be excluded from the establishment if they have symptoms that include diarrhea, vomiting, or jaundice. Managers must restrict them from working with or around food if they have a sore throat with fever.

Management plays a critical role in the effectiveness of a personal hygiene program. By establishing a program that includes specific policies and by training and enforcing those policies, managers can minimize the risk of causing a foodborne illness. Most important, managers must set a good example by modeling proper personal hygiene practices.

Apply Your Knowledge

1. Does this situation represent a threat to food safety?
2. Explain what Chris did right.
3. Explain what Chris did wrong.

A Case in Point 1

Chris works at a quick-service restaurant. She is suffering from seasonal allergies, so she carries a small pack of tissues with her. Her assigned responsibility is to make salads. She washes her hands properly and puts on single-use gloves before she starts her task. When Chris needs to sneeze, she steps away from the food-preparation area, pulls a clean tissue out of her pocket, sneezes into it, then discards it. Because her medication gives her a dry mouth, Chris keeps a glass of water at her station. Her clearly labeled bottle of allergy medication has to be kept cold, so she stores it in the walk-in refrigerator.

For answers, please turn to the Answer Key.

Apply Your Knowledge

What could have been done to prevent this outbreak?

A Case in Point 2

Sixteen people became sick and one was hospitalized after drinking milk shakes contaminated with shiga toxin-producing *E. coli* at a drive-in restaurant. The outbreak was started by an employee who came to work despite being ill. The employee made a milk-shake mix that was served over a five-day period. She had diarrhea and severe abdominal cramps prior to making the shake mix. However, the employee continued to work until she was found to have a possible case of *E. coli* five days later.

Fourteen of the sixteen people who became ill had consumed milk shakes. Three people were briefly hospitalized. A fourth, a fifteen-year-old girl, required dialysis due to kidney failure resulting from the illness. The drive-in was temporarily closed, but it was cleaned and reopened a few days later.

For answers, please turn to the Answer Key.

Apply Your Knowledge

Use these questions to review the concepts presented in this chapter.

Discussion Questions

1. What are some basic work-attire requirements for employees?

2. What personal behaviors can contaminate food?

3. What is the proper procedure for handling employee cuts, wounds, and sores?

4. What procedures must foodhandlers follow when using gloves?

5. What employee health problems pose a possible threat to food safety? What are the appropriate actions that should be taken?

For answers, please turn to the Answer Key.

Apply Your Knowledge

Use these questions to test your knowledge of the concepts presented in this chapter.

Multiple-Choice Study Questions

1. Which personal behavior can contaminate food?
   - A. Touching a pimple
   - B. Touching hair
   - C. Nose picking
   - D. All of the above

2. After you have washed your hands, which item should be used to dry them?
   - A. Your apron
   - B. Wiping cloth
   - C. Common cloth
   - D. Single-use paper towels

3. A deli worker stops making sandwiches to use the restroom. She must first
   - A. wash her hands.
   - B. take off her hat.
   - C. take off her apron and properly store it.
   - D. change her uniform.

4. Which item can contaminate food?
   - A. Rings
   - B. Watch
   - C. Bracelet
   - D. All of the above

5. What is the proper procedure for washing your hands?
   - A. Wet hands with water at least 100°F (38°C). Apply soap. Vigorously scrub hands and arms. Apply a hand antiseptic. Dry hands.
   - B. Wet hands with water at least 100°F (38°C). Apply soap. Vigorously scrub hands and arms. Rinse hands. Dry hands.
   - C. Wet hands with water at least 41°F (5°C). Apply soap. Vigorously scrub hands and arms. Rinse hands. Dry hands.
   - D. Wet hands with water at least 41°F (5°C). Apply soap. Vigorously scrub hands and arms. Apply a hand antiseptic. Dry hands.

6. Establishments must only use hand antiseptics that
   - A. dry quickly.
   - B. can be dispensed in a liquid.
   - C. are FDA compliant.
   - D. can be applied before handwashing.

Continued on the next page...

Apply Your Knowledge

Multiple-Choice Study Questions *continued*

7. Which foodhandler is least likely to contaminate the food she will handle?
   A. A foodhandler who keeps her fingernails long
   B. A foodhandler who keeps her fingernails short
   C. A foodhandler who wears false fingernails
   D. A foodhandler who wears nail polish

8. Kim wore disposable gloves while she formed raw ground beef into patties. When she was finished, she continued to wear the gloves while she sliced hamburger buns. What mistake did Kim make?
   A. She failed to wash her hands and put on new gloves after handling raw meat and before handling the ready-to-eat buns.
   B. She failed to wash her hands before wearing the same gloves to slice the buns.
   C. She failed to wash and sanitize her gloves before handling the buns.
   D. She failed to wear reusable gloves.

9. A foodhandler who has been diagnosed with shigellosis should be
   A. told to stay home.
   B. told to wear gloves while working with food.
   C. told to wash his or her hands every fifteen minutes.
   D. assigned to a nonfoodhandling position until he or she is feeling better.

10. Employees must be excluded from the establishment if they have been diagnosed with a foodborne illness resulting from which pathogen?
    A. *Shigella* spp.
    B. *Vibrio vulnificus*
    C. *Clostridium perfringens*
    D. *Clostridium botulinum*

11. Foodhandlers should be restricted from working with or around food if they are experiencing which symptom?
    A. Headache with soreness
    B. Sore throat with fever
    C. Thirst with itching
    D. Soreness with fatigue

Apply Your Knowledge

Multiple-Choice Study Questions

12. Which policy should be implemented at establishments?
    - A. Employees must not smoke while preparing or serving food.
    - B. Employees must not eat while in food-preparation areas.
    - C. Employees must not chew gum or tobacco while preparing or serving food.
    - D. All of the above

13. Stephanie has a small cut on her finger and is about to prepare chicken salad. How should Stephanie's manager respond to the situation?
    - A. Send Stephanie home immediately.
    - B. Cover the hand with a glove or finger cot.
    - C. Cover the cut with a bandage.
    - D. Cover the cut with a bandage and a glove or finger cot.

14. Hands should be washed after which activity?
    - A. Touching your hair
    - B. Eating
    - C. Using a tissue
    - D. All of the above

15. Al, the prep cook at the Great Lakes Senior Citizen Home, calls his manager and tells her that he has a bad headache, nausea, and vomiting. What is the manager required to do with Al?
    - A. Tell him to rest for a couple hours and then come to work.
    - B. Tell him to go to the doctor and then immediately come to work.
    - C. Tell him that he cannot come to work and that he should see a doctor.
    - D. Tell him that he can come in for a couple of hours and then go home.

For answers, please turn to the Answer Key.

Take It Back*

The following food safety concepts from this chapter should be taught to your employees:

- Proper handwashing
- Proper hand care
- Proper glove use
- Proper work attire
- Employee illness
- Personal practices that can lead to contamination

The tools below can be used to teach these concepts in fifteen minutes or less using the directions below. Each tool includes content and language appropriate for employees. Choose the tool or tools that work best.

Tool #1: ServSafe Video 3: *Personal Hygiene*	**Tool #2:** *ServSafe Employee Guide*	**Tool #3:** ServSafe Posters and Quiz Sheets	**Tool #4:** ServSafe Fact Sheets and Optional Activities

Proper Handwashing

Video segment on proper handwashing	**Section 2** **When and How to Wash Your Hands**	**Poster: When and How to Wash Your Hands**	**Proper Handwashing Fact Sheet**
1 Show employees the segment. 2 Ask employees to identify when hands must be washed and describe the proper procedure for handwashing.	1 Discuss with employees when hands must be washed and the proper procedure for handwashing. 2 Complete the Washing Order activity.	1 Discuss with employees when hands must be washed and the proper procedure for handwashing as presented in the poster. 2 Have employees complete the Quiz Sheet: When and How to Wash Your Hands.	1 Pass out a fact sheet to each employee. Explain when hands must be washed and the proper procedure for handwashing using the fact sheet. 2 Have employees complete one or two of the optional activities.

* Visit the Food Safety Resource Center at *www.ServSafe.com/FoodSafety/resource* to download free posters, quiz sheets, fact sheets, and optional activities and to learn how to obtain *Employee Guides* and videos/DVDs.

Take It Back*

Tool #1: ServSafe Video 3: *Personal Hygiene*	**Tool #2:** *ServSafe Employee Guide*	**Tool #3:** ServSafe Posters and Quiz Sheets	**Tool #4:** ServSafe Fact Sheets and Optional Activities
Proper Hand Care			
Video segment on proper hand care 1 Show employees the segment. 2 Ask employees to identify proper hand care practices that should be followed before coming to work.	**Section 2** **Good Personal Hygiene** Discuss with employees proper hand-care practices.	**Poster: Before You Come to Work…** 1 Discuss with employees hand care practices as presented in the poster. 2 Have employees complete the Quiz Sheet: Before You Come To Work…	**Hand Care Fact Sheet** 1 Pass out a fact sheet to each employee. Explain hand care requirements using the fact sheet. 2 Have employees complete the optional activity.
Proper Glove Use			
Video segment on using gloves properly 1 Show employees the segment. 2 Ask employees to identify when gloves should be changed.	**Section 2** **How to Use Gloves Properly** Discuss with employees the requirements for proper glove use.		**Proper Glove Use Fact Sheet** 1 Pass out a fact sheet to each employee. Explain proper glove use requirements using the fact sheet. 2 Have employees complete one or two of the optional activities.

* Visit the Food Safety Resource Center at *www.ServSafe.com/FoodSafety/resource* to download free posters, quiz sheets, fact sheets, and optional activities and to learn how to obtain *Employee Guides* and videos/DVDs.

Take It Back*

Tool #1: ServSafe Video 3: *Personal Hygiene*	Tool #2: *ServSafe Employee Guide*	Tool #3: ServSafe Posters and Quiz Sheets	Tool #4: ServSafe Fact Sheets and Optional Activities
Proper Work Attire			
Video segment on personal cleanliness and attire 1 Show employees the segment. 2 Ask employees to identify what should and should not be worn at work.	**Section 2** **Good Personal Hygiene Practices** Discuss with employees proper work attire.	**Poster: Before You Come to Work…** 1 Discuss with employees proper work attire as presented in the poster. 2 Have employees complete the Quiz Sheet: Before You Come to Work…	**Personal Cleanliness and Proper Attire Fact Sheet** 1 Pass out a fact sheet to each employee. Explain work attire requirements using the fact sheet. 2 Have employees complete the optional activity.
Employee Illness			
Video segment on reporting illness and injury 1 Show employees the segment. 2 Ask employees to identify what symptoms of illness must be reported to the manager.	**Section 2** **Good Personal Hygiene Practices** Discuss with employees symptoms of illness that must be reported to the manager.		**Employee Illness Fact Sheet** 1 Pass out a fact sheet to each employee. Explain the symptoms of illness that must be reported to the manager using the fact sheet. 2 Have employees complete the optional activity.

*** Visit the Food Safety Resource Center at *www.ServSafe.com/FoodSafety/resource* to download free posters, quiz sheets, fact sheets, and optional activities and to learn how to obtain *Employee Guides* and videos/DVDs.**

Take It Back*

Tool #1: ServSafe Video 3: *Personal Hygiene*	**Tool #2:** *ServSafe Employee Guide*	**Tool #3:** ServSafe Posters and Quiz Sheets	**Tool #4:** ServSafe Fact Sheets and Optional Activities
Personal Practices That Can Lead to Contamination			
	Section 2 **How Food Can Become Contaminated** Discuss with employees practices that can lead to contamination.	**Poster: How Food Can Become Contaminated** 1 Discuss with employees practices that can lead to contamination as presented in the poster. 2 Have employees complete the Quiz Sheet: How Food Can Become Contaminated.	

* Visit the Food Safety Resource Center at *www.ServSafe.com/FoodSafety/resource* to download free posters, quiz sheets, fact sheets, and optional activities and to learn how to obtain *Employee Guides* and videos/DVDs.

ADDITIONAL RESOURCES

Articles and Texts

Fendler, Eleanor J., Michael J. Dolan and Ronald A. Williams. 1998. Handwashing and Gloving for Food Protection Part I: Examination of the Evidence. *Dairy, Food and Environmental Sanitation.* 18 (12): 814.

Fendler, Eleanor J., Michael J. Dolan, Ronald A. Williams, and Paulson, Daryl S. 1998. Handwashing and Gloving for Food Protection Part II: Effectiveness. *Dairy, Food and Environmental Sanitation.* 18 (12): 824.

Larson, Elaine. 2001. Hygiene of the Skin: When is Clean Too Clean? *Emerging Infectious Diseases.* 7 (2): 2001.

Paulson, Daryl S. 2000. Handwashing, Gloving, and Disease Transmission by the Food Preparer. *Dairy, Food and Environmental Sanitation.* 20 (11): 838.

Taylor, Anne K. 2000. Food Protection: New Developments in Handwashing. *Dairy, Food and Environmental Sanitation.* 20 (2): 114.

Web Sites

Centers for Disease Control and Prevention

www.cdc.gov

The Centers for Disease Control and Prevention (CDC) are collectively the nation's health promotion, prevention, and preparedness agency and a global leader in public health. The CDC apply research and findings to improve people's daily lives and respond to health emergencies. This Web site provides information on disease and disease prevention, including foodborne illness, its causes, public-health impact, and methods for control.

FDA Center for Food Safety and Applied Nutrition

www.cfsan.fda.gov/list.html

As the center within the Food and Drug Administration (FDA) responsible for food safety, the Center for Food Safety and Applied Nutrition (CFSAN) promotes and protects the public health by researching and implementing guidelines, policies, and standards to ensure that food is safe, nutritious, wholesome, and properly labeled. This Web site provides information relevant to all aspects of food safety and security, including corresponding guidelines, policies, and standards.

National Center for Infectious Diseases

www.cdc.gov/ncidod/index.htm

The National Center for Infectious Disease (NCID), one of the Centers for Disease Control and Prevention, works in partnership with local and state public-health officials, other federal agencies, medical and public-health professional associations, infectious-disease experts from academic and clinical practice, and international and public-service organizations to fulfill its mission to prevent illness, disability, and death caused by infectious diseases in the United States. This Web site houses information on bacterial, viral, and parasitic diseases.

National Institute for Occupational Safety and Health

www.cdc.gov/niosh/homepage.html

The National Institute for Occupational Safety and Health (NIOSH) is part of the Centers for Disease Control and Prevention (CDC) and the federal agency responsible for helping to ensure and healthful working conditions for working men and women. NIOSH does this by providing research, information, education, and training in the field of occupational safety and health. This Web site provides access to information about NIOSH programs and training sessions; guidance documents on illness; accident prevention and safety, including proper handling of chemicals; and preventing latex allergies.

NSF International

www.nsf.org

NSF International is a nonprofit, nongovernmental organization focused on the development of standards in the areas of food, water, indoor air, and the environment. The organization also certifies products against these standards. Visit this Web site for NSF International's food-equipment standards and a listing of certified food-equipment.

U.S. Department of Health and Human Services

www.hhs.gov

The Department of Health and Human Services (HHS) is the federal government's principle agency for protecting American health. The department oversees more than three hundred programs executed by such agencies as the Food and Drug Administration (FDA) and the Centers for Disease Control and Prevention (CDC). Visit this Web site to access food safety information provided by the various agencies and departments overseen by HHS.

Gateway to Government Food Safety Information

www.FoodSafety.gov

This Web site provides links to selected government food safety-related information.

Documents and Other Resources

Emerging Infectious Diseases

www.cdc.gov/ncidod/eid/index.htm

An online journal, *Emerging Infectious Diseases* represents the scientific communications component of the Centers for Disease Control and Prevention's (CDC) efforts against the threat of emerging infections. The journal relies on a broad, international authorship base to publish reports of interest to researchers in infectious diseases and related sciences. It also reports on laboratory and epidemiologic findings within a broader public-health perspective. Visit this Web site for articles published about foodborne disease in the United States and around the world.

FDA Foodborne Illness

www.cfsan.fda.gov/~mow/foodborn.html

The Food and Drug Administration's Center for Food Safety and Applied Nutrition (FDA CFSAN) provides the FDA Foodborne Illness Web site containing resources related to foodborne illness. Visit this site to access information about HACCP, foodborne pathogens, specific government food safety initiatives, and other government sources of foodborne-illness information.

2005 *FDA Food Code*

www.cfsan.fda.gov/~dms/fc05-toc.html

The Food and Drug Administration (FDA) publishes the *FDA Food Code,* a scientifically sound technical and legal document that serves as a model for regulating the retail and foodservice industry at the federal, state, and local level. Updates to the code are issued every other year on the odd-numbered years. The *FDA Food Code* provides a system of safeguards designed to minimize foodborne illness and ensure employee health, food protection manager knowledge, safe food, nontoxic and cleanable equipment, and appropriate sanitation of the food establishment. It is used as the basis for information in this textbook.

Morbidity and Mortality Weekly Report

www.cdc.gov/mmwr

The *Morbidity and Mortality Weekly Report* (*MMWR*), prepared by the Centers for Disease Control and Prevention (CDC), is available free of charge on this Web site. The report contains data based on weekly accounts of reportable diseases from state health departments. Visit this site to search report databases for synopses of actual foodborne-illness outbreaks.

80
100
120
60
140
40
20
180
0
CT220
200
220
°F
MIDDLEFIELD, CT U.S.A.

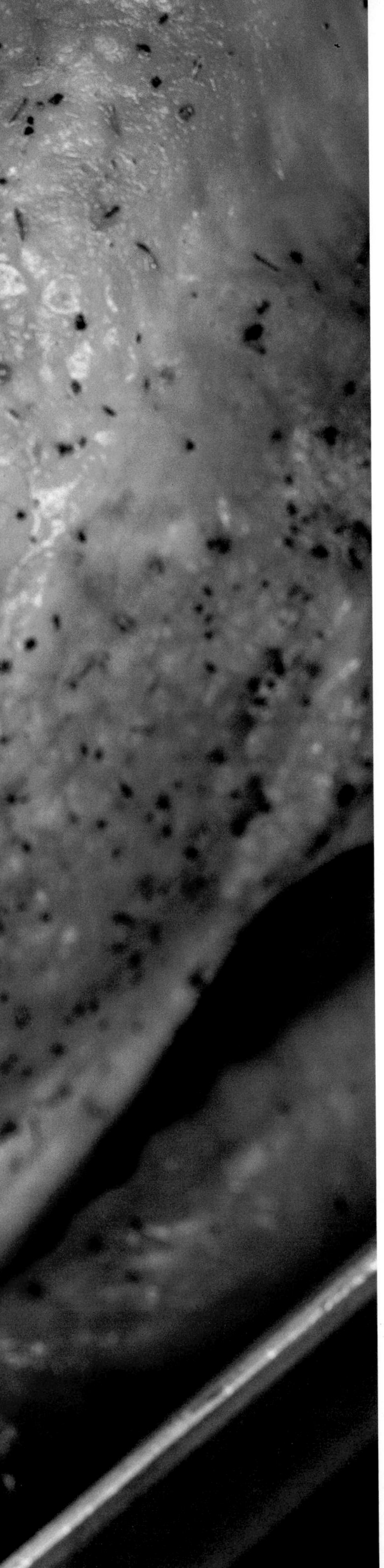

Unit 2

The Flow of Food Through the Operation

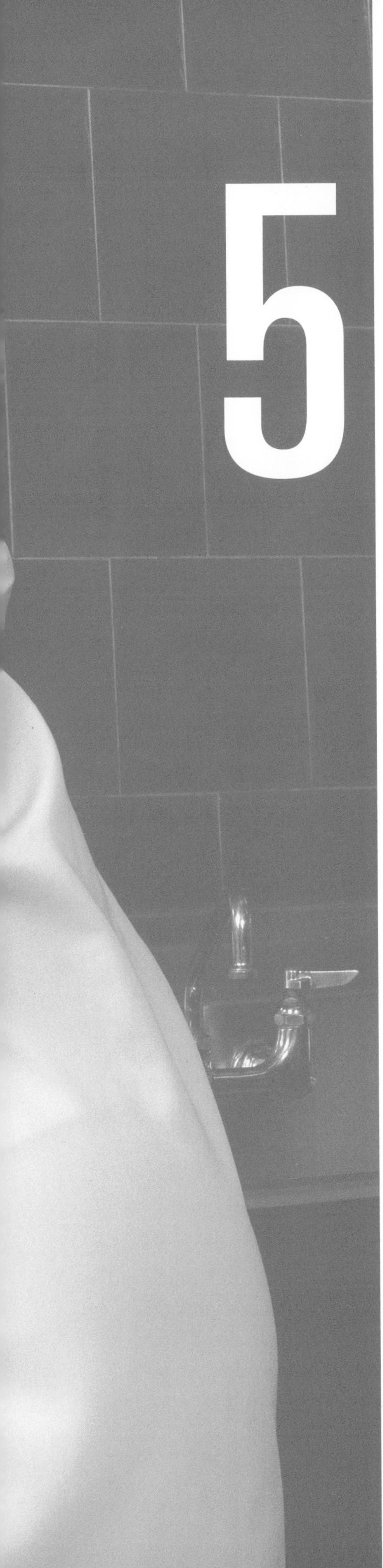

The Flow of Food: An Introduction

Inside this chapter:

- Preventing Cross-Contamination
- Time and Temperature Control

After completing this chapter, you should be able to:

- Identify methods for preventing cross-contamination.
- Identify methods for preventing time-temperature abuse.
- Identify different types of temperature-measuring devices and their uses.
- Calibrate and maintain different temperature-measuring devices.
- Properly measure the temperature of food at each point in the flow of food.

Key Terms

- Flow of food
- Bimetallic stemmed thermometer
- Time-temperature indicator (TTI)
- Calibration

Apply Your Knowledge

Check to see how much you know about the concepts in this chapter. Use the page references provided with each question to explore the topic.

Test Your Food Safety Knowledge

1. **True or False:** Chicken held at an internal temperature of 125°F (52°C) has been temperature abused. *(See page 5-4.)*
2. **True or False:** Infrared thermometers are best for measuring the internal temperature of food. *(See page 5-7.)*
3. **True or False:** When checking the temperature of a roast using a bimetallic stemmed thermometer, only the tip of the thermometer stem should be inserted into the product. *(See page 5-10.)*
4. **True or False:** A thermometer calibrated by the boiling-point method must be set to 135°F (57°C), after being placed in the boiling water. *(See page 5-8.)*
5. **True or False:** Washing and rinsing a cutting board will prevent it from cross-contaminating the next product placed on it. *(See page 5-3.)*

For answers, please turn to the Answer Key.

Exhibit 5a

The Flow of Food

Purchasing
Receiving
Storing
Preparing
Cooking
Holding
Cooling
Reheating
Serving
The Flow of Food

INTRODUCTION

Your responsibility for the safety of the food in your establishment starts long before any food is actually served to the customer. Many things can happen to a product on its path through the establishment, from purchasing and receiving, through storing, preparing, cooking, holding, cooling, reheating, and serving. This path is known as the **flow of food.** (See *Exhibit 5a.*) A frozen product that leaves the processor's plant in good condition, for example, may thaw on its way to the distributor's warehouse and go unnoticed during receiving. Once in your establishment, the product might not be stored properly or cooked to the proper internal temperature. These mistakes could cause a foodborne illness.

The safety of the food served at your establishment will depend largely on how well you apply food safety concepts presented in this program throughout the flow of food. To be effective, you must have a good understanding of how to prevent cross-contamination and time-temperature abuse. You must also develop a system that prioritizes, monitors, and verifies the most important food safety practices. This will be discussed in Chapter 10.

Exhibit 5b

Cross-Contamination

Cross-contamination is the transfer of microorganisms from one food or surface to another.

Exhibit 5c

Color-Coded Equipment

Color-coded equipment can help prevent cross-contamination by making it easier to assign specific equipment to specific food.

PREVENTING CROSS-CONTAMINATION

A major hazard in the flow of food is cross-contamination, which is the transfer of microorganisms from one food or surface to another. (See *Exhibit 5b.*) Microorganisms move around easily in a kitchen. They can be transferred from food or unwashed hands to prep tables, equipment, utensils, cutting boards, or other food.

Cross-contamination can occur at almost any point in an operation. When you know how and where microorganisms can be transferred, cross-contamination is fairly simple to prevent. It starts with the creation of barriers between food products. These barriers can be physical or procedural.

Physical Barriers for Preventing Cross-Contamination

- **Assign specific equipment to each type of food product.** For example, use one set of cutting boards, utensils, and containers for poultry; another set for meat; and a third set for produce. Some manufacturers make colored cutting boards and utensils with colored handles. Color-coding can tell employees which equipment to use with what products, such as green for produce, yellow for chicken, and red for meat. (See *Exhibit 5c.*) Although color-coding minimizes the risk of cross-contamination, it does not eliminate the need to practice other methods for preventing it.

- **Clean and sanitize all work surfaces, equipment, and utensils after each task.** After cutting up raw chicken, for example, it is not enough to simply rinse the cutting board. Wash, rinse, *and* sanitize cutting boards and utensils in a three-compartment sink, or run them through a dishwashing machine. Make sure employees know which cleaners and sanitizers to use for each job. Sanitizers used on food-contact surfaces must meet local or state codes conforming to the Code of Federal Regulations 40CFR180.940. (See Chapter 12 for more information on cleaning and sanitizing.)

Procedural Barriers for Preventing Cross-Contamination

- **When using the same prep table, prepare raw meat, fish and poultry and ready-to-eat food at different times.** For example, establishments with limited prep space can prepare lunch salads in the morning, clean and sanitize the utensils and surfaces, and then debone chicken for dinner entrees in the same space in the afternoon.
- **Purchase ingredients that require minimal preparation.** For example, an establishment can switch from buying raw chicken breasts to purchasing precooked frozen chicken breasts.

Exhibit 5d

Temperature and Bacterial Growth

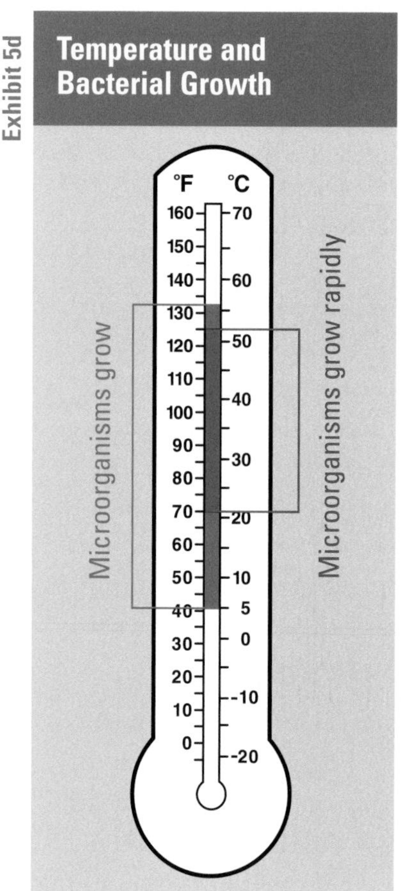

Foodborne microorganisms grow most rapidly at temperatures between 70°F and 125°F (21°C and 52°C).

TIME AND TEMPERATURE CONTROL

One of the biggest factors responsible for foodborne-illness outbreaks is time-temperature abuse. Remember, food has been time-temperature abused when it has been allowed to remain at temperatures favorable to the growth of microorganisms. Foodborne microorganisms grow at temperatures between 41°F and 135°F (5°C and 57°C), which is why this range is known as the temperature danger zone. Microorganisms grow much faster in the middle of the zone, at temperatures between 70°F and 125°F (21°C and 52°C). (See *Exhibit 5d.*) Whenever food is held in the temperature danger zone, it is being abused.

As you learned in Chapter 2, time also plays a critical role in food safety. The longer food stays in the temperature danger zone, the more time microorganisms have to grow and make food unsafe. To keep food safe throughout the flow of food, you must minimize the amount of time it spends in the temperature danger zone. (See *Exhibit 5e.*) If food is held in this dangerous range for more than four hours, you must throw it out.

Potentially hazardous food can be time-temperature abused as it flows through your establishment. This can occur when food is *not:*

- Cooked to the required minimum internal temperature
- Cooled properly
- Reheated properly
- Held at the proper temperature

Exhibit 5e

Preventing Time-Temperature Abuse

To keep food safe, you must minimize the amount of time it spends in the temperature danger zone.

Exhibit 5f

Time and Temperature Control

Print simple forms employees can use to record temperatures.

Reprinted with permission from Roger Bonifield and Dingbats Restaurants

The best way to avoid time-temperature abuse is to establish procedures employees must follow and then monitor them. Make time and temperature control part of every employee's job. To be successful, you should:

- **Determine the best way to monitor time and temperature in your establishment.** Determine which food items should be monitored, how often, and by whom. Then assign responsibility to employees in each area. Make sure they understand exactly what you want them to do, how to do it, and why it is important.
- **Make sure the establishment has the right kind of thermometers available in the right places.** Give employees their own calibrated thermometers. Have them use timers in prep areas to monitor how long food is being kept in the temperature danger zone.
- **Make sure employees regularly record temperatures and the times they are taken.** (See *Exhibit 5f.*) Print simple forms employees can use to record times and temperatures throughout the shift. Post these forms on clipboards outside of refrigerators and freezers, near prep tables, and next to cooking and holding equipment.
- **Incorporate time and temperature controls into standard operating procedures for employees.** This might include:
  - Removing from the refrigerator only the amount of food that can be prepared in a short period of time
  - Refrigerating utensils and ingredients before preparing certain recipes, such as tuna or chicken salad
  - Cooking potentially hazardous food to required minimum internal temperatures
- **Develop a set of corrective actions.** Decide what action should be taken if time and temperature standards are not met.

Choosing the Right Thermometer

To manage both time and temperature, you need to monitor and control them. The thermometer may be the single most important tool you have to protect your food.

There are many types of thermometers used in an establishment. Each is designed for a specific purpose. Some are used to measure the temperature of refrigerated or frozen storage areas. Others measure the temperature of equipment, such as ovens, hot-holding cabinets, and dishwashing machines. Perhaps the most important type are thermometers that measure the temperature of food. The most common types used in establishments are the bimetallic stemmed thermometer, the thermocouple, and the thermistor. Infrared thermometers are also becoming increasingly popular.

Bimetallic Stemmed Thermometers

A common type of thermometer used in the restaurant and foodservice industry is the **bimetallic stemmed thermometer.** This type of thermometer measures temperature through a metal probe with a sensor toward the end. Bimetallic stemmed thermometers often have scales measuring temperatures from 0°F to 220°F (–18°C to 104°C). This makes them useful for measuring the temperatures of everything from incoming shipments to the internal temperature of food in hot-holding units. If you select this type of thermometer, make sure it has (see *Exhibit 5g*):

- An adjustable calibration nut to keep it accurate
- Easy-to-read, numbered temperature markings
- A dimple to mark the end of the sensing area (which begins at the tip)
- Accuracy to within ±2°F (±1°C)

Exhibit 5g

Components of a Bimetallic Stemmed Thermometer

Indicator head
Calibration nut
Holding clip
Stem
Sensing area
Dimple

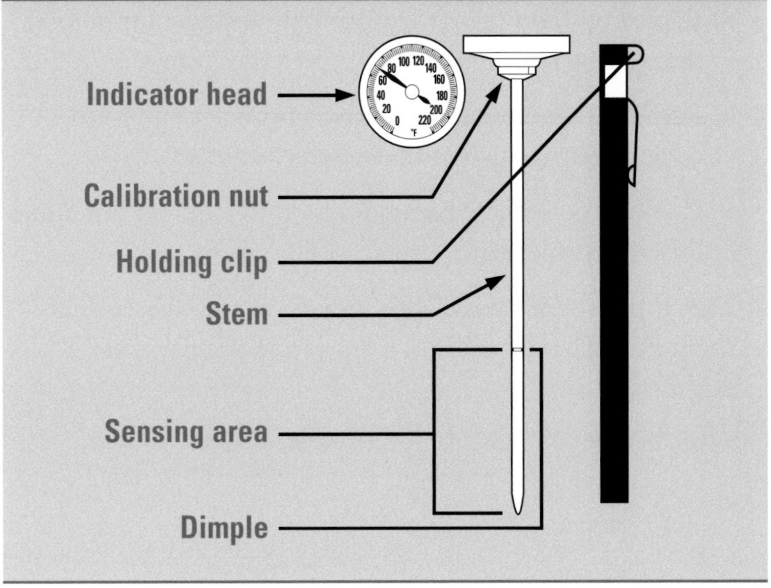

Exhibit 5h

Thermocouple

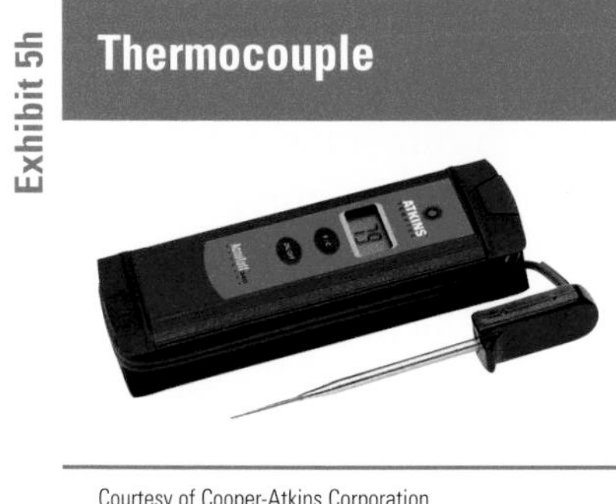

Courtesy of Cooper-Atkins Corporation

Thermocouples and Thermistors

Thermocouples and thermistors measure temperatures through a metal probe or sensing area and display results on a digital readout. They come in a wide variety of styles and sizes, from small pocket models to panel-mounted displays. (See *Exhibit 5h.*) Many come with interchangeable temperature probes designed to measure the temperature of equipment and food.

Basic types of probes include immersion, surface, penetration, and air probes. (See *Exhibit 5i.*) Immersion probes are designed to measure temperatures of liquids, such as soups, sauces, or frying oil. Surface probes measure temperatures of flat cooking equipment like griddles. Penetration probes are used to measure the internal temperature of food. Air temperature probes measure temperatures inside refrigerators or ovens.

Exhibit 5i

Types of Temperature Probes

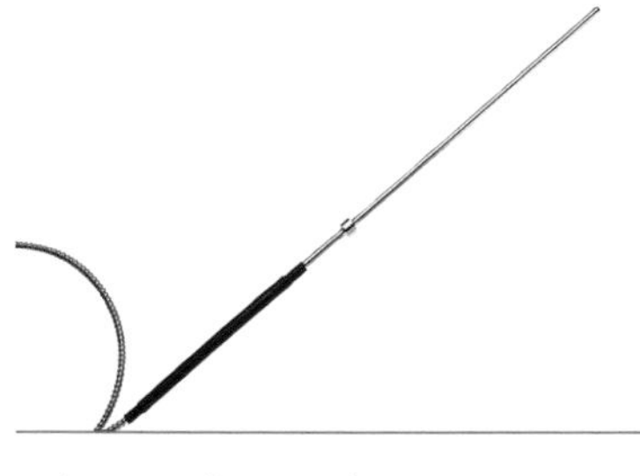

Immersion probe

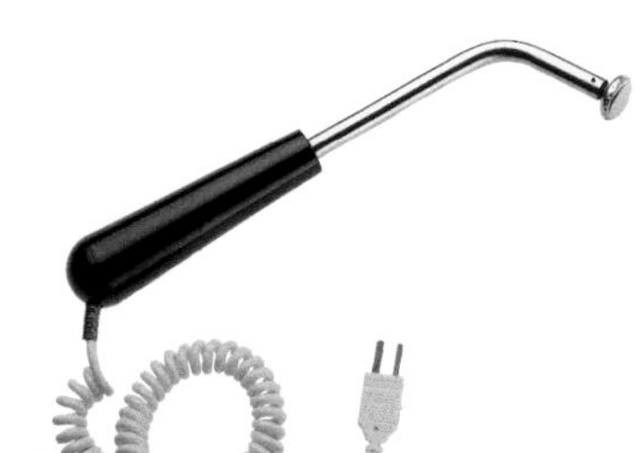

Surface probe

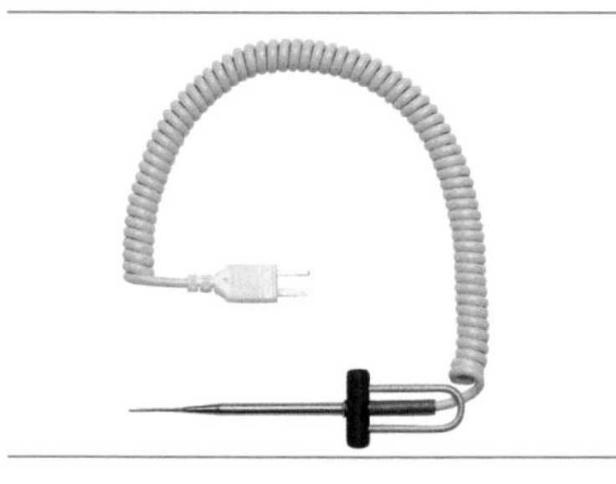

Penetration probe

Courtesy of Cooper-Atkins Corporation

Infrared (Laser) Thermometers

Infrared thermometers use infrared technology to produce accurate temperature readings of food and equipment surfaces. They are quick and easy to use. Infrared thermometers can reduce the risk of cross-contamination and damage to food products because they do not require contact with food. However, they should not be used to measure air temperature or the internal temperature of food.

When using infrared thermometers, remember the following:

- **Hold the thermometer as close as possible to the product without touching it.**
- **Remove any barriers between the thermometer and the product being checked.** Do not take temperature measurements through glass or shiny or polished-metal surfaces, such as stainless steel or aluminum.
- **Always follow the manufacturers' guidelines.** They can provide tips on obtaining the most accurate temperature reading with the infrared thermometer you are using.

Time-Temperature Indicators (TTI) and Other Time-Temperature Recording Devices

Some instruments are designed to monitor both time and product temperature. The **time-temperature indicator (TTI)** is one example. Some suppliers attach these self-adhesive tags to a food shipment

Exhibit 5j

Time-Temperature Indicator (TTI)

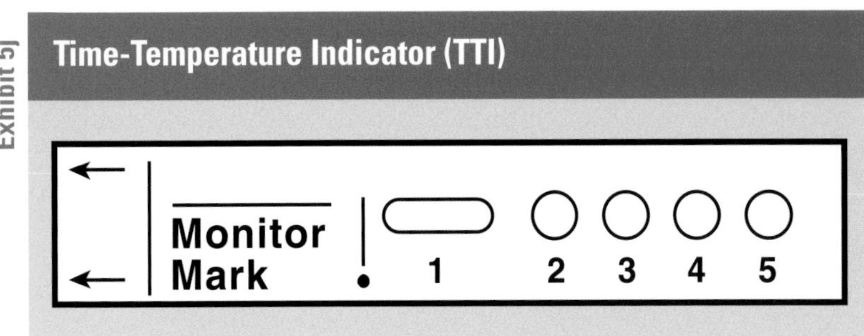

A change in color in the windows of this TTI alerts the receiver that time-temperature abuse has occurred.

to determine if the temperature has exceeded safe limits during shipment or later storage. If it has, the TTI provides an irreversible record of the incident. A change in color inside the TTI windows notifies the employee receiving the shipment that the product has been time-temperature abused. (See *Exhibit 5j*.)

More suppliers are using recording devices that continuously monitor temperatures in their delivery trucks. If delivered products appear to have been time-temperature abused, the recording device can be checked to see if the temperature in the delivery truck changed at any time during transit.

How to Calibrate Thermometers

A thermometer must be adjusted in order to give an accurate reading. This adjustment is called **calibration.** Thermometers can be calibrated by adjusting them to the boiling point of water or to the point at which water turns to ice—the ice point.

When calibrating thermometers using the boiling-point method, follow these steps:

1. **Bring clean tap water to a boil in a deep pan.**
2. **Put the thermometer stem or probe into the boiling water so the sensing area is completely submerged.** Wait thirty seconds, or until the indicator stops moving. The thermometer stem or probe must remain in the boiling water, but do not let it touch the pan's bottom or sides.
3. **Hold the calibration nut securely with a wrench or other tool and rotate the head of the thermometer until it reads 212°F (100°C) or the appropriate boiling-point temperature for your elevation.** The boiling point of water is about 1°F (about 0.5°C) lower for every 550 feet (168 meters) you are above sea level. On some thermocouples or thermistors, it may be possible to press a reset button to adjust the readout.

Thermometers are typically calibrated using the ice-point method. See *Exhibit 5k* for the proper procedure for calibrating thermometers using this method.

Exhibit 5k

Ice-Point Method for Calibrating a Thermometer

1 **Fill a large container with crushed ice.** Add clean tap water until the container is full.

Note: Stir the mixture well.

2 **Put the thermometer stem or probe into the ice water so the sensing area is completely submerged.** Wait thirty seconds or until the indicator stops moving.

Note: Do not let the stem or probe touch the container's bottom or sides. The thermometer stem or probe must remain in the ice water.

3 **Hold the calibration nut securely with a wrench or other tool and rotate the head of the thermometer until it reads 32°F (0°C).**

Note: On some thermocouples or thermistors, it may be possible to press a reset button to adjust the readout.

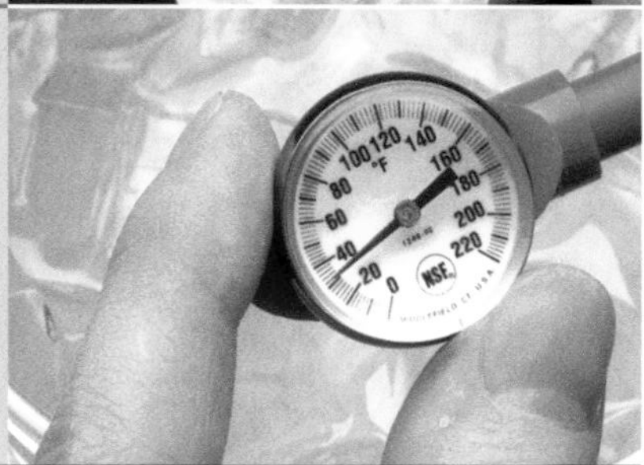

General Thermometer Guidelines

It is important to know how to use and care for each type of thermometer found in your operation. When it comes to maintenance, always follow manufacturers' recommendations. Here are a few simple guidelines for using thermometers.

- **Keep thermometers and their storage cases clean.** Thermometers should be washed, rinsed, sanitized, and air-dried before and after each use to prevent cross-contamination. Use an approved food-contact-surface sanitizing solution to sanitize them. Always have an adequate supply of clean and sanitized thermometers on hand.

- **Calibrate thermometers regularly to ensure accuracy.** This should be done before each shift or before each day's deliveries. Thermometers should also be recalibrated any time they suffer a severe shock—for example, after being dropped or after an extreme change in temperature. Thermometers that hang or sit in refrigerators or freezers can be damaged easily. To make sure these thermometers are accurate, use a thermocouple with an air probe to check the air temperature in the unit and compare it to the thermometer readout. Hanging thermometers usually cannot be recalibrated and must be replaced if they are not accurate.
- **Never use glass thermometers to monitor the temperature of food.** If they break, they can pose a serious danger to employees and customers.
- **Measure internal temperatures of food by inserting the thermometer stem or probe into the thickest part of the product (usually the center).** It is a good practice to take at least two readings in different locations because product temperatures may vary across the food portion. When checking the internal temperature of food using a bimetallic stemmed thermometer, insert the stem into the product so that it is immersed from the tip to the end of the sensing area. When measuring the internal temperature of thin food, such as meat or fish patties, small diameter probes should be used.
- **Wait for the thermometer reading to steady before recording the temperature of a food item.** Wait at least fifteen seconds from the time the thermometer stem or probe is inserted into the food.

SUMMARY

The flow of food is the path that food takes through your establishment from purchasing and receiving through storing, preparing, cooking, holding, cooling, reheating, and serving. Many things can happen to food as it flows through the establishment, but the hazards that the manager has most control over are cross-contamination and time-temperature abuse.

Cross-contamination is the transfer of microorganisms from one food or surface to another. Prevention starts with the creation of physical or procedural barriers between food products. Physical barriers include assigning specific equipment to each type of food product, and cleaning and sanitizing all work surfaces, equipment, and utensils after each task. Procedural barriers include purchasing ingredients that require minimal preparation, and preparing raw meat, fish, and poultry and ready-to-eat food at different times.

Food has been time-temperature abused when it has been allowed to remain at temperatures between 41°F and 135°F (5°C and 57°C). This temperature range is known as the temperature danger zone. To keep food safe, you must minimize the amount of time food spends in this dangerous range. To prevent time-temperature abuse in your establishment, you should incorporate time and temperature controls into your standard operating procedures, make thermometers available to your employees, and regularly record temperatures and the times they were taken.

Thermometers are the most important tools managers have to prevent time-temperature abuse. Managers should make sure employees know what different thermometers are used for and how to calibrate and use them properly. Thermometers can be calibrated using either the boiling-point or the ice-point method. The boiling-point method requires the thermometer to be adjusted to 212°F (100°C) or the appropriate boiling point for the establishment's elevation, after the stem or probe is placed in boiling water. Using the ice-point method, the thermometer is submerged in ice water and adjusted to 32°F (0°C).

Thermometers should be washed, rinsed, sanitized, and air-dried before and after each use to prevent cross-contamination. They should also be calibrated regularly to ensure accuracy. When measuring the internal temperature of food, the thermometer stem or probe should be inserted in the thickest part of the product (usually the center). When using a bimetallic stemmed thermometer, the stem should be immersed in the product from the tip to the end of the sensing area. When measuring the internal temperature of thin food, use a small diameter probe. Always wait for the thermometer reading to steady before recording the temperature of the food. Never use glass thermometers to measure food temperatures.

Apply Your Knowledge

Use these questions to review the concepts presented in this chapter.

Discussion Questions

1. What are some ways food can be time-temperature abused?

2. How can cross-contamination be prevented in the establishment?

3. How is a thermometer calibrated using the ice-point method?

For answers, please turn to the Answer Key.

Apply Your Knowledge

Use these questions to test your knowledge of the concepts presented in this chapter.

Multiple-Choice Study Questions

1. An employee has just trimmed raw chicken on a cutting board and must now use that board to prepare vegetables. What should the employee do with the board before preparing the vegetables?
   A. Wash, rinse, and sanitize the cutting board.
   B. Dry it with a paper towel.
   C. Rinse it under very hot water.
   D. Turn it over and use the reverse side.

2. Which practice will *not* prevent cross-contamination?
   A. Preparing meat separately from ready-to-eat food
   B. Assigning specific equipment for preparing specific food
   C. Rinsing cutting boards between preparing raw food and ready-to-eat food
   D. Using specific storage containers for specific food

3. Infrared thermometers should be used to measure the
   A. air temperature of a refrigerator.
   B. internal temperature of a cooked turkey.
   C. surface temperature of a steak.
   D. internal temperature of a batch of soup.

4. Which practice will *not* prevent time-temperature abuse?
   A. Storing milk at 41°F (5°C)
   B. Holding chicken noodle soup at 120°F (49°C)
   C. Reheating chili to 165°F (74°C) for fifteen seconds within two hours
   D. Holding the ingredients for tuna salad at 39°F (4°C)

5. You have a thermocouple with several different types of probes. Which probe should you use to check the temperature of a large stockpot of soup?
   A. Penetration probe
   B. Surface probe
   C. Air probe
   D. Immersion probe

Continued on the next page...

Apply Your Knowledge

Multiple-Choice Study Questions *continued*

6. Which step for calibrating a thermometer using the ice-point method is *incorrect?*
   A. First, insert the thermometer stem or probe into a container of ice water.
   B. Second, wait thirty seconds or until the temperature indicator stops moving.
   C. Third, remove the thermometer stem or probe from the ice water.
   D. Finally, adjust the thermometer until it reads 32°F (0°C).
7. You must purchase a new thermometer for the restaurant. Which type would *not* be a proper choice?
   A. Bimetallic stemmed thermometer
   B. Thermistor
   C. Thermocouple
   D. Glass thermometer

For answers, please turn to the Answer Key.

Apply Your Knowledge

Notes

Take It Back*

The following food safety concepts from this chapter should be taught to your employees:

- Preventing time-temperature abuse
- Preventing cross-contamination
- Calibrating a thermometer

The tools below can be used to teach these concepts in fifteen minutes or less using the directions below. Each tool includes content and language appropriate for employees. Choose the tool or tools that work best.

Tool #1: ServSafe Video 4: *Purchasing, Receiving, and Storage*	**Tool #2:** *ServSafe Employee Guide*	**Tool #3:** ServSafe Posters and Quiz Sheets	**Tool #4:** ServSafe Fact Sheets and Optional Activities
		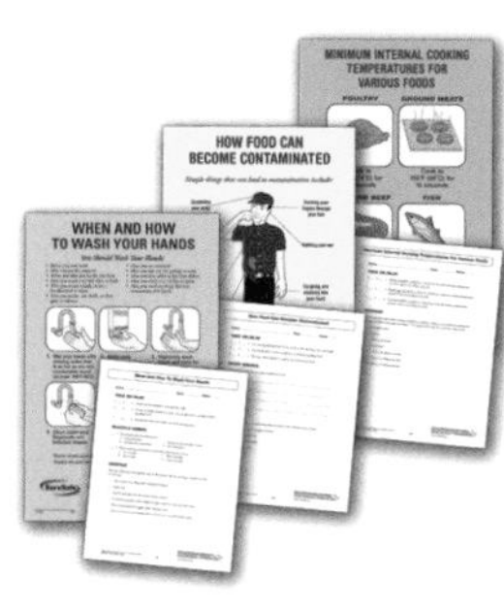	

Preventing Time-Temperature Abuse

Video segment on controlling time and temperature 1. Show employees the segment. 2. Ask employees to define time-temperature abuse and explain why it is important to keep food out of the temperature danger zone.	**Section 1** **How Food Can Become Unsafe** Discuss with employees time-temperature abuse.	**Poster: How Food Can Become Unsafe** 1. Discuss with employees time-temperature abuse as presented in the poster. 2. Have employees complete the Quiz Sheet: How Food Can Become Unsafe.	

* **Visit the Food Safety Resource Center at *www.ServSafe.com/FoodSafety/resource* to download free posters, quiz sheets, fact sheets, and optional activities and to learn how to obtain *Employee Guides* and videos/DVDs.**

Take It Back*

Tool #1: ServSafe Video 4: *Purchasing, Receiving, and Storage*	Tool #2: *ServSafe Employee Guide*	Tool #3: ServSafe Posters and Quiz Sheets	Tool #4: ServSafe Fact Sheets and Optional Activities
Preventing Cross-Contamination			
Video segment on preventing cross-contamination 1 Show employees the segment. 2 Ask employees to define cross-contamination and identify ways it can be prevented.	**Section 1** **How Food Can Become Unsafe** Discuss with employees cross-contamination.	**Poster: How Food Can Become Unsafe** 1 Discuss with employees cross-contamination as presented in the poster. 2 Have employees complete the Quiz Sheet: How Food Can Become Unsafe.	

*** Visit the Food Safety Resource Center at *www.ServSafe.com/FoodSafety/resource* to download free posters, quiz sheets, fact sheets, and optional activities and to learn how to obtain *Employee Guides* and videos/DVDs.**

Take It Back*

Tool #1: ServSafe Video 4: *Purchasing, Receiving, and Storage*	**Tool #2:** *ServSafe Employee Guide*	**Tool #3:** ServSafe Posters and Quiz Sheets	**Tool #4:** ServSafe Fact Sheets and Optional Activities
		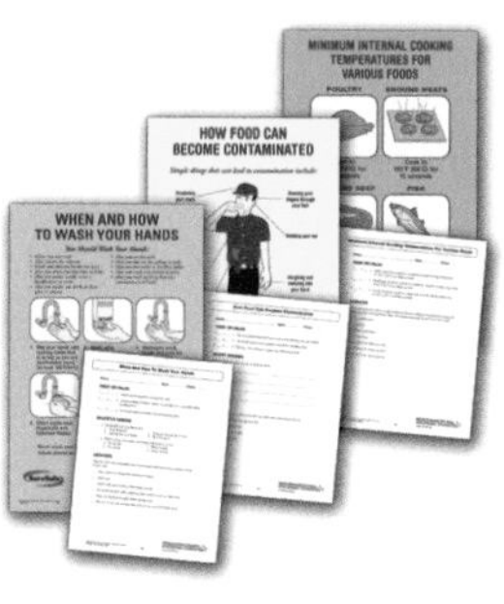	
Calibrating a Thermometer			
	Section 3 **How to Calibrate a Thermometer** 1 Discuss with employees the procedure for calibrating a thermometer using the ice-point method. 2 Complete the Calibrate This! activity.	**Poster: How to Calibrate a Thermometer** 1 Discuss with employees the procedure for calibrating a thermometer using the ice-point method as presented in the poster. 2 Have employees complete the Quiz Sheet: How To Calibrate A Thermometer.	

* Visit the Food Safety Resource Center at *www.ServSafe.com/FoodSafety/resource* to download free posters, quiz sheets, fact sheets, and optional activities and to learn how to obtain *Employee Guides* and videos/DVDs.

Notes

ADDITIONAL RESOURCES

Web Sites

Chlorine Chemistry Council®

www.c3.org

The Chlorine Chemistry Council® is a national trade association representing the manufacturers and users of chlorine and chlorine-related products. Visit this Web site for facts on how chlorine contributes to the safety of food served in restaurants.

FDA Center for Food Safety and Applied Nutrition

www.cfsan.fda.gov/list.html

As the center within FDA responsible for food safety, the Center for Food Safety and Applied Nutrition (CFSAN) promotes and protects the public health by researching and implementing guidelines, policies, and standards to ensure that food is safe, nutritious, wholesome, and properly labeled. This Web site provides information relevant to all aspects of food safety and security, including corresponding guidelines, policies, and standards.

Gateway to Government Food Safety Information

www.foodsafety.gov

This Web site provides links to selected government food safety-related information.

National Frozen & Refrigerated Foods Association

www.nfraweb.org

The mission of the National Frozen & Refrigerated Foods Association is to promote the sales and consumption of frozen and refrigerated food through education, training, research, sales planning, and menu development. Visit this Web site for information on the proper receiving, storage, and preparation of frozen and refrigerated food, including shelf-life charts, tips, and recommended food and equipment storage temperatures.

National Cattlemen's Beef Association

www.beef.org

The National Cattlemen's Beef Association (NCBA) is the national trade association representing U.S. cattle producers. The organization works to advance the economic, political, and social interests of the U.S. cattle industry. This Web site links to other NCBA-associated sites addressing topics such as nutrition, research, food safety, and marketing. A separate site, ***www.beeffoodservice.com,*** provides recipes, facts on foodservice cuts, and food safety information specifically for restaurant and foodservice operators.

NSF International

www.nsf.org

NSF International is a nonprofit, nongovernmental organization focused on the development of standards in the areas of food, water, indoor air, and the environment. This organization also certifies products against these standards. Visit this Web site for NSF International's food-equipment standards and a listing of certified food equipment.

Underwriters Laboratories, Inc.

www.ul.com

Underwriters Laboratories, Inc. (UL) is an independent, nonprofit testing and certification organization for product safety. UL both develops and tests products against standards essential to helping ensure public safety and confidence, reducing costs, and improving quality. This Web site provides access to UL's Standards for Safety and listings of UL-certified food equipment.

Other Documents and Resources

Code of Federal Regulations

www.access.gpo.gov/nara/cfr/cfr-table-search.html

The Code of Federal Regulations (CFR) is the codification of the general and permanent rules published in the Federal Register by the executive departments and agencies of the federal government. It is divided into fifty titles that represent broad areas subject to federal regulation. Those titles relevant to the restaurant and foodservice industry include 7, 21, 40, and 42.

2005 *FDA Food Code*

www.cfsan.fda.gov/~dms/fc05-toc.html

The Food and Drug Administration (FDA) publishes the *FDA Food Code,* a scientifically sound technical and legal document that serves as a model for regulating the retail and foodservice industry at the federal, state, and local level. Updates to the code are issued on the odd-numbered years. The *FDA Food Code* provides a system of safeguards designed to minimize foodborne illness and ensure employee health, food protection manager knowledge, safe food, nontoxic and cleanable equipment, and appropriate sanitation of the food establishment. It is used as the basis for information in this textbook.

Notes

ENRICO'S
ITALIAN DINING

The Flow of Food: Purchasing and Receiving

Inside this chapter:

- Choosing a Supplier
- Inspection Procedures
- Receiving and Inspecting Specific Food

After completing this chapter, you should be able to:

- Identify an approved food source.
- Identify accept and reject criteria for:
  - Meat and poultry
  - Seafood
  - Milk and dairy products
  - Eggs
  - Fruit and vegetables
  - Ready-to-eat food
  - Canned goods and other dry food
  - Frozen food
  - Bakery goods

Key Terms

- Shellstock identification tags
- Modified atmosphere packaging (MAP)
- Vacuum-packed food
- *Sous vide* food
- Ultra-high temperature (UHT) pasteurization
- Aseptically packaged food

Apply Your Knowledge

Check to see how much you know about the concepts in this chapter. Use the page references provided with each question to explore the topic.

Test Your Food Safety Knowledge

1 **True or False:** A delivery of fresh fish should be received at an internal temperature of 41°F (5°C) or lower. *(See page 6-9.)*

2 **True or False:** Turkey should be rejected if the texture is firm and springs back when touched. *(See page 6-13.)*

3 **True or False:** You should reject a delivery of frozen steaks covered in large ice crystals. *(See page 6-17.)*

4 **True or False:** If a sack of flour is dry upon delivery, the contents may still be contaminated. *(See page 6-19.)*

5 **True or False:** A supplier that has been inspected is in compliance with local, state, and federal law can be considered an approved source. *(See page 6-2.)*

For answers, please turn to the Answer Key.

INTRODUCTION

The final responsibility for the safety of food entering your establishment rests with you. You can avoid many potential food safety hazards by using approved suppliers and properly inspecting products when they are delivered.

CHOOSING A SUPPLIER

A number of factors go into selecting the right suppliers. While service needs to be considered, choosing a supplier who can deliver safe food is the ultimate goal. Make sure suppliers can meet or exceed your standards. Keep the following criteria in mind when making your selection:

- **Make sure the supplier is getting products from approved sources.** An approved source is one that has been inspected and is in compliance with applicable local, state, and federal law. It is your responsibility to ensure that food you purchase comes from suppliers (distributors) and sources (points of origin) that have been approved.

Key Point

Make sure suppliers are getting products from approved sources that have a documented GMP program in place.

- **Make sure the supplier and its sources have a documented Good Manufacturing Practices (GMP) program in place.** GMPs are the Food and Drug Administration's (FDA) minimum sanitary and processing requirements for producing safe food. They describe the methods, equipment, facilities, and controls used to process food. Both suppliers and their sources are subject to GMP inspections.
- **Make sure the supplier is reputable.** Ask other operators what their experience has been with the supplier. Find out if they have ever received product that was handled carelessly (e.g., broken boxes, leaky packages, dented cans, etc.).
- **Inspect the supplier's warehouse or plant if possible.** See if it is clean and well run.
- **Ask the supplier if they have a HACCP program in place.** If not, ask what precautions or procedures they take to ensure food safety. (See Chapter 10.)
- **Find out if the supplier's employees are trained in food safety.**
- **Check the condition of the supplier's delivery trucks.** Are they clean and well maintained? Do they hold refrigerated or frozen products at the proper temperatures? Are raw products separated from processed food and fresh produce?
- **Ask the supplier if they can deliver products when your staff have time to receive them.**

Produce Suppliers

It is critical that your produce suppliers get their product from sources with good food safety practices in place. These sources, which include farms, growers, packers, and shippers, should have a GMP program in place. These programs should be documented and audited each season. GAP (Good Agricultural Practices) and GMP audits measure critical areas in which food safety could be compromised. These audits typically focus on:

- Water supply and sources
- Soil, ground history, and nearby land use

- Employee handwashing, personal hygiene, illness, and toilet facilities
- Facility and equipment sanitation
- Fertilizer use
- Training and inspection documentation
- Safe practices for employees working in the fields
- Proper labeling and traceability
- Pest control
- Storage and warehousing practices

When choosing a produce supplier, ask them how well their sources performed on their GAP or GMP audit.

INSPECTION PROCEDURES

If you establish procedures for inspecting products, you can reduce hazards before they enter your establishment. Here are some general guidelines that can help you improve the way you receive deliveries:

- **Train employees to inspect deliveries properly.** Ideally, you should assign the responsibility for receiving and inspecting deliveries to specific employees. These employees should be trained to check products for proper temperatures, expired code dates, signs of thawing and refreezing, pest damage, and so on. They also should be authorized to accept, reject, and sign for deliveries.
- **Plan ahead for shipments.** Have clean hand trucks, carts, dollies, and containers available in the receiving area. Make sure enough space is available in walk-ins and storerooms prior to receiving a shipment. Some operations use a refrigerator and freezer in the receiving area for temporary storage. If products need to be washed or broken down and rewrapped, make workspace available as close to the receiving area as possible. This will prevent dirt and pests from being brought into storage areas or into the kitchen.

- **Schedule deliveries during off-peak hours.** Arrange deliveries so products arrive during times when employees have adequate time to inspect them.
- **Plan a backup menu in case you have to return food items.** If food is not safe or does not meet your standards, you may have to take an item off the menu, substitute another menu item, or try to arrange delivery from another supplier.
- **If possible, receive only one delivery at a time.** Inspect and store each delivery before accepting another one to avoid product abuse in the receiving area.
- **Have the right information available.** Receivers should have a purchase order or order sheet ready to check against the supplier's invoice. The sheet should list quantities, quality specifications, and agreed-upon prices. It should also have room to record the date and time of delivery, product temperatures, and other notes.

Exhibit 6a

Checking Temperatures during Receiving

Take sample temperatures of all refrigerated food.

- **Inspect deliveries immediately.** Do a thorough visual inspection to count quantities, check for damaged products, and look for items that might have been repacked or mishandled. Spot-check weights and take sample temperatures of all refrigerated food. (See *Exhibit 6a.*) There is always a possibility that food—even government-inspected products—may have been mishandled during shipment.

- **Correct mistakes immediately.** If any products are damaged, are not at the correct temperature, or have not been delivered to specifications, do not accept them.
- **Put products away as quickly as possible, especially products requiring refrigeration.**
- **Keep the receiving area clean and well lighted to discourage pests.**

Key Point

You have the right to refuse any delivery that does not meet your standards.

Rejecting Shipments

You have the right to refuse any delivery that does not meet your standards. You should have a company policy about returns, and your suppliers should be aware of it and agree to it. When food does not meet your standards, your employees should know what to do.

To reject a product or shipment:

- **Set the rejected product aside.** Keep it separate from other food and supplies.
- **Tell the delivery person exactly what is wrong with the rejected product.** Use your purchase agreement and company standards to back up your decision to reject the product.
- **Get a signed adjustment or credit slip from the delivery person before throwing the product away or letting the delivery person remove it.**
- **Log the incident on the invoice or receiving document.** Note the food involved, including lot number and expiration date if appropriate, the standard that was not met, and the corrective action taken.

How to Check Temperatures of Deliveries

Use the following guidelines when checking the temperature of various types of food (see *Exhibit 6b* on the next page):

- **Meat, poultry, and fish:** Insert the thermometer stem or probe into the thickest part of the product (usually the center). Additionally, you may want to check the surface temperature using an appropriate thermometer since it can be a better indicator of potential temperature abuse.
- **Reduced oxygen packaged (ROP) and bulk food:** Insert the thermometer stem or probe between two packages. As an alternative, it may be possible to check product temperature by folding the packaging around the thermometer stem or probe. Be careful not to puncture the packaging.
- **Other packaged food:** Open the package and insert the thermometer stem or probe into the product. The sensing area must be fully immersed in the product. The stem or probe must not touch the sides or bottom of the container.
- **Live, molluscan shellfish:** Insert the thermometer stem or probe into the middle of the carton or case—between the shellfish—for an air temperature reading. Check the temperature of shucked shellfish by inserting the stem or probe into the container until the sensing area is immersed.
- **Eggs:** Check the air temperature of the delivery truck, as well as the truck's temperature chart recorder for extreme temperature fluctuations during transport. Temperature fluctuations, high humidity, and warm temperatures may result in the growth of harmful microorganisms.

Always be sure to use a clean, sanitized thermometer each time you check a temperature. If you do not have extra thermometers, clean and sanitize your thermometer after each use. Keep a bucket of sanitizing solution in the receiving area, or use approved sanitizing wipes.

Exhibit 6b

Checking the Temperature of Various Types of Food

Meat, Poultry, Fish

Insert the thermometer stem or probe directly into the thickest part of the product (usually the center).

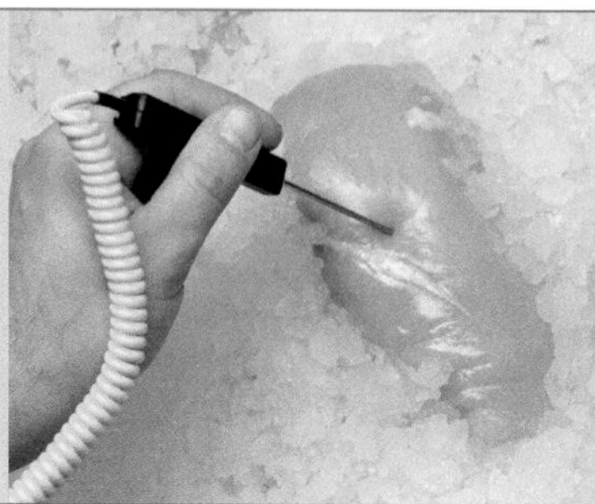

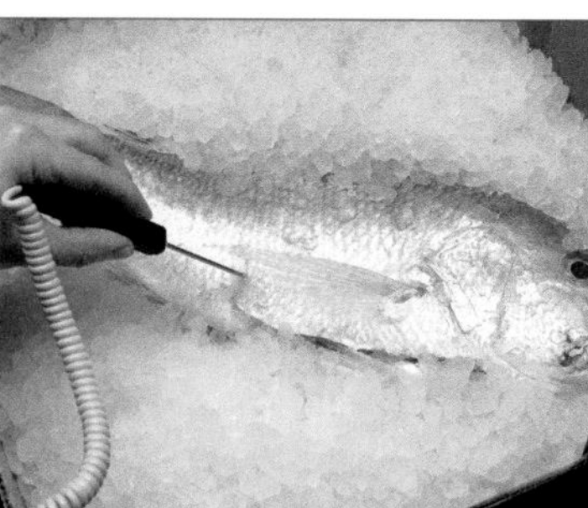

ROP and Bulk Food

Insert the thermometer stem or probe between two packages, or fold the packaging around it.

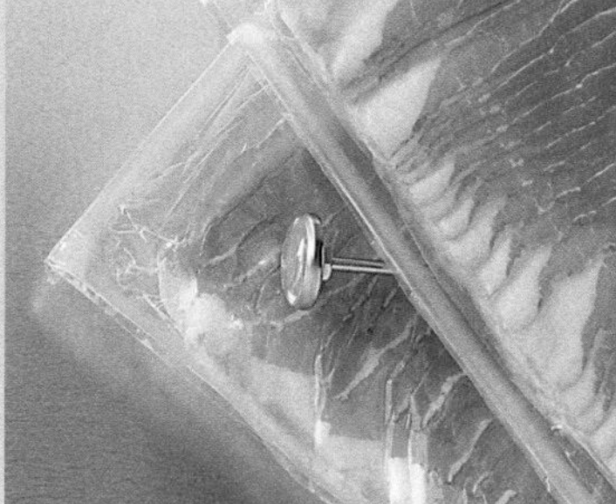

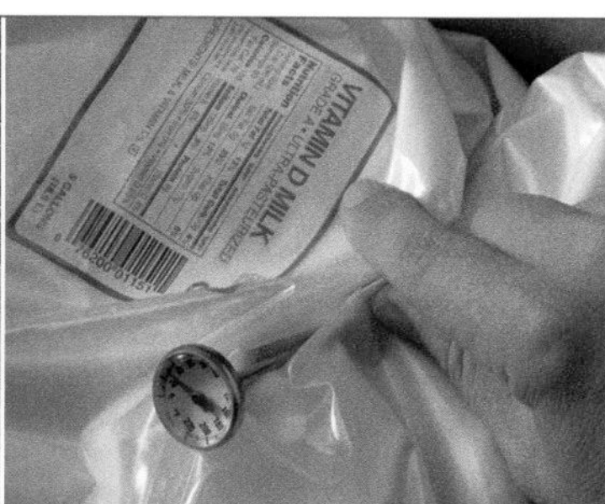

Other Packaged Food

Open the package and insert the thermometer stem or probe into the product.

RECEIVING AND INSPECTING SPECIFIC FOOD

Every food product delivered to your establishment should be inspected carefully for damage, potential contamination, and proper temperature. While receiving temperatures for fresh food are product specific, frozen food should always be received frozen. In addition, note each food's appearance, texture, smell, and in some cases, taste. An in-depth look at receiving criteria for specific products follows in the next several pages.

Fish

Fresh fish is very sensitive to time-temperature abuse and can deteriorate quickly if handled improperly. It should be packed in self-draining, crushed or flaked ice. Upon delivery, it should be received at a temperature of 41°F (5°C) or lower.

Exhibit 6c

Acceptable versus Unacceptable Fish

Acceptable

Unacceptable

To be acceptable, fresh fish must also meet the following criteria:

- **Color:** bright red gills; bright shiny skin
- **Texture:** firm flesh that springs back when touched
- **Odor:** mild ocean or seaweed smell
- **Eyes:** bright, clear, and full
- **Packaging:** product should be surrounded by crushed, self-draining ice

The following criteria are grounds for rejecting fish:

- **Color:** dull gray gills; dull dry skin
- **Texture:** soft flesh that leaves an imprint when touched
- **Odor:** strong fishy or ammonia smell
- **Eyes:** cloudy, red-rimmed, sunken
- **Product:** tumors, abscesses, or cysts on the skin

See *Exhibit 6c* for examples of acceptable and unacceptable fish.

Frozen fish should be received frozen. If there is any indication it has been allowed to thaw, do not accept it. Fish that has thawed and then refrozen before reaching your establishment may have a sour odor and be off color. Fillets often turn brown at the edges when they have been refrozen. Other signs include large amounts of ice or liquid in the bottom of the shipping box and moist, discolored, or slimy wrapping paper.

Fish that will be served raw or partially cooked should be frozen by the processor for a specific amount of time in order to kill parasites. The lower the temperature to which the fish are frozen, the shorter the freezing time. Fish that will be served raw or partially cooked should be frozen to one of the following temperatures prior to shipment:

- –4°F (–20°C) or lower for seven days (168 hours) in a storage freezer
- –31°F (–35°C) or lower until solid and then stored at –31°F (–35°C) for fifteen hours
- –31°F (–35°C) or lower until solid and then stored at –4°F (–20°C) or lower for a minimum of twenty-four hours

Certain species of tuna and some farm-raised fish can be served raw or partially cooked without being frozen to eliminate parasites. To be served this way, farm-raised fish must be raised in a controlled environment with feed that is parasite free. The supplier should provide documentation verifying this and the establishment must keep the documentation on file for ninety days.

Shellfish

Shellfish varieties include mollusks, such as clams, oysters, and mussels. They can be shipped live, frozen, in the shell, or shucked. Interstate shipping is monitored by the state, the FDA, and the shellfish industry. Shellfish must be purchased from suppliers listed in the Interstate Certified Shellfish Shippers List.

Shucked shellfish must be packaged in nonreturnable containers clearly labeled with the name, address, and certification number of the packer. Packages containing less than one-half gallon must have a sell-by date. Packages containing more than one-half gallon should list the date the shellfish were shucked.

Exhibit 6d

Acceptable versus Unacceptable Shellfish

Acceptable

Unacceptable

Live shellfish must be received on ice or at an air temperature of 45°F (7°C) or lower, while shucked product must be received at an internal temperature of 45°F (7°C) or lower. Live shellfish must be delivered alive in nonreturnable containers. The FDA requires that they carry **shellstock identification tags**—tags that document where the shellfish was harvested. The ID tag should remain attached to the container the shellfish came in until all of the shellfish have been used. Operators must keep the tags on file for ninety days from the harvest date of the shellfish.

To be acceptable, shellfish must also meet the following criteria:

- **Odor:** mild ocean or seaweed smell
- **Shells:** closed and unbroken (indicates shellfish are alive) (see *Exhibit 6d*)
- **Condition:** if fresh, they are received alive

The following criteria are grounds for rejecting shellfish:

- **Texture:** slimy, sticky, or dry
- **Odor:** strong fishy smell
- **Shells:** broken shells (see *Exhibit 6d*)
- **Condition:** dead on arrival (open shells that do not close when tapped)

Exhibit 6e

Unacceptable Lobster

If a lobster's tail does not curl when picked up, the lobster is dead and should be rejected.

Crustaceans

Crustaceans include shrimp, crab, and lobster. All processed crustaceans must be received at an internal temperature of 41°F (5°C) or lower. Live lobsters and crabs must be received alive. A live lobster in good condition will show signs of movement and will curl its tail when picked up, while a dead one will not. (See *Exhibit 6e.*) Those showing weak signs of life should be cooked right away, while dead ones must be discarded or returned to the vendor for credit.

To be acceptable, crustaceans must also meet the following criteria:

- **Odor:** mild ocean or seaweed smell
- **Condition:** shipped alive; packed in seaweed and kept moist

The following criteria are grounds for rejecting crustaceans:

- **Odor:** strong fishy smell
- **Condition:** dead on arrival

Exhibit 6f

Inspection and Grading Stamps for Meat

USDA Inspection Stamp

USDA Grading Stamp

Meat inspection is mandatory.

Meat

Meat must be purchased from plants inspected by the U.S. Department of Agriculture (USDA) or state department of agriculture. During the mandatory inspection process, USDA inspectors examine the animal carcass and viscera for possible signs of illness and check processing plants for sanitary conditions. *Inspected* does not mean the product is free of microorganisms but that the product and processing plant have met certain standards. Meat products that have been inspected will be stamped with abbreviations for "inspected and passed" by the inspecting agency, along with a number identifying the processing plant. (See *Exhibit 6f.*) These stamps will not appear on every cut of meat, but one should be present on every inspected carcass and on packaging.

Most meat also carries a stamp indicating its "grade," or palatability, and level of quality. Grading is a voluntary service offered by the USDA and is paid for by processors and packers. USDA grades are printed inside a shield-shaped stamp. (See *Exhibit 6f.*)

Meat must be delivered at 41°F (5°C) or lower. To be acceptable, it must also meet the following criteria:

- **Color:**
  - **Beef:** bright, cherry red; aged beef may be darker in color; vacuum-packed beef will appear purplish in color
  - **Lamb:** light red
  - **Pork:** light pink meat; firm, white fat
- **Texture:** firm flesh that springs back when touched

- **Odor:** no odor
- **Packaging:** intact and clean

The following criteria are grounds for rejecting meat:

- **Color:**
  - **Beef:** brown or green
  - **Lamb:** brown, whitish surface covering the lean meat
  - **Pork:** excessively dark color, soft or rancid fat
- **Texture:** slimy, sticky, or dry
- **Odor:** sour odor
- **Packaging:** broken cartons, dirty wrappers, torn packaging, vacuum packaging with broken seals

Exhibit 6g

Inspection and Grading Stamps for Poultry

USDA Inspection Stamp

USDA Grading Stamp

Poultry inspection is mandatory.

Poultry

Poultry is inspected by the USDA or state department of agriculture in much the same way as meat. As with meat, grading is voluntary and paid for by processors. See *Exhibit 6g* for examples of poultry inspection and grading stamps.

Fresh poultry should be shipped in self-draining, crushed ice and delivered at a temperature of 41°F (5°C) or lower, or chill packed.

To be acceptable, poultry must also meet the following criteria:

- **Color:** no discoloration
- **Texture:** firm flesh that springs back when touched
- **Odor:** no odor
- **Packaging:** product should be surrounded by crushed, self-draining ice

The following criteria are grounds for rejecting poultry:

- **Color:** purple or green discoloration around the neck; dark wing tips (red tips are acceptable)
- **Texture:** stickiness under the wings and around joints
- **Odor:** abnormal, unpleasant odor

Exhibit 6h

Inspection and Grading Stamps for Eggs

USDA Inspection Stamp

USDA Grading Stamp

Egg inspection is mandatory.

Eggs

Eggs must be purchased from approved, government-inspected suppliers. The mandatory USDA inspection stamp on egg cartons indicates federal regulations are enforced to maintain quality and reduce contamination. As with meat and poultry, grading is voluntary and is provided by the USDA. The grading stamp certifies that eggs have been graded for quality under federal and/or state supervision. *Exhibit 6h* provides examples of inspection and grading stamps for eggs.

You should use suppliers who can deliver eggs within a few days of the packing date. Eggs must be delivered in refrigerated trucks capable of documenting air temperature during transport. When the eggs arrive, the truck's air temperature should be 45°F (7°C) or lower, and the eggs must be stored immediately in refrigeration units that will hold them at an air temperature of 45°F (7°C) or lower.

Liquid, frozen, and dehydrated eggs must be pasteurized as required by law and bear the USDA inspection mark. When delivered, liquid and frozen eggs should be refrigerated or frozen at the proper temperature. Check packages for damage or signs of refreezing, and note use-by dates to be sure the product is in good condition and still usable. Cases or cartons of eggs for direct sale to the consumer must display safe-handling instructions on them.

To be acceptable, shell eggs must also meet the following criteria:

- **Odor:** no odor
- **Shells:** clean and unbroken

The following criteria are grounds for rejecting shell eggs:

- **Odor:** sulfur smell or off odor
- **Shells:** dirty or cracked (see *Exhibit 6i*)

Exhibit 6i

Unacceptable Shell Eggs

Eggs that are dirty or cracked should be rejected.

Exhibit 6j

Acceptable versus Unacceptable Cheese

Acceptable

Unacceptable

Dairy Products

Purchase only pasteurized dairy products since unpasteurized products are potential sources of foodborne pathogens such as *Listeria monocytogenes* and *Salmonella* spp. All milk and milk products should be labeled "Grade A." This means they meet standards for quality and sanitary processing methods set by the FDA and the U.S. Public Health Service. Dairy products with the Grade A label—such as cream, cottage cheese, butter and ice cream—are made with pasteurized milk.

Milk and dairy products should be received at 41°F (5°C) or lower unless otherwise specified by law. Check with the proper regulatory agency for temperature requirements.

In the United States, all cheese must meet certain standards of identity. In order for the product to be called "Cheddar" or "mozzarella," for example, the government specifies ingredients that must be used, maximum moisture content, minimum fat content, and general characteristics.

To be acceptable, dairy products must also meet the following criteria:

- **Milk:** sweetish flavor
- **Butter:** sweet flavor, uniform color, firm texture
- **Cheese:** typical flavor and texture; uniform color; clean and unbroken rind (see *Exhibit 6j*)

The following criteria are grounds for rejecting dairy products:

- **Milk:** sour, bitter, or moldy taste; off odor; expired sell-by date
- **Butter:** sour, bitter, or moldy taste; uneven color; soft texture; contains foreign matter
- **Cheese:** abnormal flavor or texture; uneven color; unnatural mold; unclean or broken rind (see *Exhibit 6j*)

Fresh Produce

Fresh fruit and vegetables have different temperature requirements for transportation and storage. No specific temperature is mandated by regulation with the exception of cut melons, which must be received and stored at 41°F (5°C)

Exhibit 6k

Unacceptable Produce

Reject any produce with signs of spoilage.

or lower. When receiving fresh-cut produce, reject any items that have passed their expiration date.

Reject deliveries if there is evidence of mishandling or insect infestation (including insect eggs and egg cases). Produce should also be rejected if there are signs of spoilage, including:

- Mold (see *Exhibit 6k*)
- Cuts
- Wilting and mushiness
- Discoloration and dull appearance
- Unpleasant odors and tastes

Keep in mind that what applies to one item may not apply to another. For example, peaches with cuts in them could be considered poor quality, but potatoes and carrots with cuts would be considered acceptable. Discoloration in produce may vary as well. For example, oranges may actually revert to a green color without affecting the quality of the orange or its juice.

If produce items are pinched, squeezed, or roughly handled, they will bruise and spoil more quickly. Bruised produce can pose a hazard because the bruises provide a potential entry point for pathogens.

Use smell and taste to help determine product quality. Unpleasant odors will tell you when a product is not acceptable. With fruit, sometimes taste is the best test. Outer peels or skins can be blemished without affecting flavor or quality. Be sure to wash or peel fruit and vegetables carefully before tasting them.

Since there are so many ways produce can show signs of spoilage, employees need to learn not only how to identify obviously unacceptable produce, but also how to identify produce that will spoil quickly in storage. Most produce should not sit out at room temperature. Fresh fruit and vegetables are highly perishable and should be put into storage quickly. In general, produce should not be washed before it is stored. While washing would not hurt leafy green items, many other products are likely to decay faster if washed before storage. This is especially true for mushrooms and berries. Produce should be washed right before preparing and serving it.

Exhibit 6l

Unacceptable Refrigerated Ready-to-Eat Food

Reject product with holes or tears in the packaging.

Exhibit 6m

Unacceptable Frozen Processed Food

Reject product with signs of thawing and refreezing, such as large ice crystals on the product.

Refrigerated Ready-to-Eat Food

More and more establishments are purchasing ready-to-eat food items, such as precut meats, salads containing potentially hazardous food, and refrigerated entrées that only require heating. These items must be received at 41°F (5°C) or lower unless otherwise specified by the manufacturer. Packaging must be intact and in good condition. Reject product with holes or tears in packaging (see *Exhibit 6l*) or expired use-by dates.

Frozen Processed Food

Many establishments receive processed food items that are frozen. These items should be delivered frozen, with the exception of ice cream. Ice cream may be delivered and stored at temperatures of 6°F to 10°F (–14°C to –12°C) without affecting product safety or quality.

Frozen food should be received in packaging that is intact and in good condition. It is important to check for signs of thawing and refreezing. Simply because a product is frozen upon receipt does not mean it did not thaw and refreeze during prior handling. Obvious signs are blocks of ice or liquid at the bottom of the case or large ice crystals on the product itself. (See *Exhibit 6m.*) Other signs include product discoloration or dryness and stains on the outer packaging.

Reduced Oxygen Packaged (ROP) Food

As you will recall, some bacteria need oxygen to grow. Reducing the oxygen inside food packaging can prevent this growth. This type of packaging is often referred to as ROP, or reduced oxygen packaging. There are several methods used by manufacturers to do this. These include MAP, vacuum packaging, and *sous vide.*

MAP stands for **modified atmosphere packaging.** By this method, the air inside of a package is altered using gases such as carbon dioxide and nitrogen. Many fresh-cut produce items are packaged this way.

Vacuum-packed food, such as bacon, is processed by removing the air around the product sealed in a package.

Exhibit 6n

Unacceptable ROP Food

Reject product with torn or leaking packaging.

By the ***sous vide*** method, cooked or partially cooked food is vacuum packed in individual pouches and then chilled. This food is heated for service in the establishment. Frozen, precooked meals are typically packaged this way.

ROP food that is refrigerated should be received at 41°F (5°C) or lower unless otherwise specified by the manufacturer. Product that is frozen should be received frozen.

To be acceptable, ROP food must also meet the following criteria:

- **Packaging:** intact and in good condition; valid code dates
- **Product:** acceptable color

The following criteria are grounds for rejecting ROP food:

- **Packaging:** torn or leaking packages (see *Exhibit 6n*); expired code dates
- **Product:** unacceptable color; slime, bubbles, or excessive liquid

Exhibit 6o

Unacceptable Canned Food

Swollen ends

Rust

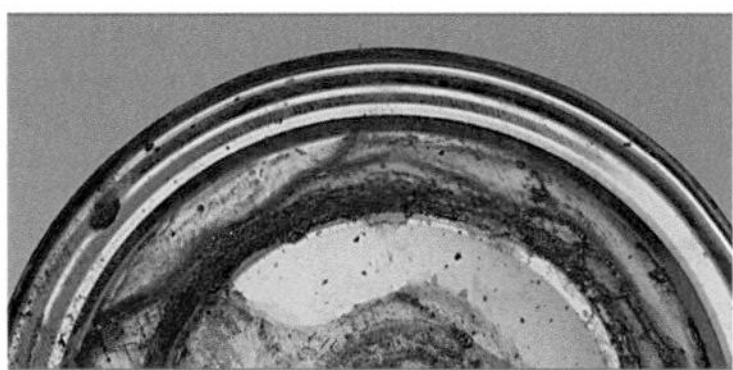

Dents

Reject cans with swollen ends, rust, or dents.

Canned Food

Canned products seem to pose little threat. Most have a fairly long shelf life and are usually used long before they have a chance to spoil. However, canned products provide a good environment for the microorganism that causes botulism. Always check cans for expired code dates and damage.

The following criteria are grounds for rejecting canned products (see *Exhibit 6o*):

- **Swollen ends.** One or both ends of a can may bulge from gas produced by the presence of chemicals or the growth of foodborne bacteria inside. If one end bulges out when the other is pressed, discard the can, because it has not gone through the proper heat-treating process to eliminate foodborne microorganisms.
- **Leaks and flawed seals.** If there are any leaks, flaws, or irregularities along the top or side seals, reject the can.
- **Rust.** If a can is rusted, it should be rejected since the contents may be too old or the rust may have eaten holes in the can. This can lead to contamination.

- **Dents.** Do not accept cans with dents along side or top seams. Reject cans with dents large enough to make it difficult to open them with a can opener since the seams may be broken. Check with your regulatory agency regarding dented cans since some jurisdictions do not allow them.
- **Missing labels.** Any cans received without labels should be rejected.

Dry Food

Most microorganisms need moisture to grow, which is why dry food has a much longer shelf life than fresh food. In order to remain safe, dry food must remain dry and should be rejected when there are signs of wetness.

Because it can be stored at room temperature, dry food might not be sealed and stored as securely as other food. For this reason, it often attracts insects and rodents. Since these pests can easily get into packages, it is critical to check for signs. You can often spot insects or insect eggs in dry products such as cereal or flour by sprinkling some of the product on brown paper.

To be acceptable, dry food must also meet the following criteria:

- **Packaging:** intact and in good condition
- **Product:** normal color and odor

The following criteria are grounds for rejecting dry food:

- **Packaging:** holes, tears, or punctures; dampness or water stains on outer cases and inner packaging, which indicates it has been wet (see *Exhibit 6p*)
- **Product:** abnormal color or odor; spots of mold or slimy appearance; contains insects, insect eggs, or rodent droppings

Exhibit 6p

Holes or tears

Moisture stain

Reject dry food when packaging has holes or tears or shows signs of prior wetness.

Exhibit 6q

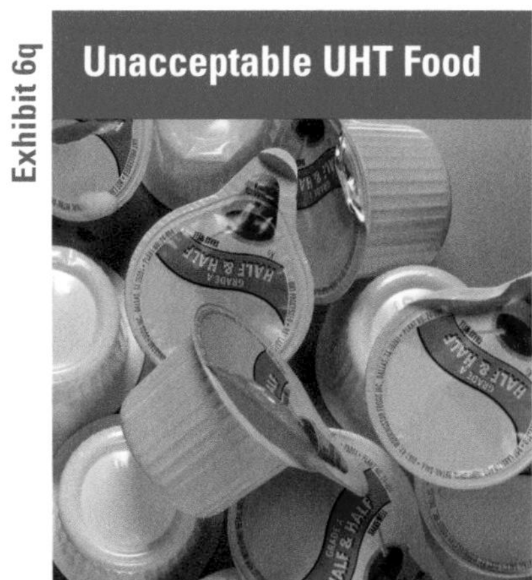

Unacceptable UHT Food

Reject UHT products if packaging is punctured or seals are broken.

Ultra-High Temperature (UHT) Pasteurized and Aseptically Packaged Food

Some food is heat-treated at very high temperatures to kill microorganisms in a process called **ultra-high temperature (UHT) pasteurization.** This food is often also **aseptically packaged**—sealed under sterile conditions to keep it from being contaminated. Examples include some puddings, juices, and creamers and milk products.

If food has been UHT pasteurized and aseptically packaged, it can be received at room temperature. If it has been UHT pasteurized but not aseptically packaged, it should be received at the temperature specified by the manufacturer or at 41°F (5°C) or lower. Packaging and seals must be intact. Reject the product if packaging is punctured or seals are broken. (See *Exhibit 6q.*)

Exhibit 6r

Unacceptable Bakery Goods

Reject bakery goods if there are signs of pests.

Bakery Goods

It is common for establishments to receive shipments of bakery items. Safe handling depends upon the items received. Bakery goods should be received at the temperature specified by the manufacturer, and packaging must be intact. Reject the product if it has passed the use-by date or there are signs of mold or pest damage. Bakery items should also be rejected if the packaging is torn or shows signs of pest damage. (See *Exhibit 6r.*)

Potentially Hazardous Hot Food

Occasionally, operators may receive a shipment of potentially hazardous hot food. It must be properly cooked as required by local or federal codes. If you purchase hot food, make sure suppliers have a HACCP plan or other means of documenting proper cooking methods and temperatures.

Potentially hazardous hot food must be received at 135°F (57°C) or higher. The shipping container must be able to maintain this temperature, and it must be undamaged. Reject the product if it does not meet these requirements.

SUMMARY

Although federal and state agencies regulate and monitor the production and transportation of food such as meat, poultry, seafood, eggs, dairy products, and canned goods, it is your responsibility to check the quality and safety of food that comes into your establishment.

Make sure suppliers are getting their products from approved sources—those that have been inspected and are in compliance with local, state, and federal law. Take steps to ensure that the suppliers you have chosen are reputable by asking other operators what their experiences have been with those suppliers.

Operators must plan delivery schedules so products can be handled promptly and correctly. Employees assigned to receive deliveries should be trained to inspect food properly as well as to distinguish between products that are acceptable and those that are not. They should also be authorized to reject products that do not meet company standards and to sign for products that do.

All products arriving at the establishment should meet agreed-upon standards. Packaging should be clean and undamaged. Use-by dates should be current. Food should not show signs of mishandling.

Products must be delivered at the proper temperature. All food items—especially meat, poultry, and fish—should be checked for proper color, texture, and odor. Live, molluscan shellfish and crustaceans must be delivered alive. Eggs should be inspected for freshness and for dirty and cracked shells. Dairy products must be checked for freshness. Produce should be fresh and undamaged. Refrigerated ready-to-eat items should be received at 41°F (5°C) or lower unless otherwise specified by the manufacturer. Packaging should be intact and in good condition. Frozen food should be inspected for signs of thawing and refreezing. ROP food should not bubble, appear slimy, or have excessive liquid. Canned food must be carefully examined for signs of damage. Dry food should be inspected for pest infestation and moisture. Bakery goods should not be moldy or show signs of pest damage, and they should not have passed their expiration date. Potentially hazardous hot food must be delivered at 135°F (57°C) or higher.

Apply Your Knowledge

A Case in Point 1

1 What was done incorrectly?

On Monday, a large food delivery arrived at the Sunnydale Nursing Home during the busy lunch hour. It included: cases of frozen ground beef patties, canned vegetables, frozen shrimp, fresh tomatoes, a case of potatoes, and fresh chicken.

Betty, the new assistant manager, thought the best thing to do was to put everything away and check it later, since she was very busy. She told Ed, in charge of receiving, to sign for the delivery and put the food into storage. Ed asked her if it would be better to ask the delivery driver to come back later. Since she needed the chicken for dinner, Betty asked Ed to accept the delivery now, and then she went back to the front of the house.

Ed put the frozen shrimp and ground beef patties in the freezer and the fresh chicken in the refrigerator. Then he put the fresh tomatoes, potatoes, and canned vegetables in dry storage. When he was finished, he went back to work in the kitchen.

For answers, please turn to the Answer Key.

Apply Your Knowledge

A Case in Point 2

1 What was done wrong?

2 What could be the result?

ABC Seafood makes its usual Thursday afternoon delivery to The Fish House. John, a prep cook, is the only person in the kitchen when the driver rings the bell at the back door. The kitchen manager, who is in charge of receiving, is in a managers' meeting. The chef is out on an errand, and the rest of the kitchen staff are on break, although some are still in the restaurant.

John follows the driver onto the dock, where the driver unloads two crates of ice-packed fresh fish, bags of live mussels, two buckets of live oysters, a case of shucked oysters in plastic containers, and a case each of frozen shrimp and frozen lobster tails. John goes back into the kitchen to get a bimetallic stemmed thermometer and remembers to look in the chef's office for a copy of the order form. He takes both out to the dock and begins to inspect the shipment.

John checks the products against both the order sheet and the invoice, then begins to check product temperatures. First, he checks the temperature of the shucked oysters by taking the cover off one container and inserting the thermometer stem into it. The thermometer reads 45°F (7°C).

After wiping the thermometer stem on his apron, John checks the internal temperature of a whole fish packed in ice. Finally, he reaches into one of the buckets of live oysters with his hand to see if it feels cold. He notices a few mussels and oysters with open and broken shells. He removes those with broken shells, knowing the chef will not use them.

John records all his findings on the order sheet he took from the chef's office, signs for the delivery, and starts putting the products away. He first puts away the live shellfish. Then he puts the shucked oysters and fresh fish in the refrigerator and the shrimp and lobster into the freezer.

For answers, please turn to the Answer Key.

Apply Your Knowledge

Use these questions to review the concepts presented in this chapter.

Discussion Questions

1. What are some general guidelines for receiving food safely?
2. What are proper methods for checking the temperature of fresh poultry delivered on ice and bulk milk? What should the temperature be for each?
3. What are three conditions that would result in rejecting a shipment of fresh poultry?
4. What types of external damage to cans are cause for rejection?

For answers, please turn to the Answer Key.

Apply Your Knowledge

Use these questions to test your knowledge of the concepts presented in this chapter.

Multiple-Choice Study Questions

1. Which is the most important factor in choosing a food supplier?
   A. Its prices are the lowest.
   B. Its warehouse is close to your establishment.
   C. It offers a convenient delivery schedule.
   D. It has been inspected and is compliant with local, state, and federal law.
2. Which shipment should be rejected?
   A. Beef that is bright, cherry red in color
   B. Lamb with flesh that is firm and springs back when touched
   C. Fish that arrives with sunken eyes
   D. Chicken received at 41°F (5°C)
3. How should whole, fresh salmon be packaged for delivery and storage?
   A. Layered with salt
   B. Vacuum sealed
   C. Wrapped in dry, clean cloth
   D. Packed in self-draining crushed ice
4. Which condition would cause you to reject a shipment of live oysters?
   A. Most of the oysters have closed shells.
   B. The oysters have a mild seaweed smell.
   C. Most of the oysters shells are open and do not close when they are tapped.
   D. The oysters are delivered with shellstock identification tags.

Continued on the next page...

Apply Your Knowledge

Multiple-Choice Study Questions *continued*

5. A box of sirloin steaks carries both a USDA inspection stamp and a USDA grading stamp. What do these stamps tell you?
   A. The farm that supplied the beef uses only USDA-certified animal feed.
   B. The meat and processing plant have met USDA standards, and the meat quality is acceptable.
   C. The meat wholesaler meets USDA quality-grading standards.
   D. The steaks are free of disease-causing microorganisms.

6. Which condition would cause you to reject a shipment of fresh chicken?
   A. The flesh is firm and springs back when touched.
   B. The wing tips are brown.
   C. The grading stamp is missing.
   D. The chicken is odorless.

7. Statements from a dairy supplier's sales brochure are listed below. Which statement should tell you not to hire this supplier?
   A. We make our cheese with only the freshest unpasteurized milk.
   B. From farm to you, our milk is kept at temperatures below 41°F (5°C).
   C. Money-back guarantee—our prices are the lowest in the area.
   D. We deliver according to your schedule and needs.

8. A shipment of eggs should be rejected for all of these reasons *except*
   A. the shells are cracked.
   B. they have a sulfur smell.
   C. they lack an inspection stamp.
   D. the air temperature of the delivery truck was 45°F (7°C).

Apply Your Knowledge

Multiple-Choice Study Questions

9. Which item does *not* have to be received at 41°F (5°C) or lower?
   A. Beef
   B. Live shellfish
   C. Pork
   D. Fish

10. All of these would be grounds for rejecting a case of frozen food *except*
    A. there are large ice crystals on the frozen food inside the case.
    B. the outside of the case is water stained.
    C. the food in the box is frozen solid.
    D. there is frozen liquid at the bottom of the case.

11. Which delivery should be rejected?
    A. Several cans in a case of peaches have torn labels.
    B. Several cans in a case of tomato soup have swollen ends.
    C. A bag of oatmeal is delivered at 60°F (16°C).
    D. A case of rice is missing a USDA inspection stamp.

12. You have just received a shipment of pumpkin pies. What would be grounds for rejecting the shipment?
    A. The pies have been received frozen.
    B. The pies have passed their expiration date.
    C. The pies contain preservatives.
    D. The pies have been received at 41°F (5°C).

For answers, please turn to the Answer Key.

Take It Back*

The following food safety concepts from this chapter should be taught to your employees:

- Checking the temperature of various types of food
- When to accept or reject a delivery

The tools below can be used to teach these concepts in fifteen minutes or less using the directions below. Each tool includes content and language appropriate for employees. Choose the tool or tools that work best.

Tool #1: ServSafe Video 4: *Purchasing, Receiving, and Storage*	**Tool #2:** *ServSafe Employee Guide*	**Tool #3:** ServSafe Posters and Quiz Sheets	**Tool #4:** ServSafe Fact Sheets and Optional Activities
			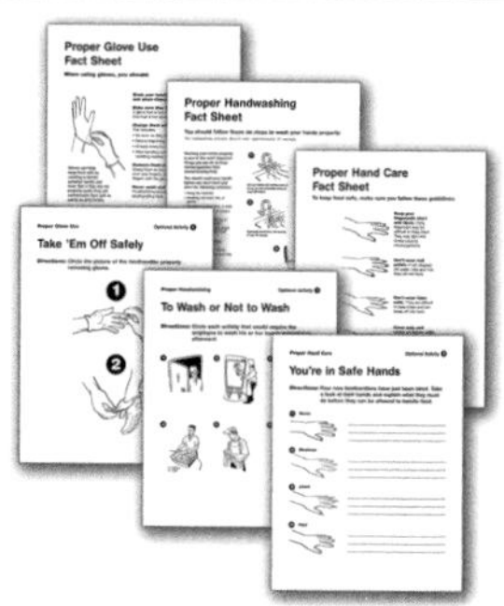
Checking the Temperature of Various Types of Food			
	Section 3 **Checking the Temperature of Various Food** Discuss with employees proper thermometer placement for checking the temperature of different food items.		

* Visit the Food Safety Resource Center at *www.ServSafe.com/FoodSafety/resource* to download free posters, quiz sheets, fact sheets, and optional activities and to learn how to obtain *Employee Guides* and videos/DVDs.

Take It Back*

Tool #1: ServSafe Video 4: *Purchasing, Receiving, and Storage*	**Tool #2:** *ServSafe Employee Guide*	**Tool #3:** ServSafe Posters and Quiz Sheets	**Tool #4:** ServSafe Fact Sheets and Optional Activities
When to Accept or Reject a Delivery			
Video segment on accepting and rejecting deliveries 1 Show employees the segment. 2 Ask employees to describe conditions that would be grounds for rejecting a shipment of meat, poultry, fish, eggs, produce, refrigerated ready-to-eat food, frozen food, and dry goods.	**Section 3** **When to Accept or Reject a Delivery** 1 Discuss with employees criteria for accepting or rejecting various food items during receiving. 2 Complete the Accept or Reject: You Make the Call activity.		

* **Visit the Food Safety Resource Center at *www.ServSafe.com/FoodSafety/resource* to download free posters, quiz sheets, fact sheets, and optional activities and to learn how to obtain *Employee Guides* and videos/DVDs.**

ADDITIONAL RESOURCES

Articles and Texts

Association of Food and Drug Officials. *A Pocket Guide to Can Defects.* Available through ***www.afdo.org/afdo/publication/index.cfm.***

Brownell, Judi and Dennis Reynolds. 2002. Actions that Make a Difference: Strengthening the F&B Purchaser-Supplier Partnership. *The Cornell Hotel and Restaurant Administration Quarterly.* 43 (6): 49.

Cadwallader, Keith R. and Hugo Weenen. *Freshness and Shelf Life of Foods.* New York: Oxford University Press, Incorporated, 2002.

Feinstein, Andrew H. and John M. Stefanelli. *Purchasing: Selection and Procurement for the Hospitality Industry.* Hoboken, NJ: John Wiley & Sons, Inc., 2002.

Han, Jung H. *Innovations in Food Packaging.* San Diego, Academic Press, 2005.

Lewis, Michael J. and Neil J. Heppell. *Continuous Thermal Processing of Foods: Pasteurization and UHT Sterilization.* New York: Springer, 2000.

Man, D.M.D. and A.A. Jones. *Shelf-Life Evaluation of Foods, 2nd edition.* New York: Springer, 2000.

Reed, Lewis. Specs: *The Foodservice and Purchasing Specification Manual.* Hoboken, NJ: John Wiley & Sons, Inc., 2005

Web Sites

Alaska Seafood Marketing Institute

www.alaskaseafood.org

The Alaska Seafood Marketing Institute (ASMI) is a public agency functioning as the state of Alaska's seafood marketing arm. ASMI's Web site provides information regarding food safety and quality assurance practices associated with Alaskan seafood.

American Egg Board

www.aeb.org

The American Egg Board (AEB) is the U.S. egg producers' link to the consumer. This Web site provides resources for the restaurant and foodservice industry, including ways to ensure safe receiving, storage and preparation of eggs, and general information about eggs and egg products.

American Lamb Board

www.americanlambboard.org

The American Lamb Board (ALB) was created by the U.S. Secretary of Agriculture to administer the Lamb Promotion, Research and Information Order. This Web site provides information about foodservice cuts, cooking temperatures, and food safety tips for lamb.

Conference for Food Protection

www.foodprotect.org

The Conference for Food Protection (CFP) is a nonprofit organization that provides a forum for regulators, industry professionals, academia, professional organizations, and consumers to identify problems, formulate recommendations, and develop and implement practices that ensure food safety. Though the conference has no formal regulatory authority, it is an organization that influences model laws and regulations among all government agencies. Visit this Web site for information regarding previously recommended changes to the *FDA Food Code,* standards for permanent outdoor cooking facilities and other guidance documents, along with how to participate in the conference.

Food and Drug Administration

www.fda.gov

The Food and Drug Administration (FDA) is responsible for promoting and protecting public health by helping safe and effective products reach consumers, monitoring products for continued safety, and helping the public obtain accurate, science-based information needed to improve health. Visit this Web site for information about all programs regulated by the FDA, including food product labeling, bottled water, and all food products except meat and poultry.

FDA Center for Food Safety and Applied Nutrition

www.cfsan.fda.gov/list.html

As the center within the Food and Drug Administration (FDA) responsible for food safety, the Center for Food Safety and Applied Nutrition (CFSAN) promotes and protects the public health by researching and implementing guidelines, policies, and standards to ensure that food is safe, nutritious, wholesome, and properly labeled. This Web site provides information relevant to all aspects of food safety and security, including corresponding guidelines, policies, and standards.

Food Products Association

www.fpa-food.org

The Food Products Association represents the food and beverage industry in the United States and worldwide. FPA's laboratory centers, scientists, and professional staff provide technical and regulatory assistance to member companies and represent the food industry on scientific and public policy issues involving food safety, food security, nutrition, consumer affairs, and international trade. This Web site posts news releases and provides the latest information and resources on food safety and security issues.

Food Safety and Inspection Service

www.fsis.usda.gov

The Food Safety and Inspection Service (FSIS) is the public health agency within the U.S. Department of Agriculture responsible for ensuring that the nation's commercial supply of meat, poultry, and egg products is safe, wholesome, and correctly labeled and packaged. Visit this Web site for food safety information and regulations related to meat, poultry, and eggs.

Gateway to Government Food Safety Information

www.foodsafety.gov

This Web site provides links to selected government food safety-related information.

International Dairy Foods Association

www.idfa.org

International Dairy Foods Association (IDFA) is the Washington, D.C.–based organization representing the nation's dairy-processing and manufacturing industries and their suppliers. IDFA is composed of three constituent organizations: the Milk Industry Foundation (MIF); the National Cheese Institute (NCI); and the International Ice Cream Association (IICA). Visit this Web site to access the Pasteurized Milk Ordinance, dairy product quality standards, and regulations impacting the dairy industry.

International Fresh-Cut Produce Association

www.fresh-cuts.org

The International Fresh-Cut Produce Association (IFPA) represents processors, distributors, retail, and foodservice buyers of fresh-cut produce and companies that supply goods and services in the fresh-cut industry. IFPA provides its members with the knowledge and technical information necessary to deliver convenient safe and wholesome food. This Web site offers a detailed list of frequently asked questions about the fresh-cut produce industry and has a contact feature in which to ask a question. Free publications on food safety and best practices for fresh-cut produce are also available.

Mushroom Council

www.mushroomcouncil.org

The Mushroom Council, composed of fresh-market producers or importers of mushrooms, administers a national promotion, research, and consumer information program to maintain and expand markets for fresh mushrooms. Information about shipping, handling, and preparation of mushrooms, as well as risk associated with the use of mushrooms harvested by inexperienced mushroom hunters, is available on this Web site.

National Cattlemen's Beef Association

www.beef.org

The National Cattlemen's Beef Association (NCBA) is the national trade association representing U.S. cattle producers. The organization works to advance the economic, political, and social interests of the U.S. cattle industry. This Web site links to other NCBA-associated sites addressing topics such as nutrition, research, food safety, and marketing. A separate site, *www.beeffoodservice.com,* provides recipes, facts on foodservice cuts, and food safety information specifically for restaurant and foodservice operators.

National Chicken Council

www.nationalchickencouncil.com

The National Chicken Council (NCC) is the national, non-profit trade association for the U.S. chicken industry representing integrated chicken producer/processors, poultry distributors, and allied firms. Visit this Web site for consumer consumption behaviors, product information, and research related to poultry-associated foodborne illnesses.

National Frozen & Refrigerated Foods Association

www.nfraweb.org

The mission of the National Frozen & Refrigerated Foods Association is to promote the sales and consumption of frozen and refrigerated foods through education, training, research, sales planning, and menu development. Visit this Web site for information on the proper receiving, storage, and preparation of frozen and refrigerated food, including shelf-life charts, tips, and recommended food and equipment storage temperatures.

National Pork Producers Council

www.nppc.org

The National Pork Producers Council (NPPC) is one of the nation's largest livestock commodity organizations. Visit the NPPC's foodservice site, *www.porkfoodservice.com,* for pork preparation tips, recipes, and consumer consumption trends.

National Turkey Federation

www.eatturkey.com

The National Turkey Federation is the national advocate for all segments of the U.S. turkey industry. This Web site contains information useful to foodservice establishments, including safe purchasing, storing, thawing, and preparing information as well as recipes, nutrition information, and promotional ideas for turkey.

Produce Marketing Association

www.pma.com

The Produce Marketing Association (PMA) is a nonprofit association serving members who market fresh fruit, vegetables, and related products worldwide. Their members represent the production, distribution, retail, and foodservice sectors of the industry. Visit this Web site for articles and publications on produce safety information within the foodservice and retail industry.

Seafood Inspection Program

www.seafood.nmfs.noaa.gov

The National Oceanic and Atmospheric Administration (NOAA), part of the U.S. Department of Commerce (USDC), oversees fisheries management in the United States and provides a voluntary seafood-inspection service to the industry, which ensures compliance with all applicable food regulations, as well as product grading and certification services. Available on this Web site are the USDC Participants List, a reference tool for determining which fishery products have been produced in fish establishments approved by the USDC, and the U.S. standards for grades of fishery products.

United Fresh Fruit & Vegetable Association

www.uffva.org

The United Fresh Fruit & Vegetable Association is a national trade organization that represents the interests of growers, shippers, brokers, wholesalers, and distributors of produce. The association promotes increased produce consumption and provides scientific and technical expertise to all segments of the industry. This Web site contains news updates and free publications, and lists upcoming events and initiatives on produce safety and quality.

Documents and Other Resources

Commodity Specific Food Safety Guidelines for the Melon Supply Chain

www.pma.com/Template.cfm?Section=Food_Safety3&CONTENTID=8727&TEMPLATE=/ContentManagement/ContentDisplay.cfm

The *Commodity Specific Food Safety Guidelines for the Melon Safety Chain,* available at this Web site, provides voluntary recommended guidelines on food safety practices that are intended to minimize the microbiological hazards associated with fresh and fresh-cut melon products. The intent of this document is to provide information on food safety and handling in a manner consistent with existing applicable regulations, standards, and guidelines.

2005 *FDA Food Code*

www.cfsan.fda.gov/~dms/fc05-toc.html

The Food and Drug Administration (FDA) publishes the *FDA Food Code,* a scientifically-sound technical and legal document that serves as a model for regulating the retail and foodservice industry at the federal, state, and local level. Updates to the code are issued on the odd-numbered years. The *FDA Food Code* provides a system of safeguards designed to minimize foodborne illness and ensure employee health, food protection manager knowledge, safe food, nontoxic and cleanable equipment, and appropriate sanitation of the food establishment. It is used as the basis for information in this textbook.

Good Agricultural Practices: A Self Audit for Growers and Handlers

www.ucce.ucdavis.edu/files/filelibrary/5453/4362.pdf

The *Good Agricultural Practice: A Self Audit for Growers and Handlers* is available on this Web site. This document is an example of a Good Agricultural Practices (GAP) audit that contains an outline of the fundamental components of a good food safety approach at the grower, shipper, and distributor levels.

Interstate Certified Shellfish Shippers List

www.cfsan.fda.gov/~ear/shellfis.html

The Food and Administration's Center for Food Safety and Applied Nutrition (FDA CFSAN) provides the Interstate Certified Shellfish Shippers list. Visit this Web site, which is updated monthly, for those shippers that have been certified by regulatory authorities in the United States, Canada, Chile, Korea, Mexico, and New Zealand under the uniform sanitation requirements of the national shellfish program.

National Shellfish Sanitation Program (NSSP) Guide for the Control of Molluscan Shellfish

www.cfsan.fda.gov/~ear/nss2-toc.html

The Food and Administration's Center for Food Safety and Applied Nutrition (FDA CFSAN) provides the NSSP Guide for the Control of Molluscan Shellfish. Available at this Web site, the guide consists of a Model Ordinance, supporting guidance documents, recommended forms, and other related materials associated with the program. The Model Ordinance includes guidelines to ensure that the shellfish produced in the United States in compliance with the guidelines are safe and sanitary. In addition, the Model Ordinance provides readily adoptable standards and administrative practices necessary for the sanitary control of molluscan shellfish.

Seafood Information and Resources

www.cfsan.fda.gov/seafood1.html

The Food and Administration's Center for Food Safety and Applied Nutrition (FDA CFSAN) provides a wealth of seafood information and resources on this Web site. Visit this site for an overview of the FDA seafood regulatory program, information on seafood-related foodborne pathogens and contaminants, and access to seafood guidance and regulation documents.

Whole Berry
CRANBERRY SAUCE
Whole Berry
CRANBERRY SAUCE
SALSA MEXICANA
Nutrition Facts
SWEET RELISH
DICED TOMATOES

The Flow of Food: Storage

Inside this chapter:

- General Storage Guidelines
- Types of Storage
- Storage Techniques
- Storing Specific Food

After completing this chapter, you should be able to:

- Properly label and date-mark refrigerated, frozen, and dry food prior to storage.
- Properly store refrigerated, frozen, dry, and canned food.
- Apply first in, first out (FIFO) practices as they relate to refrigerated, frozen, and dry-storage areas.
- Properly store raw food to prevent cross-contamination.
- Identify temperature requirements for refrigerated and dry-storage areas.
- Identify proper storage containers for refrigerated, frozen, and dry food.

Key Terms

- First in, first out (FIFO)
- Refrigerated storage
- Frozen storage
- Dry storage
- Shelf life
- Hygrometer

Apply Your Knowledge

Check to see how much you know about the concepts in this chapter. Use the page references provided with each question to explore the topic.

Test Your Food Safety Knowledge

1. **True or False:** Potato salad that has been prepared in-house and stored at 41°F (5°C) must be discarded after three days. *(See page 7-3.)*

2. **True or False:** Food can be stored near chemicals as long as the chemicals are stored in sturdy, clearly labeled containers. *(See page 7-4.)*

3. **True or False:** Storing cans of stewed tomatoes at 65°F (18°C) is acceptable. *(See page 7-12.)*

4. **True or False:** Raw chicken must be stored below ready-to-eat food, such as pumpkin pie, if it is stored in the same walk-in refrigerator. *(See page 7-6.)*

5. **True or False:** If stored food has passed its expiration date, you should cook and serve it at once. *(See page 7-3.)*

For answers, please turn to the Answer Key.

INTRODUCTION

When food is stored improperly and not used in a timely manner, quality and safety suffer. Poor storage practices can cause food to spoil quickly, with potentially serious results.

GENERAL STORAGE GUIDELINES

Every facility has a wide variety of products that need to be stored. A few general rules can be applied to most storage situations.

- **Label food.** All potentially hazardous, ready-to-eat food prepared on site that has been held for longer than twenty-four hours must be properly labeled. The label must include the name of the food and the date it should be sold, consumed, or discarded. If an item has been previously cooked and stored and is later mixed with another food item to make a new dish, the label on the new dish must indicate the discard date for the previously cooked item. For example, if ground beef has been cooked and stored at 41°F (5°C) or lower and later used

Exhibit 7a

Follow FIFO When Storing Food

One way to follow FIFO is to store products with the earliest use-by or expiration dates in front of products with later dates and use those first.

to make meat sauce, the meat sauce must be labeled with the discard date of the ground beef.

- **Rotate products to ensure that the oldest inventory is used first.** The first in, first out (FIFO) method is commonly used to ensure that refrigerated, frozen, and dry products are properly rotated during storage. By this method, a product's use-by or expiration date is first identified. The products are then stored to ensure that the oldest are used first. One way to do this is to train employees to store products with the earliest use-by or expiration dates in front of products with later dates. Once shelved, those stored in front are used first. (See *Exhibit 7a.*)
- **Discard food that has passed the manufacturers' expiration date.** All potentially hazardous, ready-to-eat food that has been prepared in-house can be stored for a maximum of seven days at 41°F (5°C) or lower before it must be thrown out.
- **Establish a schedule to ensure that stored product is depleted on a regular basis.** If a food item has not been sold or consumed by a predetermined date, throw it out, clean and sanitize the container, and refill the container with new product. For example, flour stored in plastic bins should be used within six to twelve months from the time it was placed in the bins. After that time period, the bins should be emptied, the flour discarded, and the bins cleaned and sanitized.
- **Transfer food between containers properly.** If food is removed from its original package, put it in a clean, sanitized container and cover it. The new container must be labeled with the name of the food and the original use-by or expiration date. Never use empty food containers to store chemicals or put food in empty chemical containers.
- **Keep potentially hazardous food out of the temperature danger zone.** Store deliveries as soon as they have been inspected. Take out only as much food as you can prepare at one time, and put prepared food away until needed. Properly cool and store cooked food as soon as it is no longer needed. (See Chapter 8 for more information on cooling cooked food.)

Exhibit 7b

Improper Storage

Never store food near chemicals or cleaning supplies.

Key Point

Store potentially hazardous food in refrigerators at an internal temperature of 41°F (5°C) or lower.

- **Check temperatures of stored food and storage areas.** Temperatures should be checked at the beginning of the shift. Many establishments use a preshift checklist to guide employees through this process.
- **Store food in designated storage areas.** Do not store food near chemicals or cleaning supplies. (See *Exhibit 7b.*) Avoid storing food in restrooms, locker rooms, janitor closets, furnace rooms, vestibules, or under stairways or pipes of any kind; food can easily be contaminated in any of these areas.
- **Keep all storage areas clean and dry.** Floors, walls, and shelving in refrigerators, freezers, dry storerooms, and heated holding cabinets should be properly cleaned on a regular basis. Clean up spills and leaks right away to keep them from contaminating other food.
- **Clean dollies, carts, transporters, and trays often.**

TYPES OF STORAGE

Most restaurants and foodservice establishments have several types of storage areas in their facilities. The most common include:

- **Refrigerated storage.** These areas are typically used to hold potentially hazardous food at 41°F (5°C) or lower. Refrigeration slows the growth of microorganisms and helps keep them from growing to levels high enough to cause illness. (Some jurisdictions allow food to be held at an internal temperature of 45°F [7°C] or lower. Check with the local regulatory agency for specific regulations.)
- **Frozen storage.** These areas are used to hold frozen food at temperatures that will keep it frozen. Freezing does not kill all microorganisms, but it does slow their growth substantially.
- **Dry storage.** These areas are used to hold dry and canned food. To maintain the quality of this food, dry-storage areas should be kept at the appropriate temperature and humidity levels. Storerooms should be clean, well ventilated, and well lighted.

Managers should monitor the use of storage areas since improper storage practices can affect the safety of food. For example, an overstocked refrigerator may not be able to hold the proper temperature and may not allow stock to be rotated properly.

Storage spaces should also be located to ease the flow of food through the operation and to prevent food contamination. They must be accessible to receiving, food-preparation, and cooking areas, but must be located so that food is stored away from dishwashing and garbage areas.

STORAGE TECHNIQUES

A few commonsense rules apply to each of these storage areas. Make sure employees follow these rules to keep food safe.

Exhibit 7c

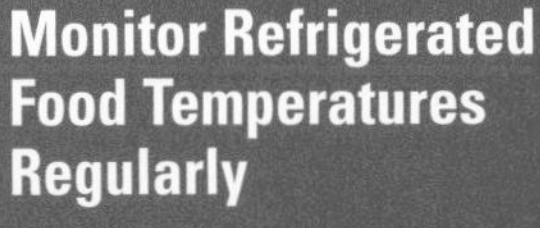

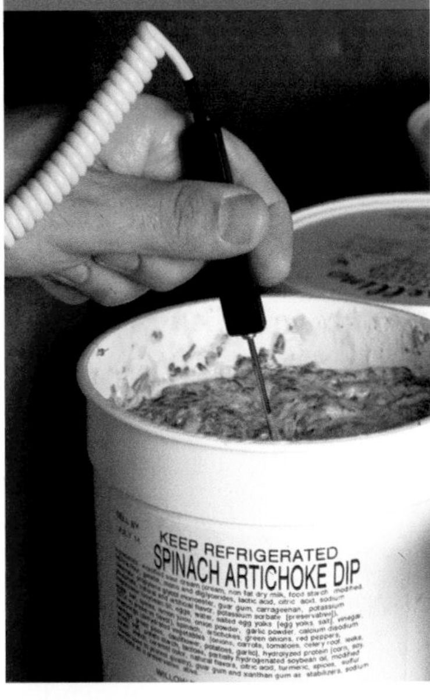

Randomly sample a food's internal temperature using a calibrated thermometer.

Refrigerated Storage

Keeping food as cold as possible without freezing extends its **shelf life,** the amount of time food will remain suitable for use. Ideal storage temperatures will vary depending on the food. Fruit and vegetables will freeze if stored at temperatures ideal for fish. Meat and poultry will have a shorter shelf life if stored at temperatures better suited for produce. If possible, store food such as meat and poultry in separate refrigerators to hold them at optimal temperatures. If this is impractical, store meat, poultry, fish, and dairy products in the coldest part of the unit, away from the door.

While there are many types of refrigeration equipment available to operators, from walk-in refrigerators to refrigerated drawers, some general guidelines apply when using all of them:

- **Set refrigerators to the proper temperature.** The setting must keep the internal temperature of the food at 41°F (5°C) or lower. At least once during each shift, check the temperature of the unit. Use hanging thermometers in the warmest part of the refrigerator. Some units have a readout panel outside to check the temperature without opening the door. These should also be checked for accuracy.
- **Monitor food temperature regularly.** Randomly sample the internal temperature of stored food using a calibrated thermometer. (See *Exhibit 7c.*)

- **Do not overload refrigerators.** Storing too many products prevents good airflow and makes units work harder to stay cold.
- **Use open shelving.** Lining shelves with aluminum foil or paper restricts circulation of cold air in the unit.
- **Never place hot food in refrigerators.** This can warm the interior enough to put other food in the temperature danger zone. (See Chapter 8 for information on properly cooling food.)
- **Keep refrigerator doors closed as much as possible.** Frequent opening lets warm air inside, which can affect food safety and make units work harder. Consider using cold curtains in walk-in refrigerators to help maintain temperatures.
- **Wrap all food properly.** Leaving food uncovered can lead to cross-contamination.
- **Store raw meat, poultry, and fish separately from cooked and ready-to-eat food to prevent cross-contamination.** If they cannot be stored separately, store cooked or ready-to-eat food above raw meat, poultry, and fish. (See *Exhibit 7d.*) This will prevent raw product juices from dripping onto the prepared food and causing a foodborne illness. It is also recommended that raw meat, poultry, and fish be stored in the following top-to-bottom order in refrigerators: whole fish, whole cuts of beef and pork, ground meat and fish, whole and ground poultry. (See *Exhibit 7e.*) This order is based on the required minimum internal cooking temperature of each food.

Exhibit 7d

Improper Storage of Different Raw Products in the Same Refrigerator

Raw meat must never be stored above cooked and ready-to-eat food.

Exhibit 7e

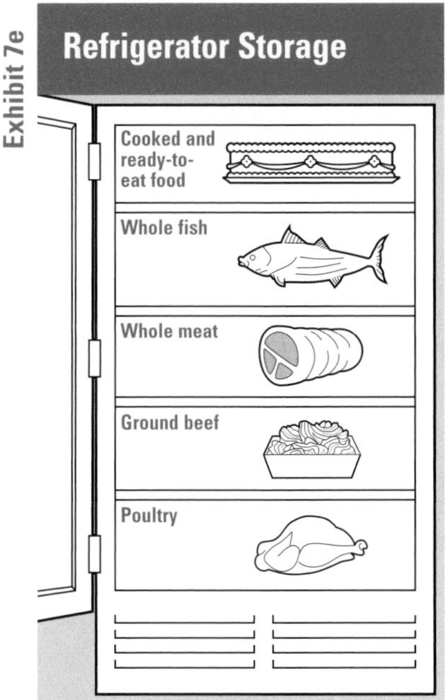

Recommended top-to-bottom order for storing different raw food in the same refrigerator

Frozen Storage

Following are some general guidelines for using freezers:

- **Keep freezers at a temperature that will keep products frozen.** This temperature will vary from product to product. A temperature that is good for one product may affect the quality of another.
- **Check freezer temperatures regularly.** Use a calibrated thermometer to check the accuracy of hanging thermometers or unit readout.
- **Place frozen food deliveries in freezers as soon as they have been inspected.** Never hold frozen food at room temperature.
- **Use caution when placing food into freezers.** Warm food can raise the temperature inside units and partially thaw the food inside. Store food to allow good air circulation. Overloading freezers makes them work harder and makes it harder to find and rotate food properly.
- **Defrost freezer units on a regular basis.** They will operate more efficiently when free of frost. Move food to another freezer while defrosting.
- **When freezing food that has been prepared on site, clearly label the food.** Identify the package's contents and use-by date, if there is one.
- **Keep the unit closed as much as possible.** Use cold curtains to help maintain temperatures.

Dry Storage

Dry food remains safe and retains quality if held in the right conditions. When storing dry food, follow these guidelines:

- **Keep storerooms cool and dry.** Moisture and heat are the biggest dangers to dry and canned food. For optimum quality and to assure safety, the temperature for the storeroom should be between 50°F and 70°F (10°C and 21°C). Keep relative humidity at 50 to 60 percent, if possible. Use a **hygrometer** to measure humidity. If high humidity is a problem, consider using a dehumidifier.

- **Make sure storerooms are well ventilated.** This will help keep temperature and humidity constant throughout the storage area.
- **Store dry food away from walls and at least six inches (fifteen centimeters) off the floor.**
- **Keep dry food out of direct sunlight.**
- **Keep the area clean and dry.**

STORING SPECIFIC FOOD

The general storage guidelines previously discussed apply to most food. However, certain types of food have special requirements.

Meat

- Store meat immediately after delivery and inspection in its own storage unit or in the coldest part of the refrigerator. Fresh meat must be held at an internal temperature of 41°F (5°C) or lower. Frozen meat should be stored at a temperature that will keep it frozen.
- Wrap meat in airtight, moisture-proof material or place it in cleaned and sanitized containers.
- Primal cuts, quarters, sides of raw meat, and slab bacon can be hung on clean, sanitized hooks or can be placed on sanitized racks. To prevent cross-contamination, do not store meat above any other food.
- Throw out any meat showing signs of spoilage.

Key Point

Store fresh raw meat, poultry, and fish at an internal temperature of 41°F (5°C) or lower.

Poultry

- Store raw, fresh poultry at an internal temperature of 41°F (5°C) or lower. Frozen poultry should be stored at temperatures that will keep it frozen. If it has been removed from its original packaging, place it in airtight containers or wrap it in airtight material.
- Ice-packed poultry can be stored in a refrigerator as is. Use containers that are self-draining. Change the ice, and sanitize the container often.

Key Point

Store live shellfish at an air temperature of 45°F (7°C) or lower.

Fish

Fresh fish is very sensitive to time-temperature abuse and can deteriorate quickly if handled improperly.

- Store fresh fish at an internal temperature of 41°F (5°C) or lower. Keep fillets and steaks in original packaging, or tightly wrap them in moisture-proof materials. Fresh, whole fish can be packed in flaked or crushed ice, and ice beds must be self-draining. Change the ice, and clean and sanitize the container regularly.
- Store frozen fish at temperatures that will keep it frozen. Wrap fish in moisture-proof packaging.

Shellfish

Store live shellfish in its original container at an air temperature of 45°F (7°C) or lower. Shellstock identification tags must be kept on file for ninety days from the harvest date.

Live, molluscan shellfish (clams, oysters, mussels, scallops) can be stored in a display tank prior to service under one of two conditions:

1. The tank carries a sign stating that the shellfish are for display only.
2. A variance is obtained from the local health department. To obtain a variance, a HACCP plan must be submitted showing that:
    - Water from other tanks will not flow into the display tank.
    - Using the display tank will not affect product quality or safety.
    - Shellstock ID tags have been retained as required.

Key Point

Store eggs at an air temperature of 45°F (7°C) or lower.

Eggs

- Eggs received at an air temperature of 45°F (7°C), in compliance with laws governing their shipment from suppliers, must be placed immediately after inspection in refrigeration equipment capable of maintaining an air temperature of 45°F (7°C) or lower. Maintain constant temperature and humidity levels in refrigerators used to store eggs.

- Do not wash eggs before storing them because they are washed and sanitized at the packing facility.
- Use the FIFO method of stock rotation. Plan to use all eggs within four to five weeks of the packing date.
- Keep shell eggs in cold storage right up until the time they are used. Take out only as many eggs as are needed for immediate use.
- Store frozen egg products at temperatures that will keep them frozen.
- Store liquid eggs according to the manufacturers' recommendations.
- Dried egg products can be stored in a cool, dry storeroom. Once they are reconstituted (mixed with water), store them in the refrigerator at 41°F (5°C) or lower. Do not reconstitute more dried egg product than is needed for immediate use.

Dairy Products

- Store dairy products at 41°F (5°C) or lower.
- Frozen dairy products such as ice cream and frozen yogurt can be stored at 6°F to 10°F (–14°C to –12°C).
- Always use the FIFO method of stock rotation. Discard products if they have passed their use-by or expiration dates.

Fresh Produce

- Fruit and vegetables have various temperature requirements for storage. While many whole, raw fruit and vegetables can be stored at 41°F (5°C) or lower, not all will be stored at these temperatures. Whole, raw produce and raw, cut vegetables—such as celery, carrots, and radishes—delivered packed in ice can be stored as is. The containers must be self-draining, and ice should be changed regularly.
- Fruit and vegetables kept in the refrigerator can dry out quickly. Keep the relative humidity at 85 to 95 percent.
- Although most produce can be stored in the refrigerator, avocados, bananas, pears, and tomatoes ripen best at room temperature.

Key Point

When soaking or storing produce in standing water or an ice-water slurry, do not mix different items or multiple batches of the same item.

Key Point

Always store ROP food at temperatures recommended by the manufacturer or at 41°F (5°C) or lower.

- Most produce should not be washed before storage. Moisture promotes the growth of mold in many instances. Instead, wash produce before preparing or serving it.
- When soaking or storing produce in standing water or an ice-water slurry, do not mix different items or multiple batches of the same item.
- Store whole citrus fruit, hard-rind squash, eggplant, and root vegetables—such as potatoes, sweet potatoes, rutabagas, and onions—in a cool, dry storeroom. Temperatures of 60°F to 70°F (16°C to 21°C) are best. Make sure containers are well ventilated. Store onions away from other vegetables that might absorb odor.

ROP Food

- Always store modified-atmosphere packaged (MAP), vacuum-packed, and *sous vide* food at temperatures recommended by the manufacturer or at 41°F (5°C) or lower. Frozen product should be stored at temperatures that will keep it frozen. Store and handle these products carefully.
- Vacuum packaging will not stop the growth of microorganisms that do not require oxygen to grow. ROP products are especially susceptible to the growth of *Clostridium botulinum.* Discard product if the package is torn or slimy, if it contains excessive liquid, or if the product bubbles, indicating the possible growth of *Clostridium botulinum.*
- Always check the expiration date before using MAP, vacuum-packed, and *sous vide* products. Labels should clearly list contents, storage temperature, preparation instructions, and a use-by date.
- Operators who produce ROP food on site must follow specific labeling rules. The Food and Drug Administration's (FDA) rules for processing ROP food on site are very strict.

UHT and Aseptically Packaged Food

- Food that has been pasteurized at ultra-high temperatures (UHT) and aseptically packaged (the packaging is free of microorganisms) can be stored at room temperature. Since much of this food is served cold, such as milk and pudding, you might want to store it in the refrigerator.
- Once opened, store UHT, aseptically packaged food in the refrigerator at 41°F (5°C) or lower.
- UHT products not aseptically packaged must be stored at an internal temperature of 41°F (5°C) or lower.

Key Point

Storage temperatures higher than 70°F (21°C) may shorten the shelf life of canned goods.

Canned Goods

- Store canned goods and other dry food at a temperature between 50°F and 70°F (10°C to 21°C). Even canned food spoils over time. Higher storage temperatures may shorten shelf life. Acidic food, such as canned tomatoes, does not last as long as food low in acid. The acid can also form pinholes in the metal over time.
- Discard damaged cans.
- Keep storerooms dry. Too much moisture will cause cans to rust.
- Wipe cans clean with a sanitized cloth before opening them to help prevent dirt from falling into the contents of the can.

Dry Food

- Keep flour, cereal, and grain products such as pasta or crackers in airtight containers. They can quickly become stale in a humid room and can become moldy if there is too much moisture.
- Before using dry food, check containers or packages for damage from insects or rodents. Cereal and grain products are favorite targets for these pests.
- Salt and sugar, if stored in the right conditions, can be held almost indefinitely.

SUMMARY

When food is stored improperly, quality and safety will suffer. Although different food has different storage needs, some common rules apply. Food should be stored in designated areas and rotated to ensure that the oldest product is used first. It should also be stored in its original packaging. If food must be removed from its original packaging, wrap it in clean, moisture-proof materials or place it in clean and sanitized containers with tight-fitting lids. Make sure all packaging and containers are labeled with the name of the food being stored.

All potentially hazardous, ready-to-eat food that is prepared on site and held for longer than twenty-four hours must be properly labeled. The label must include the name of the food and the date it should be sold, consumed, or discarded. It can be stored for a maximum of seven days at 41°F (5°C) or lower before it must be discarded. Throw out all food that has passed its manufacturers' expiration date. Check the temperatures of stored food and the storage area regularly, and keep these areas clean and dry to prevent contamination.

Refrigerators must be set to the proper temperature to slow the growth of microorganisms. The setting must keep the internal temperature of the food at 41°F (5°C) or lower. Never place hot food in refrigerators, which could raise the temperature inside. Do not line refrigerator shelves, overload units, or open doors too often. These practices make units work harder to maintain the temperature inside. If possible, store raw meat, poultry, and fish separately from cooked and ready-to-eat food to prevent cross-contamination. If not, store these items below cooked or ready-to-eat food. Product temperatures should be checked periodically.

While freezer temperatures do not kill bacteria, they do slow their growth substantially. Freezers should be kept at a temperature that will keep product frozen. Unit temperatures should be checked often.

Dry-storage areas should be kept at the appropriate temperature and humidity levels and should be clean and well ventilated to maintain food quality. Food in dry storage should be stored away from walls and at least six inches (fifteen centimeters) off the floor. Do not store food products near chemicals or cleaning supplies since food can easily become contaminated. Empty food containers should never be used to store chemicals.

Fresh meat, poultry, fish, and dairy products should be stored at 41°F (5°C) or lower. Fish and poultry can be stored under refrigeration in crushed ice as long as the containers are self-draining, the ice is changed, and the container is cleaned and sanitized regularly. Eggs should be refrigerated at an air temperature of 45°F (7°C) or lower, right up until they are used.

Live, molluscan shellfish should be stored in their original containers at an air temperature of 45°F (7°C). Fresh produce has various temperature requirements for storage. Produce should not be washed before storage because it can promote mold growth. ROP food should be stored at temperatures recommended by the manufacturer. Packages should be checked for signs of contamination, including bubbling, excessive liquid, tears, and slime. Once opened, UHT and aseptically packaged food should be stored in the refrigerator at 41°F (5°C) or lower. Dry and canned food should be stored at temperatures between 50°F and 70°F (10°C and 21°C).

Apply Your Knowledge

1. What storage errors were made?
2. What food items are at risk?

A Case in Point 1

On Monday afternoon, the kitchen staff at the Sunnydale Nursing Home were busy cleaning up after lunch and preparing for dinner. Pete, a kitchen assistant, put a large stockpot of hot, leftover vegetable soup in the refrigerator to cool. Angie, a cook, began deboning the chicken breasts that were stored earlier. When she finished, she put the chicken on an uncovered sheet pan and stored the pan in the refrigerator. She carefully placed the raw chicken on the top shelf, away from the hot soup. Next, Angie iced a carrot cake she had baked that morning. She put the carrot cake in the refrigerator on the shelf directly below the chicken breasts.

For answers, please turn to the Answer Key.

Apply Your Knowledge

A Case in Point 2

What storage errors occurred?

A shipment was delivered to Enrico's Italian Restaurant on a warm summer day. Alyce, who was in charge of receiving for the restaurant, inspected the shipment and immediately proceeded to store the items. She loaded a case of sour cream on the dolly and wheeled it over to the reach-in refrigerator. When she opened the refrigerator, she noticed that it was tightly packed; however, she was able to squeeze the case into a spot on the top shelf. Next, Alyce wheeled several cases of fresh ground beef over to the walk-in refrigerator. She noticed that the readout on the outside of the walk-in indicated 39°F (4°C). Alyce pushed through the cold curtains and bumped into a hot stockpot of soup as she moved inside. She moved the soup over and made a space next to the door for the ground beef. Alyce said hello to Mary, who had just cleaned the shelving in the unit and was lining it with new aluminum foil. Alyce returned to the receiving area and loaded several cases of pasta on the dolly. She was sweating as she stacked the boxes on the shelving unit and gave a quick glance at the thermometer in the dry-storage room, which read 85°F (29°C). When she was finished stacking the boxes, Alyce returned the dolly to the receiving area.

For answers, please turn to the Answer Key.

Apply Your Knowledge

Use these questions to review the concepts presented in this chapter.

Discussion Questions

1. What is the recommended top-to-bottom order for storing the following food in the same refrigerator: raw trout, an uncooked beef roast, raw chicken, and raw ground beef?
2. How should food that has been taken out of its original package be stored?
3. What is the proper storage temperature for frozen dairy products?
4. Explain the FIFO method of stock rotation.

For answers, please turn to the Answer Key.

Apply Your Knowledge

Use these questions to test your knowledge of the concepts presented in this chapter.

Multiple-Choice Study Questions

1. Stored ground beef would most likely be unsafe to use at which internal temperature?
   A. 0°F (–17°C)
   B. 41°F (5°C)
   C. 30°F (–1°C)
   D. 60°F (16°C)
2. Dry-storage rooms should be kept at
   A. 35°F to 41°F (2°C to 5°C).
   B. 45°F to 50°F (7°C to 10°C).
   C. 50°F to 70°F (10°C to 21°C).
   D. 70°F to 80°F (21°C to 27°C).
3. Under which condition could you use a tank to display live mussels you will be serving to customers?
   A. You have obtained a variance from the health department.
   B. You clean the display tank at least once a month.
   C. You will mix them with other mussels that have not been on display.
   D. You have removed the shellstock identification tags as required by law.
4. Which is *not* a good storage practice?
   A. Shelving food based on its expiration date
   B. Storing raw poultry at temperatures between 41°F and 135°F (5°C and 57°C)
   C. Storing live shellfish at an air temperature of 45°F (7°C) or lower
   D. Storing raw meat below ready-to-eat food
5. The FIFO method helps ensure all of these during storage *except*
   A. products are properly rotated.
   B. the oldest products are used first.
   C. prepared items are used before they expire.
   D. items that have passed their expiration date are used first.

Apply Your Knowledge

Multiple-Choice Study Questions

6. When storing products using the FIFO method, the products with the earliest use-by dates should be
   A. stored in front of products with later use-by dates.
   B. stored behind products with later use-by dates.
   C. stored alongside products with later use-by dates.
   D. stored away from products with later use-by dates.
7. A restaurant that has prepared tuna salad can store it at 41°F (5°C) for a maximum of
   A. one day.
   B. three days.
   C. seven days.
   D. fourteen days.
8. When storing potentially hazardous food, what must be included on the label?
   A. Product ingredients
   B. Nutritional information
   C. Sell-by or discard date
   D. Potential allergens
9. All of these are incorrect storage practices *except*
   A. lining refrigerator shelving with aluminum foil.
   B. cooling hot food in a refrigerator.
   C. storing products with the earliest expiration dates in front of products with older dates.
   D. storing fresh poultry over ready-to-eat food.
10. Which storage practice is incorrect?
    A. Storing fresh lamb at 41°F (5°C)
    B. Storing eggs at room temperature
    C. Storing raw clams in their shipping crate at an air temperature of 45°F (7°C)
    D. Storing raw ground beef in its original packaging

For answers, please turn to the Answer Key.

Take It Back*

The following food safety concepts from this chapter should be taught to your employees:

- General storage principles
- Preventing contamination during storage
- Preventing time-temperature abuse during storage
- Storing food at the proper temperature
- Storing food according to first in, first out (FIFO)

The tools below can be used to teach these concepts in fifteen minutes or less using the directions below. Each tool includes content and language appropriate for employees. Choose the tool or tools that work best.

Tool #1: ServSafe Video 4: *Purchasing, Receiving, and Storage*

Tool #2: *ServSafe Employee Guide*

Tool #3: ServSafe Posters and Quiz Sheets

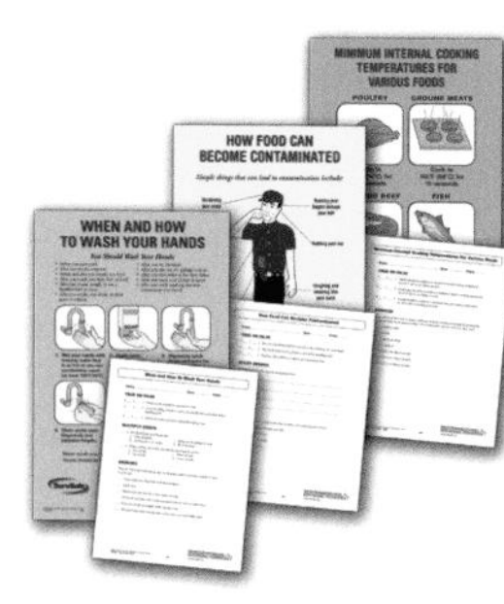

Tool #4: ServSafe Fact Sheets and Optional Activities

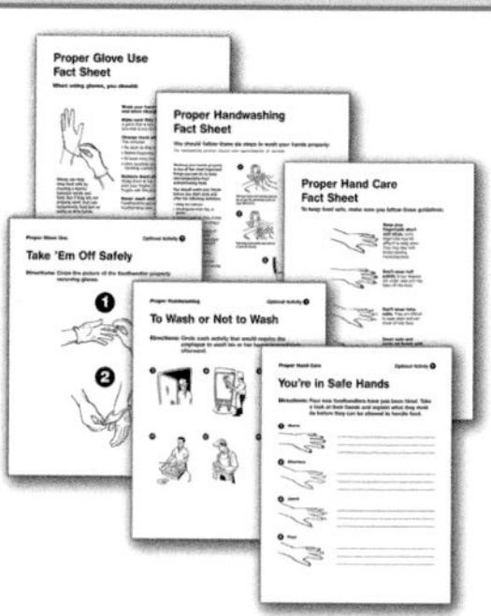

General Storage Principles

Tool #1 — Video segment on storage

1. Show employees the segment.
2. Ask employees what they can do to keep food safe during storage.

Tool #2 — Section 3

How to Properly Store Food

1. Discuss with employees general storage principles.
2. Complete the What Do You See? activity.

Tool #3 — Poster: How to Store Food Properly

1. Discuss with employees general storage principles as presented in the poster.
2. Have employees complete the Quiz Sheet: How to Store Food Properly.

Tool #4: (no entry)

* Visit the Food Safety Resource Center at *www.ServSafe.com/FoodSafety/resource* to download free posters, quiz sheets, fact sheets, and optional activities and to learn how to obtain *Employee Guides* and videos/DVDs.

Take It Back*

Tool #1: ServSafe Video 4: *Purchasing, Receiving, and Storage*	Tool #2: *ServSafe Employee Guide*	Tool #3: ServSafe Posters and Quiz Sheets	Tool #4: ServSafe Fact Sheets and Optional Activities
Preventing Contamination during Storage			
Video segment on storage 1 Show employees the segment. 2 Ask employees what they can do to prevent contamination during storage.	**Section 3** **How to Properly Store Food** Discuss with employees practices for preventing contamination during storage.	**Poster: How to Store Food Properly** 1 Discuss with employees practices for preventing contamination during storage as presented in the poster. 2 Have employees complete the Quiz Sheet: How to Store Food Properly.	**Preventing Cross-Contamination during Storage Fact Sheet** 1 Pass out a fact sheet to each employee. Discuss practices for preventing contamination during storage using the fact sheet. 2 Have employees complete one or two of the optional activities.

* Visit the Food Safety Resource Center at *www.ServSafe.com/FoodSafety/resource* to download free posters, quiz sheets, fact sheets, and optional activities and to learn how to obtain *Employee Guides* and videos/DVDs.

Take It Back*

Tool #1: ServsSafe Video 4: *Purchasing, Receiving, and Storage*	Tool #2: *ServSafe Employee Guide*	Tool #3: ServSafe Posters and Quiz Sheets	Tool #4: ServSafe Fact Sheets and Optional Activities
Preventing Time-Temperature Abuse during Storage			
Video segment on storage 1 Show employees the segment. 2 Ask employees what they can do to prevent time-temperature abuse during storage.	**Section 3** **How to Properly Store Food** Discuss with employees practices for preventing time-temperature abuse during storage.	**Poster: How to Store Food Properly** 1 Discuss with employees practices for preventing time-temperature abuse during storage as presented in the poster. 2 Have employees complete the Quiz Sheet: How To Store Food Properly.	**Preventing Time-Temperature Abuse during Storage Fact Sheet** 1 Pass out a fact sheet to each employee. Discuss practices for preventing time-temperature abuse during storage using the fact sheet. 2 Have employees complete one or two of the optional activities.
Storing Food at the Proper Temperature			
			Storing Food at the Proper Temperature Fact Sheet 1 Pass out a fact sheet to each employee. Discuss proper storage temperatures for various food items using the fact sheet. 2 Have employees complete the optional activity.

* Visit the Food Safety Resource Center at *www.ServSafe.com/FoodSafety/resource* to download free posters, quiz sheets, fact sheets, and optional activities and to learn how to obtain *Employee Guides* and videos/DVDs.

Take It Back*

Tool #1: ServSafe Video 4: *Purchasing, Receiving, and Storage*	**Tool #2:** *ServSafe Employee Guide*	**Tool #3:** ServSafe Posters and Quiz Sheets	**Tool #4:** ServSafe Fact Sheets and Optional Activities
Storing Food According to FIFO			
Video segment on storage 1 Show employees the segment. 2 Ask employees to describe how to store food using FIFO.			**Important Storage Practices Fact Sheet** 1 Pass out a fact sheet to each employee. Discuss the FIFO method of stock rotation using the fact sheet. 2 Have employees complete the optional activity.

* Visit the Food Safety Resource Center at *www.ServSafe.com/FoodSafety/resource* to download free posters, quiz sheets, fact sheets, and optional activities and to learn how to obtain *Employee Guides* and videos/DVDs.

ADDITIONAL RESOURCES

Articles and Texts

Cadwallader, Keith R. and Hugo Weenen. *Freshness and Shelf Life of Foods.* New York: Oxford University Press, Incorporated, 2002.

Man, D.M.D. and A.A. Jones. *Shelf-Life Evaluation of Foods, 2nd edition.* New York: Springer, 2000.

Todd, E.C.D. and Jeffrey M. Farber. *Safe Handling of Foods.* New York: Marcel Dekker Incorporated, 2000.

Web Sites

Alaska Seafood Marketing Institute

www.alaskaseafood.org

The Alaska Seafood Marketing Institute (ASMI) is a public agency functioning as the state of Alaska's seafood-marketing arm. ASMI's Web site provides information regarding food safety and quality-assurance practices associated with Alaskan seafood.

American Egg Board

www.aeb.org

The American Egg Board (AEB) is the U.S. egg producers' link to the consumer. This Web site provides resources for the restaurant and foodservice industry, including ways to ensure safe receiving, storage, and preparation of eggs, as well as general information about eggs and egg products.

American Lamb Board

www.americanlambboard.org

The American Lamb Board (ALB) was created by the U.S. Secretary of Agriculture to administer the Lamb Promotion, Research, and Information Order. This Web site provides information about foodservice cuts, cooking temperatures, and food safety tips for lamb.

FDA Center for Food Safety and Applied Nutrition

www.cfsan.fda.gov/list.html

As the center within the Food and Drug Administration (FDA) responsible for food safety, the Center for Food Safety and Applied Nutrition (CFSAN) promotes and protects the public health by researching and implementing guidelines, policies, and standards to ensure that food is safe, nutritious, wholesome, and properly labeled. This Web site provides information relevant to all aspects of food safety and security, including corresponding guidelines, policies, and standards.

Food Safety and Inspection Service

www.fsis.usda.gov

The Food Safety and Inspection Service (FSIS) is the public health agency within the U.S. Department of Agriculture responsible for ensuring that the nation's commercial supply of meat, poultry, and egg products is safe, wholesome, and correctly labeled and packaged. Visit this Web site for food safety information and regulations related to meat, poultry, and eggs.

International Dairy Foods Association

www.idfa.org

International Dairy Foods Association (IDFA) is the Washington, D.C.–based organization representing the nation's dairy-processing and manufacturing industries and their suppliers. IDFA is composed of three constituent organizations: the Milk Industry Foundation (MIF), the National Cheese Institute (NCI), and the International Ice Cream Association (IICA). Visit this Web site to access the Pasteurized Milk Ordinance, dairy-product quality standards, and regulations impacting the dairy industry.

International Fresh-Cut Produce Association

www.fresh-cuts.org

The International Fresh-Cut Produce Association (IFPA) represents processors, distributors, retail, and foodservice buyers of fresh-cut produce and companies that supply goods and services in the fresh-cut industry. IFPA provides its members with the knowledge and technical information necessary to deliver convenient, safe, and wholesome food. This Web site offers a detailed list of frequently asked questions about the fresh-cut produce industry and has a contact feature in which to ask a question. Free publications on food safety and best practices for fresh-cut produce are also available.

Gateway to Government Food Safety Information

www.foodsafety.gov

This Web site provides links to selected government food safety-related information.

Mushroom Council

www.mushroomcouncil.org

The Mushroom Council, composed of fresh-market producers or importers of mushrooms, administers a national promotion, research, and consumer-information program to maintain and expand markets for fresh mushrooms. Information about shipping, handling, and preparation of mushrooms, as well as risks associated with the use of mushrooms harvested by inexperienced mushroom hunters, is available on this Web site.

National Cattlemen's Beef Association

www.beef.org

The National Cattlemen's Beef Association (NCBA) is the national trade association representing U.S. cattle producers. The organization works to advance the economic, political, and social interests of the U.S. cattle industry. This Web site links to other NCBA-associated sites addressing topics such as nutrition, research, food safety, and marketing. A separate site, ***www.beeffoodservice.com,*** provides recipes, facts on foodservice cuts, and food safety information specifically for restaurant and foodservice operators.

National Chicken Council

www.nationalchickencouncil.com

The National Chicken Council (NCC) is the national, nonprofit trade association for the U.S. chicken industry, representing integrated chicken producer/processors, poultry distributors, and allied firms. Visit this Web site for consumer consumption behaviors, product information, and research related to poultry-associated foodborne illnesses.

National Frozen & Refrigerated Foods Association

www.nfraweb.org

The mission of the National Frozen & Refrigerated Foods Association is to promote the sales and consumption of frozen and refrigerated food through education, training, research, sales planning, and menu development. Visit this Web site for information on the proper receiving, storage, and preparation of frozen and refrigerated food, including shelf-life charts, tips, and recommended food and equipment storage temperatures.

National Pork Producers Council

www.nppc.org

The National Pork Producers Council (NPPC) is one of the nation's largest livestock-commodity organizations. Visit the NPPC's foodservice site, *www.porkfoodservice.com,* for pork-preparation tips, recipes, and consumer consumption trends.

National Turkey Federation

www.eatturkey.com

The National Turkey Federation is the national advocate for all segments of the U.S. turkey industry. This Web site contains information useful to foodservice establishments, including safe purchasing, storing, thawing, and preparing information, as well as recipes, nutrition information, and promotional ideas for turkey.

Produce Marketing Association

www.pma.com

The Produce Marketing Association (PMA) is a not-for-profit association serving members who market fresh fruit, vegetables, and related products. Their members represent the production, distribution, retail, and foodservice sectors of the industry. Visit this Web site for articles and publications on produce food safety information within the foodservice and retail industry.

United Fresh Fruit & Vegetable Association

www.uffva.org

The United Fresh Fruit & Vegetable Association is a national trade organization that represents the interests of growers, shippers, brokers, wholesalers, and distributors of produce. The association promotes increased produce consumption and provides scientific and technical expertise to all segments of the industry. This Web site contains news updates and free publications and lists upcoming events and initiatives on produce safety and quality.

Other Documents and Resources

2005 *FDA Food Code*

www.cfsan.fda.gov/~dms/fc05-toc.html

The Food and Drug Administration (FDA) publishes the *FDA Food Code,* a scientifically-sound technical and legal document that serves as a model for regulating the retail and foodservice industry at the federal, state, and local level. Updates to the code are issued on the odd-numbered years. The *FDA Food Code* provides a system of safeguards designed to minimize foodborne illness and ensure employee health, food protection manager knowledge, safe food, nontoxic and cleanable equipment, and appropriate sanitation of the food establishment. It is used as the basis for information in this textbook.

Food Establishment Plan Review Guide

www.cfsan.fda.gov/~dms/prev-toc.html

The *Food Establishment Plan Review Guide* is available at this Web site. This guide has been developed to provide guidance and assistance in complying with nationally recognized food safety standards. It includes design, installation, and construction recommendations regarding food equipment and facilities.

Seafood Information and Resources

www.cfsan.fda.gov/seafood1.html

The Food and Drug Administration's Center for Food Safety and Applied Nutrition (FDA CFSAN) provides a wealth of seafood information and resources on this Web site. Visit this site for an overview of the FDA's seafood regulatory program, information on seafood-related foodborne pathogens and contaminants, and access to seafood guidance and regulation documents.

The Flow of Food: Preparation

Inside this chapter:

- Thawing Food
- Preparing Specific Food
- Cooking Food
- Cooking Requirements for Specific Food
- Cooling Food
- Storing Cooled Food
- Reheating Food

After completing this chapter, you should be able to:

- Identify proper methods for thawing food.
- Identify the minimum internal cooking time and temperatures for potentially hazardous food.
- Identify the proper procedure for cooking potentially hazardous food in a microwave.
- Identify methods and time and temperature requirements for cooling cooked food.
- Identify time and temperature requirements for reheating cooked, potentially hazardous food.
- Identify methods for preventing contamination and time and temperature abuse when preparing food.
- Recognize the importance of informing consumers of risks when serving raw or undercooked food.

Key Terms

- Slacking
- Minimum internal temperature
- Ice-water bath
- Ice paddle

Apply Your Knowledge

Check to see how much you know about the concepts in this chapter. Use the page references provided with each question to explore the topic.

Test Your Food Safety Knowledge

1. **True or False:** Ground beef should be cooked to a minimum internal temperature of 140°F (60°C) for fifteen seconds. *(See page 8-13.)*
2. **True or False:** Fish cooked in a microwave must be heated to a minimum internal temperature of 145°F (63°C). *(See page 8-17.)*
3. **True or False:** Potentially hazardous food must be cooled from 135°F to 70°F (57°C to 21°C) within four hours and from 70°F to 41°F (21°C to 5°C) or lower within the next two hours. *(See page 8-17.)*
4. **True or False:** If potentially hazardous food is reheated for hot holding, the internal temperature must reach 155°F (68°C) for fifteen seconds within two hours. *(See page 8-20.)*
5. **True or False:** It is acceptable to thaw a beef roast at room temperature. *(See page 8-2.)*

For answers, please turn to the Answer Key.

INTRODUCTION

Once food has been received and stored safely, it is essential that it be prepared, cooked, cooled, and reheated with just as much care. It is at these points in the flow of food that the risk of cross-contamination and time-temperature abuse is greatest.

Exhibit 8a

Thawing Food

Food must never be thawed at room temperature since any microorganisms present can grow to dangerous levels.

THAWING FOOD

Freezing food does not kill microorganisms. If frozen food is exposed to the temperature danger zone during thawing, any foodborne microorganisms present will begin to grow. For this reason, food should never be thawed at room temperature. (See *Exhibit 8a.*)

For example, suppose a cook needs to quickly thaw a twenty-pound turkey. Because he is in a hurry, he places the turkey in a pan on a prep counter to thaw overnight. When the turkey begins to thaw, the skin and outer layers are exposed to the

Exhibit 8b

Acceptable Methods for Thawing Food

In refrigeration

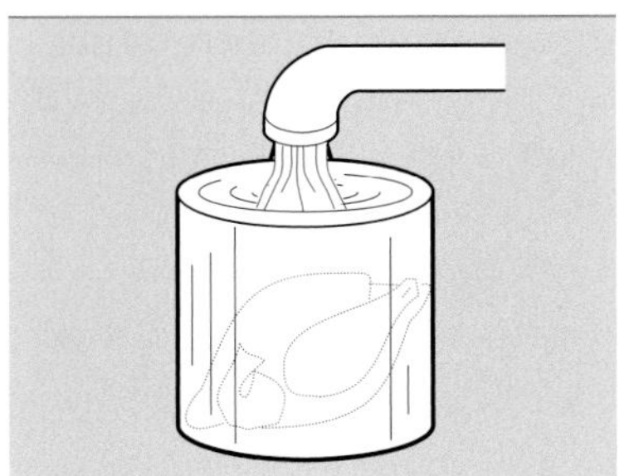

Submerged under running potable water

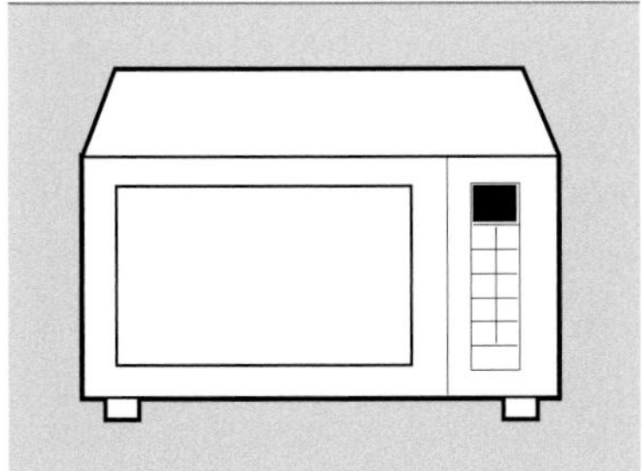

In a microwave oven

As part of cooking

temperature danger zone even though the core of the turkey is still frozen. This situation is hazardous because microorganisms present on the turkey may begin to grow.

To prevent this growth, there are only four acceptable methods for thawing potentially hazardous food (see *Exhibit 8b*):

- **Thaw food in a refrigerator at 41°F (5°C) or lower.** This method requires advance planning. Larger products, such as a turkey, can take several days to thaw completely in a refrigerator.
- **Submerge the food under running potable water at a temperature of 70°F (21°C) or lower.** Water flow must be strong enough to wash loose food particles into the overflow drain. Make sure the thawed product does not drip water onto other products or food-contact surfaces. Clean and sanitize the sink and work area before and after thawing food this way.
- **Thaw food in a microwave oven if it will be cooked immediately afterward.** Microwave thawing can actually start cooking the product, so do not use this method unless you intend to continue cooking the food immediately. Large items such as roasts or turkeys do not thaw well in the microwave.
- **Thaw food as part of the cooking process as long as the product reaches the required minimum internal cooking temperature.** Frozen hamburger patties, for example, can go straight from the freezer onto a grill without being thawed first. Frozen chicken can go straight into a deep fryer. These products cook quickly enough from the frozen state to pass through the temperature danger zone without harm. However, always make sure you verify the final internal cooking temperature with a thermometer.

Some frozen food may be slacked before cooking. **Slacking** is the process of gradually thawing frozen food in preparation for deep-frying, allowing even heating during cooking. For example, you might slack frozen, breaded chicken breasts by allowing them to warm from –10°F (–23°C) to 25°F (–4°C). Slack food just before you cook it. Do not let food get any warmer than 41°F (5°C). If slacking at room temperature is permitted in your jurisdiction, a system must be in place to ensure the product does not exceed 41°F (5°C).

PREPARING SPECIFIC FOOD

Meat, Fish, and Poultry

The sources of most cross-contamination in an operation are raw meat, poultry, and seafood. Your staff should follow safe procedures when handling these products:

- **Use clean and sanitized work areas, cutting boards, knives, and utensils.** Prepare raw meat, poultry, and seafood separately or at a different time from fresh produce.
- **Wash hands properly.** If gloves are worn, hands should be washed, and the gloves changed before starting each new task.
- **Remove from refrigerated storage only as much product as can be prepared at one time.** When cubing beef for stew, for example, take out and cube one roast and refrigerate it before taking out another roast.
- **Return raw, prepared meat to refrigeration, or cook it as quickly as possible.** Store these items properly to prevent cross-contamination.

Exhibit 8c

Salads Containing Potentially Hazardous Food

Do not let large amounts of these types of food sit out at room temperature.

Salads Containing Potentially Hazardous Food

Chicken, tuna, egg, pasta, and potato salads all have been involved in foodborne-illness outbreaks. Since these salads are not typically cooked after preparation, there is no chance to eliminate microorganisms that may have been introduced during preparation. Therefore, care must be taken when preparing these salads. Follow these preparation guidelines:

- **Make sure potentially hazardous leftovers that will be used to make salads have been handled safely.** Leftover ingredients such as pasta, chicken, and potatoes should only be used if they have been cooked, held, and cooled properly. You should also make sure they have not been stored too long. Throw out items held at 41°F (5°C) or lower after seven days. Throw them out after four days if they have been held at 45°F (7°C).
- **Prepare food in small batches, so large amounts of food do not sit out at room temperature for long periods of time.** (See *Exhibit 8c.*)

- **Consider chilling all ingredients and utensils before using them to make the salad.** For example, tuna, mayonnaise, and mixing bowls can be chilled before making tuna salad.
- **Leave food in the refrigerator until all ingredients are ready to be mixed.**

Eggs and Egg Mixtures

Historically, the contents of whole, clean, uncracked shell eggs were considered free of bacteria. It is now known that in rare cases a certain bacteria, *Salmonella* Enteritidis, can be found inside eggs. *Salmonella* Enteritidis can live inside a laying hen and can be deposited in an egg before the shell is formed. Although only a small number of eggs produced in the United States are likely to carry this type of bacteria, all untreated eggs are considered potentially hazardous food because they are able to support the rapid growth of microorganisms. When preparing eggs and egg mixtures, follow these guidelines:

Exhibit 8d

Pooled Eggs

Containers used to hold pooled eggs must be washed and sanitized before being used for a new batch.

- **Handle pooled eggs (if allowed) with special care.** Pooled eggs are eggs that are cracked open and combined in a common container. They must be handled with care because bacteria in one egg can be spread to the rest. Pooled eggs must be cooked promptly after mixing, or stored at 41°F (5°C) or lower. Containers that have been used to hold pooled eggs must be washed and sanitized before being used for a new batch. (See *Exhibit 8d.*)
- **Consider using pasteurized shell eggs or egg products when preparing egg dishes requiring little or no cooking.** These include dishes such as mayonnaise, eggnog, Caesar salad dressing, and hollandaise sauce.
- **Operations that serve high-risk populations, such as hospitals and nursing homes, must take special care when using eggs.** Pasteurized eggs or egg products must be used when dishes containing eggs will be served raw or undercooked. If shell eggs will be pooled for a recipe, they must also be pasteurized. Unpasteurized shell eggs may be used if the dish will be cooked all the way through such as in an omelet or a cake.
- **Promptly clean and sanitize all equipment and utensils used to prepare eggs.**

Key Point

Batters prepared with eggs or milk should be handled with care due to the risk of time-temperature abuse and cross-contamination.

Batters and Breading

Batters prepared with eggs or milk should be handled with care. There is a risk of time-temperature abuse and cross-contamination when batters are made with these products. Breading must also be handled with care since cross-contamination is a risk. In some cases, it may be better to buy frozen breaded items that can be taken directly from the freezer and then cooked thoroughly in the oven or fryer. If you prefer to make breaded or battered food from scratch, however, follow these guidelines:

- **Prepare batters in small batches.** Preparing small amounts prevents time-temperature abuse of both the batter and the food being coated. Store what you do not need at 41°F (5°C) or lower in a covered container.
- **When breading food that will be cooked at a later time, store it in the refrigerator as soon as possible.**
- **Throw out any unused batter or breading after each shift.** Never use a batter or breading for more than one product.
- **Cook battered and breaded food thoroughly.** The coating acts as an insulator, which can prevent food from being thoroughly cooked. When deep-frying food, make sure the temperature of the oil recovers before loading each batch. Overloading the basket also slows cooking time, which means product could be removed from the fryer before it is thoroughly cooked. Be sure to monitor oil and food temperatures using calibrated thermometers, and watch cooking time.

Cross-Contamination

When preparing raw vegetables, start with a clean, sanitized workspace. Prepare vegetables away from raw meat, poultry, and eggs, and from cooked and ready-to-eat food.

Produce

Fresh produce must be handled carefully to prevent foodborne illness. Viruses such as hepatitis A, bacteria such as *E. Coli* and parasites such as *Cryptosporidium parvum* can survive on produce, especially cut produce. The risk from such microorganisms can be minimized or eliminated by these simple preparation safeguards:

- **Make sure fruit and vegetables do not come in contact with surfaces exposed to raw meat and poultry.** Prepare produce away from raw meat, poultry, eggs, and cooked and ready-to-eat food. Clean and sanitize the work space and all utensils that will be used during preparation.

- **Wash fruit and vegetables thoroughly under running potable water to remove dirt and other contaminants before cutting, cooking, or combining with other ingredients.** The water should be slightly warmer than the temperature of the produce. Pay particular attention to leafy greens, such as lettuce and spinach. Remove the outer leaves, and pull lettuce and spinach completely apart and rinse thoroughly. Be sure to clean and sanitize surfaces that were used to prepare produce items for washing.
- **When soaking or storing produce in standing water or an ice-water slurry, do not mix different items or multiple batches of the same item.** (See *Exhibit 8e*.) Pathogens from contaminated produce can contaminate the water and ice and spread to other produce.
- **Refrigerate and hold cut melons at 41°F (5°C) or lower since they are potentially hazardous food.**
- **Do not add sulfites (preservatives that maintain freshness) to food.**
- **If your establishment primarily serves high-risk populations, do not serve raw seed sprouts.**

Exhibit 8e

Soaking or Storing Produce in Water

When soaking or storing produce in standing water or an ice-water slurry, do not mix different items or multiple batches of the same item.

Fresh Juice

Many establishments prepare fresh fruit and vegetable juices on site for their customers. However, if the juice is packaged on site for sale at a later time, the establishment must have a variance from the regulatory agency. The juice must also be treated (e.g. pasteurized) according to an approved HACCP plan. (See Chapter 10.) As an alternative, the juice can be labeled with the following phrase: *Warning: This product has not been pasteurized and therefore may contain harmful bacteria that can cause serious illness in children, the elderly, and people with weakened immune systems.* If fresh juice will be served to a high-risk population, federal public health officials require the establishment to treat the juice to eliminate pathogens that can cause a foodborne illness.

Cross-Contamination

Store ice scoops outside of the ice machine in a sanitary, protected location.

Ice

People often forget that ice is also a food. It is used to chill beverages and to chill or dilute food. In Chapter 7, you learned how to store products on ice, such as poultry, fish, and produce. Many establishments also use ice to chill food on display, such as canned beverages or fruit. Any time ice has been used to cool food in this way, you may not reuse it as a food.

Ice that will be consumed or used to chill food must be made with potable water (water that is safe to drink). Remember that ice can become contaminated just as easily as other food. When transferring ice from an ice machine to an ice bin or to a display, use a clean, sanitized scoop and container. Never transfer ice in containers that have been used to store raw meat, poultry, fish, or chemicals. Store ice scoops outside of the ice machine in a sanitary, protected location.

Preparation Practices That Require a Variance

Some preparation methods require a variance from the regulatory agency. A variance is a document that allows a requirement to be waved or modified, and is required whenever an establishment:

- Smokes food or uses food additives as a method of food preservation
- Cures food
- Custom-processes animals for personal use; for example, this may include dressing deer in the establishment for home consumption
- Packages food using a reduced-oxygen packaging method
- Serves raw or undercooked fish, eggs, shellfish, or meat (excluding steaks)
- Sprouts seeds or beans

COOKING FOOD

The only way to reduce microorganisms in food to safe levels is to cook it to the required **minimum internal temperature.** This temperature varies from product to product. Minimum standards have been developed for most food. (See *Exhibit 8f* on the next page.) These temperatures must be reached and held for the specified amount of time. A thermometer with a suitably sized probe should be used to measure these temperatures. The temperature should be checked in the thickest part of the food, and at least two readings should be taken in different locations.

Bacterial Growth

While cooking can reduce microorganisms, it will not destroy the spores or toxins they may have produced. For this reason, it is critical to handle food safely before it is cooked.

While cooking can reduce to safe levels the number of microorganisms that may be present on food, it will not destroy spores or toxins the microorganisms may have produced. For this reason, it is critical to handle food safely before it is cooked.

It is important to remember that potentially hazardous food, such as meat, eggs, and seafood, should be cooked to the minimum internal temperatures specified in this chapter unless otherwise ordered by the customer. Potentially hazardous items that have not been cooked to these temperatures, such as over-easy eggs, raw oysters, and rare hamburgers, generally do not pose a risk for foodborne illness to the healthy customer. However, if a customer is from a high-risk population, consuming raw or undercooked, potentially hazardous food could significantly increase their risk for foodborne illness. These customers should be advised of this risk when they order potentially hazardous food (or an ingredient) that is raw or not fully cooked. They may want to consult with a physician before regularly consuming these types of food. Check with your regulatory agency for specific requirements.

Here are general guidelines to follow when cooking:

- **Specify cooking time and required minimum internal cooking temperature in all recipes.**
- **Use properly calibrated thermometers with a suitably sized probe to measure food temperatures.** Check internal temperatures in several places in the thickest part of the food. Clean and sanitize the thermometer after each use.
- **Avoid overloading ovens, fryers, and other cooking equipment.** Overloading may lower the equipment or oil temperature, and the food might not cook properly.
- **Let the cooking equipment's temperature recover between batches.**

- **Use utensils or gloves to handle food after cooking.**
- **Taste food correctly to avoid cross-contamination.** The safest and most sanitary way to taste food is to ladle a small amount into a dish. Taste the food in the dish with a clean utensil. When finished, remove the dish and utensil from the area and have them cleaned and sanitized.

Exhibit 8f

Minimum Internal Cooking Temperatures

Product	Minimum Internal Cooking Temperature
Poultry (whole or ground duck, chicken, or turkey)	165°F (74°C) for 15 seconds
Stuffing and stuffed meat, fish, poultry, and pasta	165°F (74°C) for 15 seconds
Potentially hazardous food cooked in a microwave (eggs, poultry, fish, and meat)	165°F (74°C)
Ground meat (beef, pork, and other meat)	155°F (68°C) for 15 seconds
Injected meat (including brined ham and flavor-injected roasts)	155°F (68°C) for 15 seconds
Pork, beef, veal, lamb	**Steaks/Chops:** 145°F (63°C) for 15 seconds **Roasts:** 145°F (63°C) for 4 minutes
Fish	145°F (63°C) for 15 seconds
Shell eggs for immediate service	145°F (63°C) for 15 seconds
Commercially processed, ready-to-eat food (hot held for service)	135°F (57°C)

Exhibit 8g

Checking the Temperature of Poultry

Poultry should be cooked to a minimum internal temperature of 165°F (74°C) for fifteen seconds.

COOKING REQUIREMENTS FOR SPECIFIC FOOD

Poultry

Poultry should be cooked to a minimum internal temperature of 165°F (74°C) for fifteen seconds. (See *Exhibit 8g.*) Poultry has more types and higher counts of microorganisms than other meat because of the way it is processed. Therefore, it should be cooked more thoroughly.

Stuffing

Stuffing can pose a hazard, especially when it is made with potentially hazardous ingredients or when it is used in especially large birds or whole cuts of meat.

Stuffing Made with Potentially Hazardous Ingredients

Stuffing can be a potentially hazardous food when it is made with eggs, oysters, or other potentially hazardous ingredients. Since it is critical that these items are fully cooked, **stuffing made with potentially hazardous ingredients should be cooked to a minimum internal temperature of 165°F (74°C) for fifteen seconds.**

Stuffed Meat, Fish, Poultry, and Pasta

Stuffed meat, fish, poultry, and pasta should be cooked to a minimum internal temperature of 165°F (74°C) for fifteen seconds. Stuffing can be a hazard because it acts as insulation, preventing heat from reaching the center of the product. Always verify that both the stuffing and the product reach the required temperature.

Stuffing should be cooked separately, particularly when cooking whole large birds or large cuts of meat. Smaller cuts of meat, such as pork tenderloins or veal chops, may be stuffed before cooking as long as both the meat and stuffing reach the required temperature.

Dishes That Include Potentially Hazardous Ingredients

When cooking dishes that include previously cooked, potentially hazardous ingredients, such as the ground beef in a meat sauce, **these ingredients must be cooked to a minimum internal temperature of 165°F (74°C) for at least fifteen seconds within two hours.**

When cooking dishes that include raw, potentially hazardous ingredients, these ingredients must be cooked to their required minimum internal temperature. For example, when cooking jambalaya, you must ensure that the raw shrimp reaches its required minimum internal temperature of 145°F (63°C) for at least fifteen seconds.

Exhibit 8h

Checking the Temperature of Pork

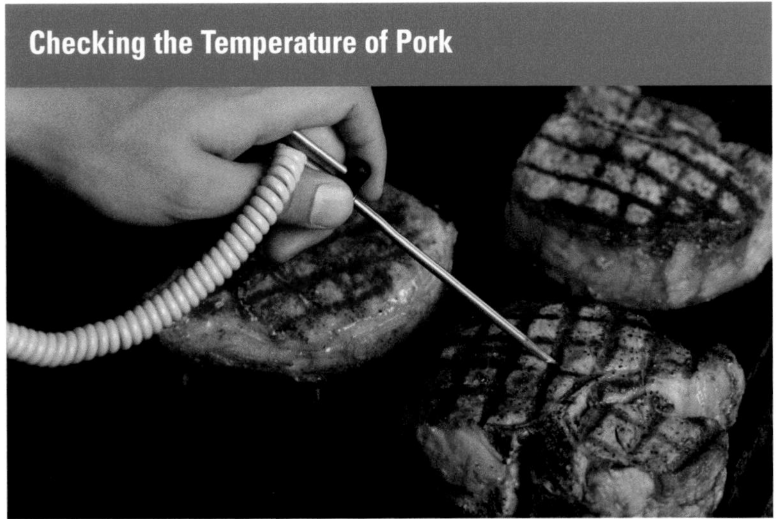

Porkchops should be cooked to a minimum internal temperature of 145°F (63°C) for fifteen seconds.

Pork

Cook pork, such as chops or medallions of tenderloin, to a minimum internal temperature of 145°F (63°C) for fifteen seconds. (See *Exhibit 8h.*) Pork roasts, on the other hand, must hold the same minimum internal temperature for at least four minutes. Depending on the type of roast used, however, pork can be cooked at different internal temperatures. (See *Exhibit 8i.*)

Exhibit 8i

Alternative Minimum Internal Temperatures for Cooking Beef and Pork Roasts

Temperature	Hold for (in minutes)
130°F (54°C)	112
131°F (55°C)	89
133°F (56°C)	56
135°F (57°C)	36
136°F (58°C)	28
138°F (59°C)	18
140°F (60°C)	12
142°F (61°C)	8
144°F (62°C)	5
145°F (63°C)	4

Chart adapted from the *FDA Food Code*

Beef

Steaks must reach and hold a minimum internal temperature of 145°F (63°C) for fifteen seconds. Roasts, on the other hand, must hold the same internal temperature for at least four minutes. Depending on the type of roast used, however, beef can be cooked at different internal temperatures. (See *Exhibit 8i.*)

Ground Meat

Ground beef, pork, and other meat must be cooked to a minimum internal temperature of 155°F (68°C) for fifteen seconds. It may also be cooked according to the alternative cooking temperatures indicated in *Exhibit 8j* on the next page. Most whole-muscle cuts of meat are likely to have microorganisms only on the surface. When meat is ground, such as for hamburger or sausage, microorganisms on the surface are mixed throughout the product. Check with your local or state regulatory agency for additional requirements.

Exhibit 8j

Alternative Minimum Internal Temperatures for Cooking Ground and Injected Meat

Temperature	Hold for
145°F (63°C)	3 minutes
150°F (66°C)	1 minute
155°F (68°C)	15 seconds
158°F (70°C)	< 1 second

Chart adapted from the *FDA Food Code*

Photo courtesy of Cooper-Atkins Corporation

Injected Meat

When meat is injected, foodborne microorganisms on the surface can be carried into the interior. For this reason, **injected meat, such as brined ham or flavor-injected roasts, must be cooked to a minimum internal temperature of 155°F (68°C) for fifteen seconds.** It may also be cooked according to the alternative cooking temperatures indicated in *Exhibit 8j*.

Game and Ratites

Commercially raised and inspected game animals, such as elk, deer, bison, and rabbit, can be cooked to the same minimum internal temperature as beef. Small cuts, such as steaks, can be cooked to a minimum internal temperature of 145°F (63°C) for fifteen seconds. Ground meat must be cooked to a minimum internal temperature of 155°F (68°C) for fifteen seconds. Stuffed meat should be cooked to a minimum internal temperature of 165°F (74°C) for fifteen seconds. Roasts can be cooked in the same way and to the same temperatures as beef and pork roasts.

Although they are birds, ratites (ostrich, emu, and rhea) are not cooked the same way as poultry. They will have a metallic taste if cooked to internal temperatures of 165°F (74°C) or higher. **Ratites are fully cooked when they reach a minimum internal temperature of 155°F (68°C) for fifteen seconds.** If portions or cutlets of these birds are stuffed, however, cook them to an internal temperature of 165°F (74°C) for fifteen seconds.

Exhibit 8k

Checking the Temperature of Fish

Cook fish to a minimum internal temperature of 145°F (63°C) for fifteen seconds.

Fish

Cook fish to a minimum internal temperature of 145°F (63°C) for fifteen seconds. (See *Exhibit 8k.*) If fish has been ground, chopped, or minced, it should be cooked to a minimum internal temperature of 155°F (68°C) for fifteen seconds. Most whole-muscle cuts of fish are likely to have microorganisms only on their surface. When fish is ground, microorganisms on the surface are mixed throughout the product.

Key Point

Eggs for immediate service should be cooked to a minimum internal temperature of 145°F (63°C) for fifteen seconds.

Eggs and Egg Mixtures

In general, shell eggs cooked for immediate service should be cooked to a minimum internal temperature of 145°F (63°C) for fifteen seconds. When eggs are cooked this way, the white is set and the yolk begins to thicken. Properly cooked scrambled eggs or omelets are firm with no visible liquid egg remaining. Poached or fried eggs should be cooked until the white is firm and the yolk begins to thicken. To hold eggs for later service, cook them to a minimum internal temperature of 155°F (68°C) for fifteen seconds.

When cooking eggs, remove from storage only as many eggs as you need for immediate use. Never stack egg trays (flats) near the grill or stove.

Fruit and Vegetables

Although most fruit and vegetables can be eaten raw, many are cooked before they are served. **When cooking fruit or vegetables for hot holding, cook them to a minimum internal temperature of 135°F (57°C).** Cooked fruit and vegetables must never be left out or held at room temperature.

Key Point

Never hold brewed tea at room temperature for more than one day.

Tea

Dry tea leaves contain low levels of bacteria, yeast, and mold (like most plant-derived food). Using improper brewing temperatures to prepare tea and storing it at room temperature for long periods of time can cause these microorganisms to grow to high levels. Improperly cleaned and sanitized equipment can also promote growth. When handling tea, follow these recommendations:

- Brew only as much tea as you reasonably expect to sell within a few hours.
- Never hold brewed tea at room temperature for more than one day. Discard any unused tea at the end of the day.
- To protect tea flavor and avoid microbial contamination and growth, clean and sanitize tea brewing, storage, and dispensing equipment at least once a day. Equipment should be disassembled, washed, rinsed, and sanitized. Urn spigots should be replaced at the end of each day with freshly cleaned and sanitized ones.
- For any brewing method, use a thermometer to make sure brewing water in your equipment meets one of these specified temperatures:
  - 195°F (91°C) for automatic iced tea and automatic coffee machine equipment; tea leaves should remain in contact with the water for a minimum of one minute.
  - 175°F (80°C) minimum when using the traditional steeping method; tea leaves must be exposed to the water for approximately five minutes by this method.

Key Point

Eggs, poultry, fish, and meat cooked in a microwave oven must be heated to a minimum internal temperature of 165°F (74°C).

Microwave Cooking

Microwave ovens tend to cook food more unevenly than other methods of cooking. For this reason, there are special rules for using microwave ovens to cook eggs, poultry, fish, and meat.

- Cover food to prevent the surface from drying out.
- Rotate or stir food halfway through the cooking process to distribute heat more evenly.
- Let food stand for at least two minutes after cooking to let product temperature equalize.

Eggs, poultry, fish, and meat cooked in a microwave oven must be heated to 165°F (74°C). Check the temperature of the food in several places to make sure it is cooked through.

Commercially Processed, Ready-to-Eat Food

Commercially processed, ready-to-eat food that will be hot held for service must be cooked to a minimum internal temperature of 135°F (57°C). This includes items such as cheese sticks, deep-fried vegetables, chicken wings, etc.

Exhibit 8l

Cooling Food

135°F 57°C
70°F 21°C
41°F 5°C

Food must be cooled from 135°F to 70°F (57°C to 21°C) within two hours and from 70°F to 41°F (21°C to 5°C) or lower in the next four hours.

COOLING FOOD

You have already learned how important it is to keep food out of the temperature danger zone. When cooked food will not be served immediately, it is essential to properly hold it or to cool it as quickly as possible.

Potentially hazardous food must be cooled from:

- 135°F to 70°F (57°C to 21°C) within two hours, **and then from**
- 70°F to 41°F (21°C to 5°C) or lower in the next four hours. (See *Exhibit 8l.*)

Keep in mind this is a two-stage process (two hours plus four hours). Microorganisms grow well in the temperature danger zone. However, they grow much faster at temperatures between 125°F and 70°F (52°C and 21°C). Food must pass through this temperature range quickly to minimize this growth. Because food is cooled to 70°F (21°C) within two hours, it passes quickly and safely through the most dangerous part of the temperature danger zone.

If food has not reached 70°F (21°C) within two hours, it must be thrown out or reheated and then cooled again.

Methods for Cooling Food

While common sense may suggest that the quickest way to cool food is to put it in the refrigerator, it is not. Refrigerators are designed to keep cold food cold. They usually do not have the capacity to cool hot food quickly. Never place large quantities of hot food in a refrigerator to cool.

Exhibit 8m

Reducing the Size of Food

Before cooling food, start by dividing large containers of food into smaller containers.

In general, the thickness or density of food is the biggest factor in how quickly it cools. The denser the food product, the more slowly it cools. For example, refried beans take longer to cool than vegetable broth since the beans are thicker.

The container in which food is stored also affects how fast it will cool. Stainless steel transfers heat from food faster than plastic. Shallow pans disperse heat faster than deep ones.

Before cooling food, you should start by reducing its size. This will allow it to cool faster. Cut large items into smaller pieces, or divide large containers of food into smaller containers or shallow pans. (See *Exhibit 8m.*)

There are a number of methods that can be used to cool food quickly and safely. These include (see *Exhibit 8n*):

- **Placing food in an ice-water bath.** After dividing food into smaller containers, place them into a sink or large pot filled with ice water. Stir the food frequently to cool it faster and more evenly.
- **Stirring food with an ice paddle.** Plastic paddles are available that can be filled with ice or with water and then frozen. Food stirred with these paddles will cool quickly. Food cools even faster when placed in an ice-water bath and stirred with an ice paddle.

Exhibit 8n

Safe Methods for Cooling Food

Ice-water bath | Ice paddle | Blast chiller

- **Placing food in a blast chiller or tumble chiller.** Blast chillers blast cold air across food at high speeds to remove heat. They are useful for chilling large food items such as roasts. Tumble chillers tumble bags of hot food in cold water. Tumble chillers work well on thick food such as mashed potatoes.

In addition to the methods identified in *Exhibit 8n,* food can be cooled by:

- **Adding ice or cold water as an ingredient.** This works for soups, stews, and other recipes that call for water as an ingredient. By this method, the recipe is prepared with less water than required. Cold water or ice is then added after cooking to cool the product and provide the remaining water.
- **Using a steam-jacketed kettle (if properly equipped).** Simply run cold water through the jacket to cool the food in the kettle.

STORING COOLED FOOD

Once food has cooled to at least 70°F (21°C), it can be stored on the top shelves in the refrigerator. Pans of food should be covered and then positioned so air can circulate around them. Be sure employees monitor the temperature of the food to ensure that it cools to 41°F (5°C) or lower in four hours. Follow the first in, first out (FIFO) rule when storing food.

Something to Think About...

That's Cool!

A small restaurant chain had noted a recurring problem at one of their older facilities. The restaurant was having trouble cooling chili, which they packed in five-gallon buckets before storing it in the walk-in. Despite filling the buckets just over half full and using an ice paddle, the chili simply did not cool fast enough.

Management and the head chef set to work to find a solution. They determined that the best approach was to pour the chili into shallow hotel pans before placing it into the walk-in. The new system cooled the chili to 41°F (5°C) in a little less than two hours.

Once implemented, the new system even earned praise from employees. They had been using hotel pans to reheat the chili anyway, and the new system saved scraping and washing the buckets. The pans were also much easier to lift.

REHEATING FOOD

Food reheated for immediate service to a customer, such as the beef in a roast beef sandwich, may be served at any temperature, as long as the food was properly cooked and cooled first.

Previously cooked, potentially hazardous food reheated for hot holding must be moved through the temperature danger zone as quickly as possible. Reheat it to an internal temperature of 165°F (74°C) for fifteen seconds within two hours. If the food has not reached this temperature within two hours, throw it out.

SUMMARY

To protect food during preparation, you must handle it safely. The keys are time and temperature control and the prevention of cross-contamination.

Thaw frozen food in the refrigerator, under cool running potable water, in a microwave oven, or as part of the cooking process. Never thaw food at room temperature. Have employees prepare food in small batches, use chilled utensils and bowls, and record product temperatures and preparation times.

Cooking can reduce the number of microorganisms in food to safe levels. To ensure that microorganisms are destroyed, food must be cooked to required minimum internal temperatures for a specific amount of time. These temperatures vary from product to product. Cooking does not kill the spores or toxins some microorganisms produce. That is why it is so important to handle food safely prior to cooking.

Once food is cooked, it should be served as quickly as possible. If it is going to be stored and served later, it must be cooled rapidly. Potentially hazardous food must be cooled from 135°F to 70°F (57°C to 21°C) within two hours and from 70°F to 41°F (21°C to 5°C) or lower in the next four hours.

Before large quantities of food are cooled, they should be reduced in size to allow them to cool faster; cut large food items into smaller pieces or divide large containers of food into smaller ones. There are several methods to cool food safely. They include using an ice-water bath, stirring food with ice paddles, or using a blast or tumble chiller. Once food is cool enough, it should be stored properly in the refrigerator.

Previously cooked, potentially hazardous food that will be hot held must be reheated to an internal temperature of 165°F (74°C) for fifteen seconds within two hours before it can be served.

Apply Your Knowledge

A Case in Point 1

1 What did John do wrong?

On Friday, John went to work at The Fish House knowing he had a lot to do. After changing clothes and punching in, he took a case of frozen raw shrimp out of the freezer. To thaw it quickly, he put the frozen shrimp into the prep sink and turned on the hot water. While waiting for the shrimp to thaw, John took several fresh, whole fish out of the walk-in refrigerator. He brought them back to the prep area and began to clean and fillet them. When he finished, he put the fillets in a pan and returned them to the walk-in refrigerator. He rinsed off the boning knife and cutting board in the sink, and wiped off the worktable with a dishtowel.

Next, John transferred the shrimp from the sink to the worktable using a large colander. On the cutting board, he peeled, deveined, and butterflied the shrimp using the boning knife. He put the prepared shrimp in a covered container in the refrigerator, then started preparing fresh produce.

For answers, please turn to the Answer Key.

Apply Your Knowledge

A Case in Point 2

1 What did Angie do wrong?

By 7:30 p.m., all the residents at Sunnydale Nursing Home had eaten dinner. As she began cleaning up, Angie realized she had a lot of chicken breasts left over. Betty, the new assistant manager, had forgotten to inform Angie that several residents were going to a local festival and would miss dinner.

"No problem," Angie thought. "We can use the leftover chicken to make chicken salad."

Angie left the chicken breasts in a pan on the prep table while she started putting other food away and cleaning up the kitchen. At 9:45 p.m., when everything else was clean, she put her hand over the pan of chicken breasts and decided they were cool enough to handle. She covered the pan with plastic wrap, and put it in the refrigerator.

Three days later, Angie came in to work on the early shift. She decided to make chicken salad from the leftover chicken breasts. After she hung up her coat and put on her apron, Angie took all the ingredients she needed for chicken salad out of the refrigerator and put them on a worktable. Then she started breakfast.

First, she cracked three dozen eggs into a large bowl, added some milk, and set the bowl near the stove. Then she took bacon out of the refrigerator and put it on the worktable next to the chicken salad ingredients. She peeled off strips of bacon onto a sheet pan and put the pan into the oven. After wiping her hands on her apron, she went back to the stove to whisk the eggs and pour them onto the griddle. When they were almost done, Angie scooped the scrambled eggs into a hotel pan and put it in the steam table.

As soon as breakfast was cooked, Angie went back to the prep table to wash and cut up celery and cut up the chicken for chicken salad.

For answers, please turn to the Answer Key.

Apply Your Knowledge

Use these questions to review the concepts presented in this chapter.

Discussion Questions

1. What are the required minimum internal cooking temperatures for poultry, fish, pork, and ground beef?
2. What are four proper methods for thawing food?
3. What methods can be used to cool cooked food?
4. What are the steps for properly cooking food in a microwave oven?

For answers, please turn to the Answer Key.

Apply Your Knowledge

Use these questions to test your knowledge of the concepts presented in this chapter.

Multiple-Choice Study Questions

1. Beef stew must be cooled from 135°F to 70°F (57°C to 21°C) within ____ hours and from 70°F to 41°F (21°C to 5°C) or lower in the next ____ hours.
   A. four, two
   B. two, four
   C. three, two
   D. two, three

2. Which is *not* a safe method for thawing frozen food?
   A. Thawing it by submerging it under running potable water at 70°F (21°C) or lower
   B. Thawing it in the microwave and cooking it immediately afterward
   C. Thawing it at room temperature
   D. Thawing it in the refrigerator overnight

3. Stuffed pork chops must be cooked to a minimum internal temperature of
   A. 135°F (57°C) for fifteen seconds.
   B. 145°F (63°C) for fifteen seconds.
   C. 155°F (68°C) for fifteen seconds.
   D. 165°F (74°C) for fifteen seconds.

4. When reheating potentially hazardous food for hot holding, reheat the food to
   A. 135°F (57°C) for fifteen seconds within two hours.
   B. 145°F (63°C) for fifteen seconds within two hours.
   C. 155°F (68°C) for fifteen seconds within two hours.
   D. 165°F (74°C) for fifteen seconds within two hours.

5. Meat, poultry, and fish cooked in a microwave must be heated to at least
   A. 140°F (60°C).
   B. 145°F (63°C).
   C. 155°F (68°C).
   D. 165°F (74°C).

6. Eggs that will be cooked and hot held for later service must be cooked to an internal temperature of
   A. 140°F (60°C) for fifteen seconds.
   B. 145°F (63°C) for fifteen seconds.
   C. 155°F (68°C) for fifteen seconds.
   D. 165°F (74°C) for fifteen seconds.

Apply Your Knowledge

Multiple-Choice Study Questions

7. All of these practices can help prevent cross-contamination during food preparation *except*
   A. preparing food in small batches.
   B. throwing out unused batter or breading after each shift.
   C. preparing raw meat at a different time than fresh produce.
   D. cleaning and sanitizing pooled egg containers between batches.

8. What is the proper way to cool a large stockpot of clam chowder?
   A. Allow the stockpot to cool at room temperature.
   B. Put the hot stockpot into the walk-in refrigerator to cool.
   C. Divide the clam chowder into smaller containers and place them in an ice-water bath.
   D. Put the hot stockpot into the walk-in freezer to cool.

9. All of these practices can help prevent time and temperature abuse *except*
   A. thawing food in a refrigerator at 41°F (5°C).
   B. chilling all ingredients used to make tuna salad.
   C. leaving food in the refrigerator until all ingredients are ready to be mixed.
   D. thawing steaks in a microwave and promptly refrigerating them for later use.

10. Which food item has been safely cooked?
    A. Hamburger cooked to an internal temperature of 135°F (57°C) for fifteen seconds
    B. Pork chops cooked to an internal temperature of 145°F (63°C) for fifteen seconds
    C. Whole turkey cooked to an internal temperature of 155°F (68°C) for fifteen seconds
    D. Fish cooked to an internal temperature of 135°F (57°C) for fifteen seconds

For answers, please turn to the Answer Key.

Take It Back*

The following food safety concepts from this chapter should be taught to your employees:

- Thawing food properly
- Preparing food safely
- Preparing specific food
- Minimum internal cooking temperatures
- Cooling and reheating food safely

The tools below can be used to teach these concepts in fifteen minutes or less using the directions below. Each tool includes content and language appropriate for employees. Choose the tool or tools that work best.

Tool #1: ServSafe Video 5: *Preparation, Cooking, and Serving*	Tool #2: *ServSafe Employee Guide*	Tool #3: ServSafe Posters and Quiz Sheets	Tool #4: ServSafe Fact Sheets and Optional Activities
		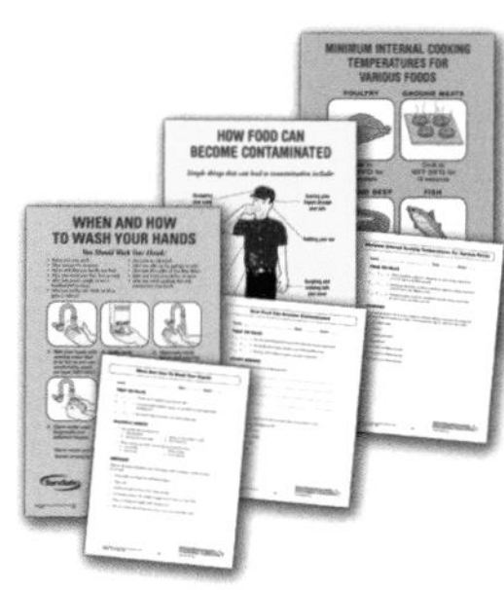	

Thawing Food Properly

Video segment on thawing food 1. Show employees the segment. 2. Ask employees to list the four acceptable methods for thawing food.	**Section 4** **The Four Methods for Thawing Food** Discuss with employees the four acceptable methods for thawing food.		**Thawing Food Properly Fact Sheet** 1. Pass out a fact sheet to each employee. Explain the four acceptable methods for thawing food using the fact sheet. 2. Have employees complete one or two of the optional activities.

* Visit the Food Safety Resource Center at *www.ServSafe.com/FoodSafety/resource* to download free posters, quiz sheets, fact sheets, and optional activities and to learn how to obtain *Employee Guides* and videos/DVDs.

Take It Back*

Tool #1: ServSafe Video 5: *Preparation, Cooking, and Serving*	**Tool #2:** *ServSafe Employee Guide*	**Tool #3:** ServSafe Posters and Quiz Sheets	**Tool #4:** ServSafe Fact Sheets and Optional Activities
Preparing Food Safely (Preventing time-temperature abuse and cross-contamination)			
			Preparing Food Safely Fact Sheet 1 Pass out a fact sheet to each employee. Discuss practices for preventing contamination and temperature abuse during preparation using the fact sheet. 2 Have employees complete one or two of the optional activities.
Preparing Specific Food			
Video segment on preparation 1 Show employees the segment. 2 Ask employees what they should do to keep salads, eggs and egg mixtures, and produce safe during preparation.			**Preparing Specific Types of Food Fact Sheet** 1 Pass out a fact sheet to each employee. Discuss the requirements for preparing specific types of food using the fact sheet. 2 Have employees complete one or two of the optional activities.

* Visit the Food Safety Resource Center at *www.ServSafe.com/FoodSafety/resource* to download free posters, quiz sheets, fact sheets, and optional activities and to learn how to obtain *Employee Guides* and videos/DVDs.

Take It Back* *continued*

Tool #1: ServSafe Video 5: *Preparation, Cooking, and Serving*	**Tool #2:** *ServSafe Employee Guide*	**Tool #3:** ServSafe Posters and Quiz Sheets	**Tool #4:** ServSafe Fact Sheets and Optional Activities
Minimum Internal Cooking Temperatures			
Video segment on specific cooking requirements 1 Show employees the segment. 2 Ask employees to identify minimum internal cooking temperatures for various food items.	**Section 4** **Minimum Internal Cooking Temperatures for Various Types of Food** 1 Discuss with employees minimum internal cooking temperatures for various food items. 2 Complete the Match Game activity.	**Poster: Minimum Internal Cooking Temperatures** 1 Discuss with employees minimum internal cooking temperatures for various food items as presented in the poster. 2 Have employees complete the Quiz Sheet: Minimum Internal Cooking Temperatures.	**Minimum Internal Cooking Temperatures Fact Sheet** 1 Pass out a fact sheet to each employee. Discuss minimum internal cooking temperatures for various food items using the fact sheet. 2 Have employees complete one or two of the optional activities.
Cooling and Reheating Food Safely			
Video segments on cooling food and reheating food 1 Show employees the segments. 2 Ask employees to list acceptable methods for cooling food. Ask employees to identify the requirement for reheating food properly.	**Section 4** **Proper Ways to Cool Food** **The Proper Way to Reheat Food** 1 Discuss with employees proper methods for cooling food with employees. 2 Discuss with employees the proper way to reheat food.	**Poster: Proper Ways to Cool Food** 1 Discuss the proper methods for cooling and reheating food as presented in the poster. 2 Have employees complete the Quiz Sheet: Proper Ways to Cool and Reheat Food.	**Cooling and Reheating Food Fact Sheet** 1 Pass out a fact sheet to each employee. Discuss the requirements for cooling and reheating food using the fact sheet. 2 Have employees complete one or two of the optional activities.

* Visit the Food Safety Resource Center at *www.ServSafe.com/FoodSafety/resource* to download free posters, quiz sheets, fact sheets, and optional activities and to learn how to obtain *Employee Guides* and videos/DVDs.

Notes

ADDITIONAL RESOURCES

Web Sites

Alaska Seafood Marketing Institute

www.alaskaseafood.org

The Alaska Seafood Marketing Institute (ASMI) is a public agency functioning as the state of Alaska's seafood marketing arm. ASMI's Web site provides information regarding food safety and quality assurance practices associated with Alaskan seafood.

American Egg Board

www.aeb.org

The American Egg Board (AEB) is the U.S. egg producers' link to the consumer. This Web site provides resources for the restaurant and foodservice industry, including ways to ensure safe receiving, storage and preparation of eggs, and general information about eggs and egg products.

American Lamb Board

www.americanlambboard.org

The American Lamb Board (ALB) was created by the U.S. Secretary of Agriculture to administer the Lamb Promotion, Research and Information Order. This Web site provides information about foodservice cuts, cooking temperatures, and food safety tips for lamb.

FDA Center for Food Safety and Applied Nutrition

www.cfsan.fda.gov/list.html

As the center within the Food and Drug Administration responsible for food safety, the Center for Food Safety and Applied Nutrition (FDA CFSAN) promotes and protects the public health by researching and implementing guidelines, policies, and standards to ensure that food is safe, nutritious, wholesome, and properly labeled. This Web site provides information relevant to all aspects of food safety and security, including corresponding guidelines, policies, and standards.

Food Safety and Inspection Service

www.fsis.usda.gov

The Food Safety and Inspection Service (FSIS) is the public health agency within the U.S. Department of Agriculture responsible for ensuring that the nation's commercial supply of meat, poultry, and egg products is safe, wholesome, and correctly labeled and packaged. Visit this Web site for food safety information and regulations related to meat, poultry, and eggs.

International Dairy Foods Association

www.idfa.org

International Dairy Foods Association (IDFA) is the Washington, D.C.–based organization representing the nation's dairy-processing and manufacturing industries and their suppliers. IDFA is composed of three constituent organizations: the Milk Industry Foundation (MIF); the National Cheese Institute (NCI); and the International Ice Cream Association (IICA). Visit this Web site to access the Pasteurized Milk Ordinance, dairy-product quality standards, and regulations impacting the dairy industry.

International Fresh-Cut Produce Association

www.fresh-cuts.org

The International Fresh-Cut Produce Association (IFPA) represents processors, distributors, retail, and foodservice buyers of fresh-cut produce and companies that supply goods and services in the fresh-cut industry. IFPA provides its members with the knowledge and technical information necessary to deliver convenient safe and wholesome food. This Web site offers a detailed list of frequently asked questions about the fresh-cut produce industry and has a contact feature in which to ask a question. Free publications on food safety and best practices for fresh-cut produce are also available.

Gateway to Government Food Safety Information

www.foodsafety.gov

This Web site provides links to selected government food safety-related information.

Mushroom Council

www.mushroomcouncil.org

The Mushroom Council, composed of fresh-market producers or importers of mushrooms, administers a national promotion, research, and consumer information program to maintain and expand markets for fresh mushrooms. Information about shipping, handling, and preparation of mushrooms, as well as risk associated with the use of mushrooms harvested by inexperienced mushroom hunters, is available on this Web site.

National Cattlemen's Beef Association

www.beef.org

The National Cattlemen's Beef Association (NCBA) is the national trade association representing U.S. cattle producers. The organization works to advance the economic, political, and social interests of the U.S. cattle industry. This Web site links to other NCBA-associated sites addressing topics such as nutrition, research, food safety, and marketing. A separate site, ***www.beeffoodservice.com,*** provides recipes, facts on foodservice cuts, and food safety information specifically for restaurant and foodservice operators.

National Chicken Council

www.nationalchickencouncil.com

The National Chicken Council (NCC) is the national, nonprofit trade association for the U.S. chicken industry representing integrated chicken producer/processors, poultry distributors, and allied firms. Visit this Web site for consumer consumption behaviors, product information, and research related to poultry-associated foodborne illnesses.

National Frozen & Refrigerated Foods Association

www.nfraweb.org

The mission of the National Frozen & Refrigerated Foods Association is to promote the sales and consumption of frozen and refrigerated foods through education, training, research, sales planning, and menu development. Visit this Web site for information on the proper receiving, storage, and preparation of frozen and refrigerated food, including shelf-life charts, tips, and recommended food and equipment storage temperatures.

National Pork Producers Council

www.nppc.org

The National Pork Producers Council (NPPC) is one of the nation's largest livestock commodity organizations. Visit the NPPC's foodservice site, ***www.porkfoodservice.com,*** for pork preparation tips, recipes, and consumer consumption trends.

National Turkey Federation

www.eatturkey.com

The National Turkey Federation is the national advocate for all segments of the U.S. turkey industry. This Web site contains information useful to foodservice establishments, including safe purchasing, storing, thawing, and preparing information as well as recipes, nutrition information, and promotional ideas for turkey.

Produce Marketing Association

www.pma.com

The Produce Marketing Association (PMA) is a nonprofit association serving members who market fresh fruit, vegetables, and related products worldwide. Their members represent the production, distribution, retail, and foodservice sectors of the industry. Visit this Web site for articles and publications on produce safety information within the foodservice and retail industry.

United Fresh Fruit & Vegetable Association

www.uffva.org

The United Fresh Fruit & Vegetable Association is a national trade organization that represents the interests of growers, shippers, brokers, wholesalers, and distributors of produce. The association promotes increased produce consumption and provides scientific and technical expertise to all segments of the industry. This Web site contains news updates and free publications and lists upcoming events and initiatives on produce safety and quality.

Documents and Other Resources

2005 *FDA Food Code*

www.cfsan.fda.gov/~dms/fc05-toc.html

The Food and Drug Administration (FDA) publishes the *FDA Food Code,* a scientifically sound technical and legal document that serves as a model for regulating the retail and foodservice industry at the federal, state, and local level. Updates to the code are issued every other year on the odd-numbered years. The *FDA Food Code* provides a system of safeguards designed to minimize foodborne illness and ensure employee health, food protection manager knowledge, safe food, nontoxic and cleanable equipment, and appropriate sanitation of the food establishment. It is used as the basis for information in this textbook.

Seafood Information and Resources

www.cfsan.fda.gov/seafood1.html

The Food and Drug Administration's Center for Food Safety and Applied Nutrition (FDA CFSAN) provides a wealth of seafood information and resources on this Web site. Visit this site for an overview of the FDA seafood regulatory program, information on seafood-related foodborne pathogens and contaminants, and access to seafood guidance and regulation documents.

Notes

SERIES
396K
50209-K
501999

The Flow of Food: Service

Inside this chapter:

- Holding Food for Service
- Serving Food Safely
- Off-Site Service

After completing this chapter, you should be able to:

- Identify time and temperature requirements for holding hot and cold, potentially hazardous food.
- Identify procedures for preventing time-temperature abuse and cross-contamination when displaying and serving food.
- Identify the requirements for using time rather than temperature as the only method of control when holding ready-to-eat food.
- Implement methods for minimizing bare-hand contact with ready-to-eat food.
- Identify hazards associated with the transportation of food and methods for preventing them.
- Identify hazards associated with the service of food off site and methods for preventing them.
- Identify hazards associated with vending food and methods for preventing them.
- Prevent customers from contaminating self-service areas.
- Prevent employees from contaminating food.

Key Terms

- Hot-holding equipment
- Cold-holding equipment
- Food bar
- Sneeze guard
- Off-site service
- Single-use item
- Mobile unit
- Temporary unit
- Vending machine

Apply Your Knowledge	Test Your Food Safety Knowledge
Check to see how much you know about the concepts in this chapter. Use the page references provided with each question to explore the topic.	1 **True or False:** Cold, potentially hazardous food must be held at an internal temperature of 41°F (5°C) or lower. *(See page 9-4.)* 2 **True or False:** Hot, potentially hazardous food must be held at an internal temperature of 120°F (49°C) or higher. *(See page 9-4.)* 3 **True or False:** Chicken salad can be held at room temperature, if it has a label specifying it must be discarded after eight hours. *(See page 9-4.)* 4 **True or False:** When holding potentially hazardous food for service, the internal temperature must be checked at least every four hours. *(See page 9-3.)* 5 **True or False:** Servers can contaminate food simply by handling the food-contact surface of a plate. *(See page 9-6.)* **For answers, please turn to the Answer Key.**

INTRODUCTION

The job of protecting food continues even after the food has been prepared and cooked properly, since microorganisms can still contaminate it before it is eaten. The key to serving safe food is to prevent time-temperature abuse and cross-contamination. Hold, display, and serve food at the correct temperature, and handle it safely. People do many things without knowing their actions can lead to contamination. Train employees to serve food properly, and make sure food safety rules are followed.

HOLDING FOOD FOR SERVICE

In many establishments, food is cooked to order. If it has been stored, prepared, and cooked properly and then served immediately, it is less likely to cause illness. Even in facilities that cook food to order, many menu items are cooked and held for service. A coffee shop might hold soup in a warming kettle.

A steakhouse might keep prime rib warm on a steam table. Many establishments, such as cafeterias and buffets, hold almost all food they serve.

Kitchen staff might be tempted to hold hot food at a lower temperature than required to maintain quality. However, employees must remember that microorganisms can grow at temperatures between 41°F and 135°F (5°C and 57°C). To ensure the safety of food that is held hot or cold, specific procedures must be followed.

General Rules For Holding Food

Follow these guidelines when holding food:

- **Check the internal temperature of food using a thermometer.** The temperature gauge on a holding unit may not provide an accurate indication of a food's internal temperature. Therefore, it is critical to use a thermometer to check temperature.
- **Check the temperature of food at least every four hours.** (See *Exhibit 9a.*) Throw out food that is not at 135°F (57°C) or higher or at 41°F (5°C) or lower. As an alternative, check the temperature every two hours to leave time for corrective action.
- **Establish a policy to ensure that food being held for service will be discarded after a predetermined amount of time.** For example, a policy may state that a pan of veal on a buffet can be replenished all day as long as it is discarded at the end of the day.
- **Cover food and provide sneeze guards to protect food from contamination.** Covers also maintain the internal temperature of food.
- **Prepare food in small batches so it will be used faster.** Do not prepare food any further in advance than necessary to minimize the potential for time-temperature abuse.

Exhibit 9a

Checking Temperatures during Holding

Check the temperature of food at least every four hours.

Exhibit 9b

Hot Holding

Potentially hazardous hot food must be held at an internal temperature of 135°F (57°C) or higher.

Exhibit 9c

Cold Holding

Potentially hazardous cold food must be held at an internal temperature of 41°F (5°C) or lower.

Hot Food

- **Potentially hazardous hot food must be held at an internal temperature of 135°F (57°C) or higher.** (See *Exhibit 9b.*)
- **Only use hot-holding equipment that can keep food at the proper temperature.**
- **Never use hot-holding equipment to reheat food if it is not designed to do so.** Most hot-holding equipment is not designed to pass food through the temperature danger zone quickly enough. Food should be properly reheated first, and then transferred to the holding unit.
- **Stir food at regular intervals to distribute heat evenly.**

Cold Food

- **Potentially hazardous cold food must be held at an internal temperature of 41°F (5°C) or lower.** (See *Exhibit 9c.*)
- **Only use cold-holding equipment that can keep food at the proper temperature.**
- **Do not store food directly on ice.** Whole fruit and vegetables and raw, cut vegetables are the only exceptions. Place all other food in pans or on plates first.

Holding Food Without Temperature Control

Ready-to-eat, potentially hazardous food can be displayed or held for consumption without temperature control under certain conditions. **Before using time as a method of control, check with your regulatory agency for specific requirements.**

Cold Food

Cold food can be held without temperature control for up to **six hours** if:

- **It was held at 41°F (5°C) or lower prior to removing it from refrigeration.**
- **It does not exceed 70°F (21°C) during the six hours.** Throw out any food that exceeds this temperature.
- **It has a label that specifies both the time it was removed from refrigeration and the time it must be thrown out.** The label must indicate a discard time that is six hours from

Exhibit 9d

Holding Cold Food without Temperature Control

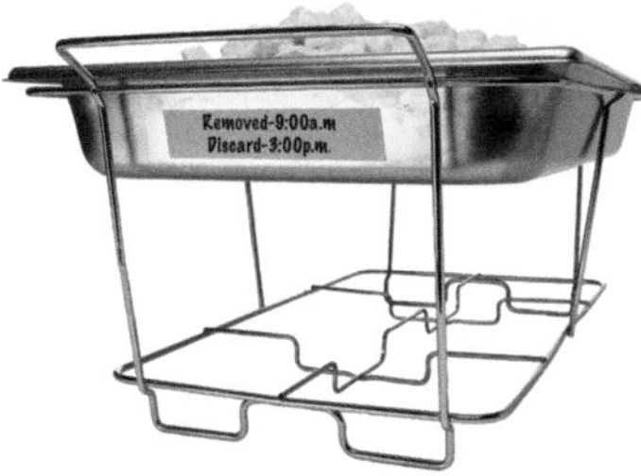

In some jurisdictions, cold, potentially hazardous food can be held without temperature control for up to six hours under certain conditions.

the point the food was removed from refrigeration. For example, if potato salad served at a picnic was removed from refrigeration at 9:00 a.m., the discard time on the label should indicate 3:00 p.m., which is six hours from the time it was removed from refrigeration. (See *Exhibit 9d*.)

- **It is sold, served, or discarded within six hours.**

Hot Food

Hot food can be held without temperature control for up to **four hours** if:

- **It was held at 135°F (57°C) or higher prior to removing it from temperature control.**
- **It has a label that specifies when the item must be thrown out.**
- **It is sold, served, or discarded within four hours.**

SERVING FOOD SAFELY

After handling food safely and cooking it properly, you do not want to risk contamination when serving it.

Exhibit 9e

Properly Stored Utensils

If stored in food, utensils should be stored with the handle extended above the rim of the container.

Kitchen Staff

Train your kitchen staff to follow these procedures for serving food safely.

- **Use clean and sanitized utensils for serving.** Use separate utensils for each food item, and properly clean and sanitize them after each serving task. Utensils should be cleaned and sanitized at least once every four hours during continuous use.
- **Use serving utensils with long handles.** Long-handled utensils keep the server's hands away from food.
- **Store serving utensils properly.** Serving utensils can be stored in the food with the handle extended above the rim of the container. (See *Exhibit 9e*.) They can also be placed on a clean, sanitized food-contact surface. Spoons or scoops used to serve food such as ice cream or mashed potatoes can be stored under running water.

Exhibit 9f

Handling Cooked and Ready-to-Eat Food

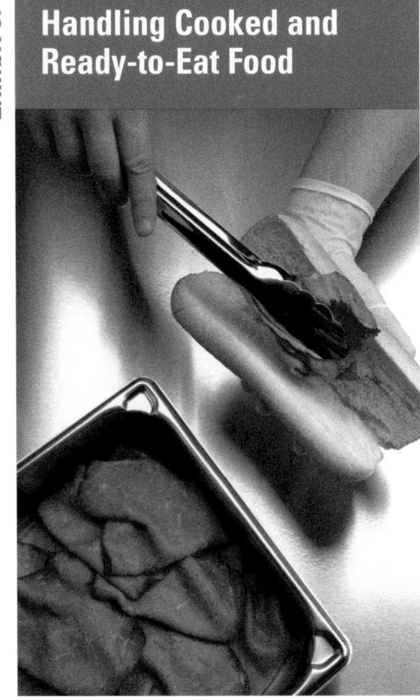

Minimize bare-hand contact with cooked and ready-to-eat food by using gloves or tongs to serve it.

- **Minimize bare-hand contact with food that is cooked or ready to eat.** Handle food with tongs, deli sheets, or gloves. (See *Exhibit 9f.*) Bare-hand contact is allowed in some jurisdictions if the establishment has written policies and procedures on employee health, handwashing, and other hygiene practices. Check with your regulatory agency for requirements in your jurisdiction.
- **Practice good personal hygiene.** Proper handwashing is essential to keep food safe.

Servers

Food servers can contaminate food simply by handling the food-contact surfaces of glassware, dishes, and utensils. The following guidelines should be followed when serving food (see *Exhibit 9g*):

- **Glassware and dishes should be handled properly.** The food-contact area of plates, bowls, glasses, or cups should not be touched. Dishes should be held by the bottom or the edge. Cups should be held by their handles, and glassware should be held by the middle, bottom, or stem.
- **Glassware and dishes should not be stacked when serving.** The rim or surface of one item can be contaminated by the one above it.
- **Flatware and utensils should be held at the handle.** Store flatware so servers grasp handles, not food-contact surfaces.
- **Minimize bare-hand contact with food that is cooked or ready to eat.**
- **Use ice scoops or tongs to get ice.** Servers should never scoop ice with their bare hands or use a glass since it may chip or break. Ice scoops should always be stored in a sanitary location—not in the ice bin.
- **Practice good personal hygiene.**
- **Never use cloths meant for cleaning food spills for any other purpose.** When tables are cleaned between guest seatings, spills should be wiped up with a disposable, dry cloth. The table should then be cleaned with a moist cloth that has been stored in a fresh sanitizer solution.

Exhibit 9g

Right and Wrong Ways to Handle Food, Glassware, Dishes, and Utensils

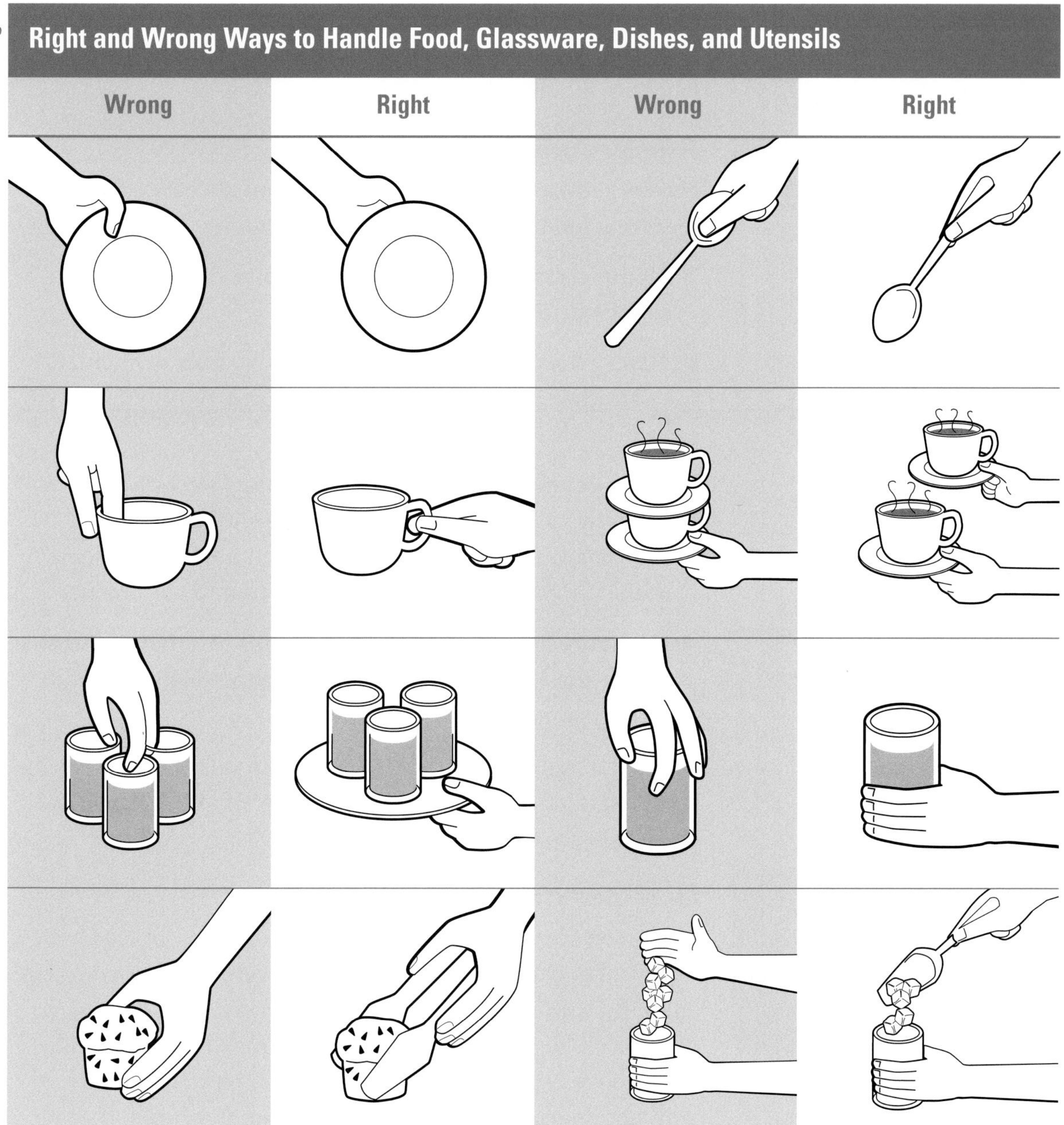

Division of Labor

To prevent cross-contamination, it is a good idea to schedule staff so they are not assigned to do more than one job during a shift. Serving food, setting tables, and bussing dirty dishes are separate tasks with different responsibilities. Since this division of labor is difficult to manage in most establishments, it is important for servers and bussers who do double duty to wash their hands often and handle food safely. After wiping tables or bussing dirty dishes, servers must wash their hands before handling food or place settings.

Exhibit 9h

Reserving Food

Never re-serve plate garnishes, uncovered condiments, or uneaten bread.

Re-serving Food Safely

Servers and kitchen staff should also know the rules about re-serving food previously served to a customer. (See *Exhibit 9h.*)

- **Menu items returned by one customer cannot be re-served to another customer.**
- **Never re-serve plate garnishes, such as fruit or pickles, to another customer.** Served but unused garnishes must be discarded.
- **Never re-serve uncovered condiments.** Do not combine leftovers with fresh food. Opened portions of salsa, mayonnaise, mustard, butter, and other condiments should be thrown away after being served to customers.
- **Do not re-serve uneaten bread or rolls to other customers.** Linens used to line bread baskets must be changed after each customer.

In general, only unopened, prepackaged food, such as condiment packets, wrapped crackers, or wrapped breadsticks, can be re-served.

Exhibit 9i

Self-Service Areas

Assign a trained staff member to monitor food bars and buffets.

Self-Service Areas

Customers choosing food from self-service areas, or **food bars,** often unknowingly serve themselves in ways that can put them and other customers in danger. A customer may eat from his plate or nibble from the food bar while moving through the line. Another might pick up carrot sticks, pickles, and olives with her fingers, or dip a finger into salad dressing to taste it. Another might return unwanted food items, use a soiled plate for a second helping, or put his head under the sneeze guard to reach items in the back of the display.

To prevent contamination, self-service areas should be monitored closely by employees trained in food safety. Assign a staff member to replenish food-bar items and to hand out fresh plates for return visits. (See *Exhibit 9i.*) Post signs with polite tips about food-bar etiquette. These practices will go a long way toward keeping self-service areas more sanitary.

Here are some additional rules for food bars:

- **Maintain proper food temperatures.** Keep hot food hot—135°F (57°C) or higher, and cold food cold—41°F (5°C) or lower.
- **Replenish food on a timely basis.** Stock and replenish small amounts at a time. Practice the first in, first out (FIFO) method of product rotation.
- **Keep raw meat, fish, and poultry separate from cooked and ready-to-eat food.** Use separate displays or food bars for raw and cooked food (for example, a Mongolian barbecue) to reduce the chance of cross-contamination.
- **Protect food on display with sneeze guards or food shields.** Sneeze guards should be fourteen inches (thirty-six centimeters) above the food counter, and shields should extend seven inches (eighteen centimeters) beyond the food. (See *Exhibit 9j.*)

Exhibit 9j

Sneeze Guards

14″

7″

Sneeze guards should be fourteen inches (thirty-six centimeters) above the food counter. The shield should extend seven inches (eighteen centimeters) beyond the food.

- **Identify all food items.** Label containers on the food bar. Place names of salad dressings on ladle handles. (See *Exhibit 9k.*)
- **Do not let customers refill soiled plates or use soiled silverware at the food bar.** Encourage customers to take a clean plate for return trips to the food bar. Customers can use glassware for refills as long as beverage-dispensing equipment does not come in contact with the rim or interior of the glass.

Exhibit 9k

Identifying Food Items

Label all items on a food bar.

OFF-SITE SERVICE

Off-site services, including delivery, mobile/temporary kitchens, and vending machines, all present special challenges. Those operating these services must follow the same food safety rules as permanent establishments. Food must be protected from contamination and time-temperature abuse, and facilities and equipment used to prepare food must be clean and sanitary. Menu items must lend themselves to safe service. Food must be handled safely.

Exhibit 9l

Delivery Containers

When transporting food, use rigid, insulated containers capable of maintaining the proper temperature.

Delivery

Many establishments such as schools, hospitals, caterers, and even restaurants may prepare food at one location and then deliver it to remote sites. The greater the time and distance between the point of preparation and the point of consumption, the greater the risk that food will be exposed to contamination or time-temperature abuse. Equipment used to transport food—both containers and vehicles—must be designed to maintain safe food temperatures and be easy to keep clean.

When transporting food, the following safety procedures should be followed:

- **Use rigid, insulated food containers capable of maintaining food at 135°F (57°C) or higher or 41°F (5°C) or lower.** (See *Exhibit 9l.*) Containers should be sectioned so that food does not mix, leak, or spill. They must also allow air circulation to keep temperatures even and should be kept clean and sanitized.
- **Clean the inside of delivery vehicles regularly.**
- **Practice good personal hygiene when distributing food.**
- **Check internal food temperatures regularly.** Take corrective action if food is not at the proper temperature. If containers or delivery vehicles are not maintaining proper food temperatures, reevaluate the length of the delivery route or the efficiency of the equipment being used.
- **Label food with storage, shelf-life, and reheating instructions for employees at off-site locations.**
- **Consider providing food safety guidelines for consumers.**

Catering

Caterers must follow the same food safety rules as permanent establishments. Food must be protected from contamination and time-temperature abuse. Facilities must be clean and sanitary. Food must be prepared and served safely, and employees must follow good personal hygiene practices.

Caterers must meet the special challenges of off-site foodhandling. They must make sure there is safe drinking water for cooking, dishwashing, and handwashing, as well as adequate power for holding and cooking equipment.

Outdoor catering for barbecues and cookouts may require special arrangements. When power or running water is not available, caterers may have to change their foodhandling procedures.

- **Use insulated containers to hold potentially hazardous food.** Raw meat should be wrapped and stored on ice. Deliver milk and dairy products in a refrigerated vehicle or on ice.
- **Serve cold food in containers on ice or in chilled, gel-filled containers.** If that is not desirable, the food may be held without temperature control according to the guidelines specified in this chapter.
- **Store raw and ready-to-eat products separately.** For example, raw chicken should be stored separately from ready-to-eat salads.
- **Use single-use items.** Make sure customers get a new set of disposable tableware for refills.
- **If leftovers are given to customers, provide instructions on how they should be handled.** Information such as a discard date and the food's storage and reheating instructions should be clearly labeled on the container.
- **Place garbage-disposal containers away from food-preparation and serving areas.**

Key Point

Mobile kitchens serving potentially hazardous food must follow the same rules as permanent kitchens.

Mobile Units

Mobile units are portable facilities ranging from concession vans to elaborate field kitchens. Those serving only frozen novelties, candy, packaged snacks, and soft drinks have to meet basic sanitation requirements. Mobile kitchens preparing and serving potentially hazardous food, however, must follow the same rules required of permanent foodservice kitchens. Both might be required to apply for a special permit or license from the local regulatory agency.

Like permanent establishments, mobile kitchens must have adequate cooking equipment, cold-storage and hot-holding units, and dishwashing and handwashing sinks with hot and cold potable water under pressure. Mobile kitchens must provide adequate ventilation, garbage storage and disposal facilities, and pest control. Operators should clean and maintain mobile units.

Exhibit 9m

Temporary Unit

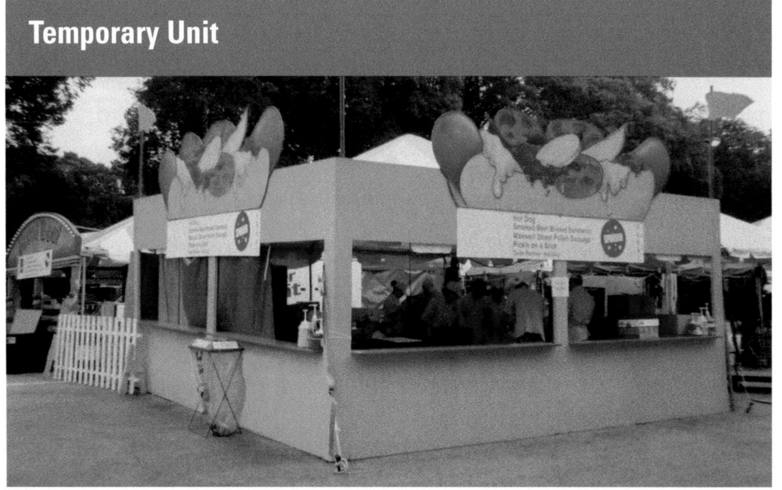

It is best to keep the menu simple to limit the amount of on-site food preparation.

Courtesy of the Illinois Restaurant Association

Temporary Units

Temporary units typically operate in one location for less than fourteen days. Foodservice tents or kiosks set up for food fairs, special celebrations, or sporting events may be temporary units. (See *Exhibit 9m.*) In some areas, the definition also extends to units set up for longer periods of time. Temporary units usually serve prepackaged food or food requiring limited preparation, such as hot dogs. It is best to keep the menu simple to limit the amount of on-site food preparation. Check with your local regulatory agency for operating requirements.

Temporary units should be constructed to keep dirt and pests out. If floors are made of dirt or gravel, cover them with mats or platforms to control dust and mud. Construct walls and the ceiling with materials that will protect food from weather and windblown dust.

In addition, the same safe-handling rules previously discussed apply to food preparation in temporary units. If food is prepared on site, the unit must have adequate cooking, cold storage, and hot-holding equipment. If food is prepared off site, it must be transported and held at 135°F (57°C) or higher or 41°F (5°C) or lower.

Safe drinking water must be available for cleaning, sanitizing, and handwashing. Since dishwashing facilities will most likely be limited, it is best to use disposable, single-use items.

Exhibit 9n

Replace food with expired code dates, and discard food if it is not used within seven days of preparation.

Vending Machines

Food prepared and packaged for **vending machines** must be handled with the same care as any other food served to a customer. Vending operators also have to protect food from contamination and time-temperature abuse during transport, delivery, and service.

To keep vended food safe, it must be held at the proper temperature. Machines that vend potentially hazardous food must hold it at an internal temperature of 135°F (57°C) or higher, or 41°F (5°C) or lower. These machines must also have controls that prevent the food from being dispensed if the temperature stays in the danger zone for a specified amount of time.

Check product shelf life daily. If a food item's code date has expired, discard it immediately. If refrigerated potentially hazardous food is not used within seven days of preparation, it must be discarded. (See *Exhibit 9n.*) Dispense potentially hazardous food, such as milk, in its original container. Fresh fruit with an edible peel should be washed and wrapped before being put into a machine.

Place machines in appropriate locations, away from garbage containers, sewage drains, and overhead pipes. Make sure the vending area is clean and well lighted. Supply safe drinking water for beverage machines.

Clean and service vending machines regularly. Sanitize food-contact surfaces in machines each time food is replenished. Employees must also wash their hands before and after servicing or refilling machines.

SUMMARY

Safe foodhandling does not stop once food is properly prepared and cooked. You must continue to protect it from time-temperature abuse and contamination until it is eaten. When holding potentially hazardous food for service, keep hot food hot—at 135°F (57°C) or higher, and cold food cold—at 41°F (5°C) or lower. Check the internal temperature of food being held at least every four hours and discard it if it is not at the proper temperature. Protect food from contaminants with covers and lids, and establish policies to ensure that food being held for service will be discarded after a predetermined amount of time. If food can be held without temperature control in your jurisdiction, follow requirements closely.

Use clean, sanitized utensils to serve food and minimize bare-hand contact with cooked and ready-to-eat food.

Make sure all employees practice good personal hygiene. Train them to avoid cross-contamination when handling service items and tableware. Teach them about the potential hazards posed by re-serving food such as plate garnishes, breads, or open dishes of condiments.

Customers can unknowingly contaminate food in self-service areas. Post signs to communicate self-service rules, and station employees in these areas to ensure compliance. Protect food in food bars and buffets with sneeze guards and make sure equipment can hold food at the proper temperature.

Take special precautions when preparing, delivering, or serving food off site. Catering, mobile kitchens, temporary units, and vending machines pose unique challenges to food safety. Learn and follow all regulations in your jurisdiction. Prepare backup plans in case off-site facilities are not adequate or equipment breaks down.

Apply Your Knowledge

1. What did Jill do wrong?
2. What should have been done?

A Case in Point 1

Jill, a line cook on the morning shift at Memorial Hospital, was busy helping the kitchen staff put food on display for lunch in the hospital cafeteria. Ann, the kitchen manager who usually supervised lunch in the cafeteria, was at an all-day seminar on food safety. Jill was responsible for making sure meals were trayed and put into food carts for transport to the patients' rooms. The staff also packed two-dozen meals each day for a neighborhood group that delivered them to homebound elderly people.

After Jill helped the kitchen staff, she looked for insulated food containers for the delivery meals. When she could not find them, she loaded the meals into cardboard boxes she found near the back door, knowing the driver would arrive soon to pick them up. To help the cafeteria staff, Jill filled a baine with soup by dipping a two-quart measuring cup into the stockpot and pouring it into the baine. She carried the baine out to the cafeteria, put it into the steam table, and turned it on low.

The lunch hour was hectic. The cafeteria was busy, and the staff had many patient meals to tray and deliver. Halfway through lunch, a cashier came back to the kitchen to tell Jill that the salad bar needed replenishing. Since she was busy, Jill asked a kitchen employee to take pans of prepared ingredients out of the refrigerator and put them on the salad bar. When she looked up a few moments later, she saw the kitchen employee send away two children who were eating carrot sticks from the salad bar.

With lunch almost over, Jill breathed a sigh of relief. She moved down the cafeteria serving line, checking food temperatures. One of the casseroles was about 130°F (54°C). Jill checked the water level in the steam table and turned up the thermostat, then went to clean up the kitchen and finish her shift.

For answers, please turn to the Answer Key.

Apply Your Knowledge

1. What errors did Megan make?
2. What should she have done?

A Case in Point 2

Megan, a new server at The Fish House, reported for work ten minutes early on Thursday. Excited about her new job, she made sure to shower and wash her hair before going to work. When she arrived, she changed into a clean uniform, pulled her hair back tightly into a ponytail, and checked her appearance. She had used makeup sparingly and wore only a watch and a few rings.

Her shift started off well. One of her customers ordered a menu item that Megan had not tried yet. When the order came up, Megan dipped her finger into the sauce at the edge of the plate for a taste. As the shift progressed, Megan's station got busier. When the hostess came to ask how soon one of Megan's tables could be cleared, Megan decided to do it herself. She took the dirty dishes to a bus station and then wiped down the table with a serving cloth she kept in her apron.

The busser, John, finished another task and came to help her set the table. While he put out silverware and linens, Megan filled water glasses with ice by scooping the glasses into the ice bin in the bus station. Only one slice of bread and one pat of butter were missing from the basket that had been on a previous customer's table, so she put that on a tray with the water glasses and brought it to the table.

The table was reset in record time, and Megan soon had more guests. While taking their orders, Megan reached up to scratch a sore on her neck. Then she went to the kitchen to turn in the order and pick up a dessert order for another table.

For answers, please turn to the Answer Key.

Apply Your Knowledge

Use these questions to review the concepts presented in this chapter.

Discussion Questions

1. What can be done to minimize contamination in self-service areas?
2. What hazards are associated with the transportation of food and how can they be prevented?
3. What are the requirements for using time rather than temperature as the only method of control when holding potentially hazardous, ready-to-eat food?
4. What practices should be followed to serve food safely off site?

For answers, please turn to the Answer Key.

Apply Your Knowledge

Use these questions to test your knowledge of the concepts presented in this chapter.

Multiple-Choice Study Questions

1. Which is an unsafe serving practice?
   A. Stacking plates of food before serving them to the customer
   B. Holding flatware by the handles when setting a table
   C. Serving soup with a long-handled ladle
   D. Holding glassware by the stem
2. Which statement about serving utensils is *not* true?
   A. They should be cleaned and sanitized at least once every four hours during continuous use.
   B. They can be used to handle more than one food item at a time.
   C. They can be stored in the food with the handle extended above the rim of the container.
   D. They must be cleaned and sanitized after each task.
3. To keep vended food safe, you should
   A. leave fresh fruit with edible peels unwrapped.
   B. discard potentially hazardous food within fourteen days of preparation.
   C. dispense potentially hazardous food in its original container.
   D. ensure that cold, potentially hazardous food is kept at 50°F (10°C) or lower.
4. To hold cold food safely, you should
   A. store it directly on ice.
   B. store it at 41°F (5°C) or lower.
   C. stir it regularly.
   D. leave it uncovered.
5. Which is an acceptable serving practice at a self-service bar?
   A. Holding hot, potentially hazardous food at 120°F (49°C)
   B. Storing raw meat next to ready-to-eat food
   C. Allowing customers to use the same plate for a return trip to the self-service bar
   D. Allowing customers to reuse glassware for beverage refills

Apply Your Knowledge

Multiple-Choice Study Questions

6. Hot, potentially hazardous food should be held at an internal temperature of
   A. 135°F (57°C) or higher.
   B. 130°F (54°C) or higher.
   C. 120°F (49°C) or higher.
   D. 110°F (43°C) or higher.

7. Which statement is *not* true about holding potentially hazardous cold food without temperature control?
   A. The temperature of the product must not exceed 50°F (10°C).
   B. The food must be sold, served, or discarded within six hours.
   C. The food must contain a label specifying when it must be discarded.
   D. The food must contain a label specifying when it was removed from refrigeration.

For answers, please turn to the Answer Key.

Take It Back*

The following food safety concepts from this chapter should be taught to your employees:

- Holding food at the proper temperature
- Preventing contamination when serving food

The tools below can be used to teach these concepts in fifteen minutes or less using the directions below. Each tool includes content and language appropriate for employees. Choose the tool or tools that work best.

Tool #1: ServSafe Video 5: *Preparation, Cooking, and Serving*	**Tool #2:** *ServSafe Employee Guide*	**Tool #3:** ServSafe Posters and Quiz Sheets	**Tool #4:** ServSafe Fact Sheets and Optional Activities

Holding Food at the Proper Temperature

Video segment on holding food 1. Show employees the segment. 2. Ask employees to identify the proper temperatures for holding hot and cold potentially hazardous food.	**Section 4** **Proper Temperatures for Hot and Cold Holding** 1. Discuss with employees proper temperatures for holding hot and cold, potentially hazardous food. 2. Complete the Take Your Pick activity.	**Poster: Proper Temperatures for Hot and Cold Holding** 1. Discuss with employees the proper temperatures for holding hot and cold, potentially hazardous food as presented in the poster. 2. Have employees complete the Quiz Sheet: Proper Temperatures for Hot and Cold Holding.	

* Visit the Food Safety Resource Center at *www.ServSafe.com/FoodSafety/resource* to download free posters, quiz sheets, fact sheets, and optional activities and to learn how to obtain *Employee Guides* and videos/DVDs.

Take It Back*

Tool #1: ServSafe Video 5: *Preparation, Cooking, and Serving*	Tool #2: *ServSafe Employee Guide*	Tool #3: ServSafe Posters and Quiz Sheets	Tool #4: ServSafe Fact Sheets and Optional Activities
Preventing Contamination When Serving Food			
Video segment on serving food 1 Show employees the segment. 2 Ask employees to identify the proper ways to serve food and carry utensils to prevent contamination.	**Section 4** **Proper Ways to Serve Food** Discuss with employees the proper way to serve food and carry utensils to prevent contamination.	**Poster: Proper Ways to Serve Food** 1 Discuss with employees the proper way to serve food and carry utensils as presented in the poster. 2 Have employees complete the Quiz Sheet: Proper Ways to Serve Food.	

* Visit the Food Safety Resource Center at *www.ServSafe.com/FoodSafety/resource* to download free posters, quiz sheets, fact sheets, and optional activities and to learn how to obtain *Employee Guides* and videos/DVDs.

ADDITIONAL RESOURCES

Web Sites

Alaska Seafood Marketing Institute

www.alaskaseafood.org

The Alaska Seafood Marketing Institute (ASMI) is a public agency functioning as the state of Alaska's seafood-marketing arm. ASMI's Web site provides information regarding food safety and quality assurance practices associated with Alaskan seafood.

American Egg Board

www.aeb.org

The American Egg Board (AEB) is the U.S. egg producers' link to the consumer. This Web site provides resources for the restaurant and foodservice industry, including ways to ensure safe receiving, storage, and preparation of eggs and general information about eggs and egg products.

American Lamb Board

www.americanlambboard.org

The American Lamb Board (ALB) was created by the U.S. Secretary of Agriculture to administer the Lamb Promotion, Research, and Information Order. This Web site provides information about foodservice cuts, cooking temperatures, and food safety tips for lamb.

Conference for Food Protection

www.foodprotect.org

The Conference for Food Protection (CFP) is a nonprofit organization that provides a forum for regulators, industry professionals, academia, professional organizations, and consumers to identify problems, formulate recommendations, and develop and implement practices that ensure food safety. Though the conference has no formal regulatory authority, it is an organization that influences model laws and regulations among all government agencies. Visit this Web site for information regarding previously recommended changes to the *FDA Food Code,* standards for permanent outdoor cooking facilities, and other guidance documents, along with how to participate in the conference.

FDA Center for Food Safety and Applied Nutrition

www.cfsan.fda.gov/list.html

As the center within the Food and Drug Administration (FDA) responsible for food safety, the Center for Food Safety and Applied Nutrition (CFSAN) promotes and protects the public health by researching and implementing guidelines, policies, and standards to ensure that food is safe, nutritious, wholesome, and properly labeled. This Web site provides information relevant to all aspects of food safety and security, including corresponding guidelines, policies, and standards.

Food Safety and Inspection Service

www.fsis.usda.gov

The Food Safety and Inspection Service (FSIS) is the public health agency within the U.S. Department of Agriculture responsible for ensuring that the nation's commercial supply of meat, poultry, and egg products is safe, wholesome, and correctly labeled and packaged. Visit this Web site for food safety information and regulations related to meat, poultry, and eggs.

Gateway to Government Food Safety Information

www.foodsafety.gov

This Web site provides links to selected government food safety-related information.

International Fresh-Cut Produce Association

www.fresh-cuts.org

The International Fresh-Cut Produce Association (IFPA) represents processors, distributors, retail, and foodservice buyers of fresh-cut produce and companies that supply goods and services in the fresh-cut industry. IFPA provides its members with the knowledge and technical information necessary to deliver convenient, safe, and wholesome food. This Web site offers a detailed list of frequently asked questions about the fresh-cut produce industry and has a contact feature in which to ask a question. Free publications on food safety and best practices for fresh-cut produce are also available.

National Cattlemen's Beef Association

www.beef.org

The National Cattlemen's Beef Association (NCBA) is the national trade association representing U.S. cattle producers. The organization works to advance the economic, political, and social interests of the U.S. cattle industry. This Web site links to other NCBA-associated sites addressing topics such as nutrition, research, food safety, and marketing. A separate site, ***www.beeffoodservice.com,*** provides recipes, facts on foodservice cuts, and food safety information specifically for restaurant and foodservice operators.

National Chicken Council

www.nationalchickencouncil.com

The National Chicken Council (NCC) is the national, nonprofit trade association for the U.S. chicken industry, representing integrated chicken producer/processors, poultry distributors, and allied firms. Visit this Web site for consumer consumption behaviors, product information, and research related to poultry-associated foodborne illnesses.

National Frozen & Refrigerated Foods Association

www.nfraweb.org

The mission of the National Frozen & Refrigerated Foods Association is to promote the sales and consumption of frozen and refrigerated foods through education, training, research, sales planning, and menu development. Visit this Web site for information on the proper receiving, storage, and preparation of frozen and refrigerated food, including shelf-life charts, tips, and recommended food and equipment storage temperatures.

National Pork Producers Council

www.nppc.org

The National Pork Producers Council (NPPC) is one of the nation's largest livestock-commodity organizations. Visit the NPPC's foodservice site, ***www.porkfoodservice.com,*** for pork preparation tips, recipes, and consumer consumption trends.

National Turkey Federation

www.eatturkey.com

The National Turkey Federation is the national advocate for all segments of the U.S. turkey industry. This Web site contains information useful to foodservice establishments, including safe purchasing, storing, thawing, and preparing information, as well as recipes, nutrition information, and promotional ideas related to turkey.

The Produce Marketing Association

www.pma.com

The Produce Marketing Association (PMA) is a nonprofit association serving members who market fresh fruit, vegetables, and related products worldwide. Their members represent the production, distribution, retail, and foodservice sectors of the industry. Visit this Web site for articles and publications on produce food safety information within the foodservice and retail industry.

National Automatic Merchandising Association

www.vending.org/index.php

National Automatic Merchandising Association (NAMA) is the national trade association of the food and refreshment vending, coffee-service, and foodservice management industries including on site, commissary, catering, and mobile. The basic mission of the association is to collectively advance and promote the automatic-merchandising and coffee-service industries. Visit this Web site for valuable information on health, safety, and technical issues relating to the food and beverage vending industry.

United Fresh Fruit & Vegetable Association

www.uffva.org

The United Fresh Fruit & Vegetable Association is a national trade organization that represents the interests of growers, shippers, brokers, wholesalers, and distributors of produce. The association promotes increased produce consumption and provides scientific and technical expertise to all segments of the industry. This Web site contains news updates and free publications and lists upcoming events and initiatives on produce safety and quality.

Documents and Other Resources

2005 *FDA Food Code*

www.cfsan.fda.gov/~dms/fc05-toc.html

The Food and Drug Administration (FDA) publishes the *FDA Food Code,* a scientifically sound technical and legal document that serves as a model for regulating the retail and foodservice industry at the federal, state, and local level. Updates to the *FDA Food Code* are issued on the odd-numbered years. The *FDA Food Code* provides a system of safeguards designed to minimize foodborne illness and ensure employee health, food protection manager knowledge, safe food, nontoxic and cleanable equipment, and appropriate sanitation of the food establishment. It is used as the basis for information in this textbook.

Notes

RANGE GUARD
IN CASE OF FIRE
PULL PIN

10 Food Safety Management Systems

Inside this chapter:

- Prerequisite Food Safety Programs
- Active Managerial Control
- Hazard Analysis Critical Control Point (HACCP)
- Crisis Management

After completing this chapter, you should be able to:

- Identify how active managerial control can impact food safety.
- Identify HACCP principles for preventing foodborne illness.
- Implement HACCP principles when applicable.
- Identify when a HACCP plan is required.
- Implement a crisis-management program.
- Cooperate with regulatory agencies in the event of a foodborne-illness investigation.

Key Terms

- Food safety management systems
- Active managerial control
- HACCP
- HACCP plan
- Critical control point (CCP)

Apply Your Knowledge

Check to see how much you know about the concepts in this chapter. Use the page references provided with each question to explore the topic.

Test Your Food Safety Knowledge

1. **True or False:** Active managerial control focuses on controlling the most common foodborne-illness risk factors identified by the Centers for Disease Control and Prevention (CDC). *(See page 10-3.)*

2. **True or False:** Purchasing fish directly from local fishermen would be considered a risk in an active managerial control system. *(See page 10-3.)*

3. **True or False:** A critical control point (CCP) is a point in the flow of food where a hazard can be prevented, eliminated, or reduced to safe levels. *(See page 10-7.)*

4. **True or False:** If cooking is a CCP for ground beef patties in a particular establishment, then ensuring the internal temperature reaches 155°F (68°C) for fifteen seconds would be an appropriate critical limit. *(See page 10-8.)*

5. **True or False:** An establishment that cures food must have a HACCP plan. *(See page 10-12.)*

For answers, please turn to the Answer Key.

INTRODUCTION

In Chapters 5 through 9, you learned how to handle food safely throughout the flow of food. This accumulated knowledge will help you take the next step in preventing foodborne illness—the development of a food safety management system.

A **food safety management system** is a group of programs, procedures, and measures for preventing foodborne illness by actively controlling risks and hazards throughout the flow of food. Active managerial control and Hazard Analysis Critical Control Point (HACCP) offer two systematic and proactive approaches. Before these systems can be implemented, however, some prerequisite food safety programs must first be established.

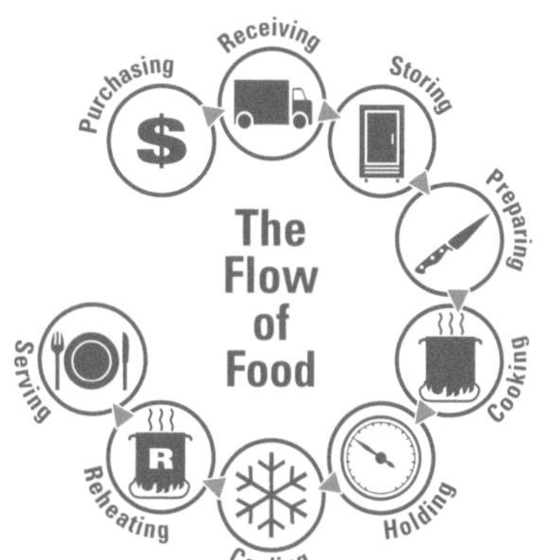

PREREQUISITE FOOD SAFETY PROGRAMS

For your food safety management system to be effective, you must first have the necessary food safety programs in place. (See *Exhibit 10a.*) The principles you have learned throughout the ServSafe program will help you develop these programs. They address the basic operational and sanitation conditions within your establishment, and can include processes, policies, and procedures.

ACTIVE MANAGERIAL CONTROL

One way to manage food safety risks in your establishment is to implement **active managerial control.** This approach focuses on controlling the five most common risk factors responsible for foodborne illness as identified by the Centers for Disease Control and Prevention (CDC). They include:

- Purchasing food from unsafe sources
- Failing to cook food adequately
- Holding food at improper temperatures
- Using contaminated equipment
- Practicing poor personal hygiene

Exhibit 10b

Active Managerial Control

Managers must monitor policies and procedures to ensure they are being followed.

The Active Managerial Control Approach

There are specific steps that should be taken when using active managerial control to manage food safety risks in your establishment.

1. **Consider the five risk factors as they apply throughout the flow of food, and identify any issues that could impact food safety.**
2. **Develop policies and procedures that address the issues that were identified.** Consider input from staff when developing them. It may be necessary to provide training on these policies and procedures.
3. **Regularly monitor the policies and procedures that have been developed.** This proactive step, which is critical to the success of an active managerial control system, can help you determine if the policies and procedures are being followed. (See *Exhibit 10b.*) If not, it may be necessary to revise them, create new ones, or retrain employees.
4. **Verify that the policies and procedures you have established are actually controlling the risk factors.** Use feedback from internal sources (records, temperature logs, and self-inspections) and external sources (health-inspection reports, customer comments, and quality assurance audits) to adjust the policies and procedures to continuously improve the system.

Example of Active Managerial Control

A seafood restaurant chain identified purchasing seafood from unsafe sources as a risk in its establishments. To avoid buying unsafe product, management developed a list of approved vendors based on a predetermined set of criteria. Next, they created a policy stating that seafood could only be purchased from vendors on this list. To ensure the policy was being followed, management decided that seafood invoices and deliveries would be monitored. On a regular basis, they looked at the criteria they had established for selecting seafood vendors to ensure they were still appropriate for controlling the risk. Management also decided to review their policy whenever a problem arose and to change it if necessary.

Something to Think About... **Get a Handle on It!**

A local health department was inspecting a store in a large quick-service chain. The inspector noticed that the grill operator who was handling raw chicken fillets also put cooked fillets in a holding drawer. A sandwich maker touched the handle of the drawer each time she retrieved a cooked fillet.

The health inspector recognized that the grill operator was contaminating the handle of the holding drawer each time he put a cooked fillet inside—since his hands had touched raw chicken. As the sandwich maker touched the contaminated handle, there was a risk that she could contaminate the sandwiches she was assembling.

While working with the unit manager to find a solution, the inspector recommended adding an extra handle to the holding drawer and designating one for each position. The chain adopted the recommendation in all of its units. They also modified standard operating procedures (SOPs) to control the risk, retrained employees, and incorporated the new SOPs in the chain's monitoring program.

HAZARD ANALYSIS CRITICAL CONTROL POINT (HACCP)

A HACCP system can also be used to control risks and hazards throughout the flow of food. **HACCP** (pronounced *HASS-ip*) is based on the idea that if significant biological, chemical, or physical hazards are identified at specific points within a product's flow through an operation, they can be prevented, eliminated, or reduced to safe levels.

To be effective, a HACCP system must be based on a written plan that is specific to each facility's menu, customers, equipment, processes, and operations. Since each **HACCP plan** is unique, a plan that works for one establishment may not work for another.

Exhibit 10c

The Seven HACCP Principles

1. Conduct a hazard analysis.
2. Determine critical control points (CCPs).
3. Establish critical limits.
4. Establish monitoring procedures.
5. Identify corrective actions.
6. Verify that the system works.
7. Establish procedures for record keeping and documentation.

The HACCP Approach

In order to focus on the critical aspects of the HACCP plan, it is essential that you have the necessary prerequisite food safety programs in place. These programs are the foundation on which an effective HACCP system is built.

A HACCP plan is based on the seven basic principles outlined by the National Advisory Committee on Microbiological Criteria for Foods. (See *Exhibit 10c.*) These principles are seven sequential steps that outline how to create a HACCP plan. Since each principle builds on the information gained from the previous principle, you must consider all seven principles in order when developing your plan.

In general terms:

- Principles One and Two help you identify and evaluate your hazards.
- Principles Three, Four, and Five help you establish ways for controlling those hazards.
- Principles Six and Seven help you maintain the HACCP plan and system and verify its effectiveness.

The Seven HACCP Principles

The information covered in the next several pages is designed to provide you with an introduction to the seven HACCP principles and an overview of the process for developing a HACCP program. A real-world example, highlighted in blue, has also been included for each principle. It documents the efforts of *Enrico's,* an Italian restaurant, as it implements a HACCP program.

Introduction to the Seven HACCP Principles

Principle One: Conduct a Hazard Analysis.

To identify and assess potential hazards in the food you serve, start by taking a look at how it is processed in your establishment. Many types of food are processed similarly. The most common processes include:

- Preparing and serving without cooking (salads, cold sandwiches, etc.)
- Preparing and cooking for same-day service (grilled chicken sandwiches, hamburgers, etc.)
- Preparing, cooking, holding, cooling, reheating, and serving (chili, soup, pasta sauce with meat, etc.)

Take a look at your menu, and identify items that are processed similarly. Next, identify those that are potentially hazardous, and determine where food safety hazards are likely to occur for each one. Hazards include contamination by:

- Bacteria, viruses, or parasites
- Cleaning compounds, sanitizers, and allergens
- General physical contaminants

The management team at *Enrico's* decided to implement a HACCP program. They began by conducting a hazard analysis.

When they reviewed their menu, they noted that several of their dishes—including the spicy, charbroiled chicken breast—were received, stored, prepared, cooked, and served the same day.

The team determined that bacteria were the most likely hazard to food prepared by this process.

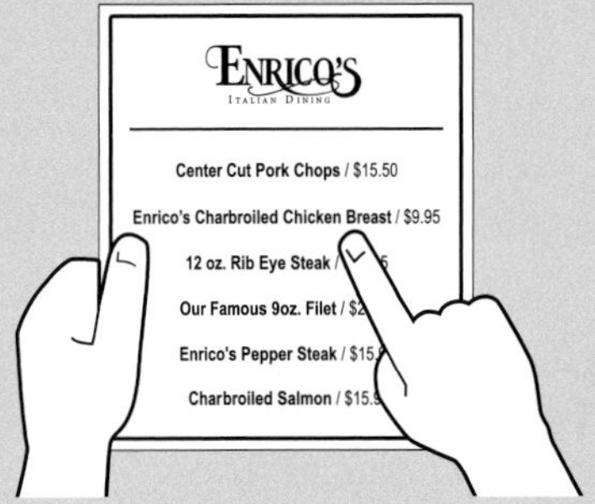

Principle Two: Determine Critical Control Points (CCPs)

Find the points in the process where the identified hazard(s) can be prevented, eliminated, or reduced to safe levels. These are the **critical control points (CCPs)**. Depending on the process, there may be more than one CCP.

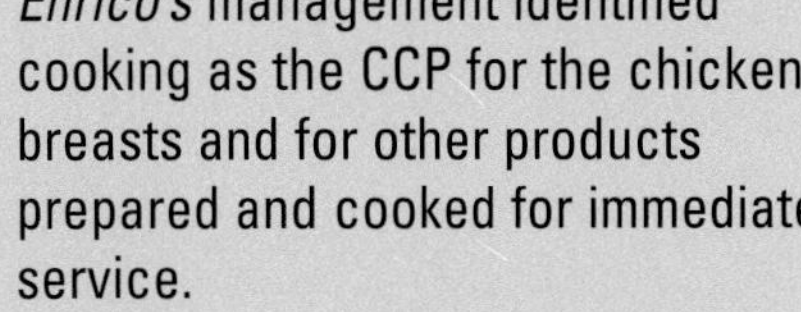

Enrico's management identified cooking as the CCP for the chicken breasts and for other products prepared and cooked for immediate service.

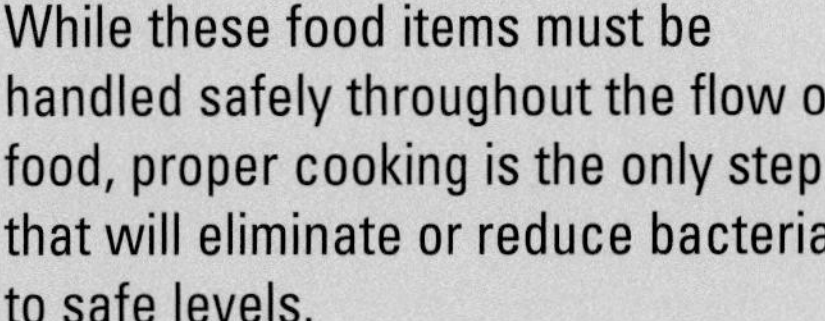

While these food items must be handled safely throughout the flow of food, proper cooking is the only step that will eliminate or reduce bacteria to safe levels.

Since the chicken breasts were prepared for immediate service, cooking was the only CCP identified.

Continued on the next page...

Introduction to the Seven HACCP Principles *continued*

Principle Three: Establish Critical Limits

For each CCP, establish minimum or maximum limits that must be met to prevent or eliminate the hazard or to reduce it to a safe level.

Since cooking was identified as the CCP for *Enrico's* chicken breasts, management determined that the critical limit would be cooking the chicken to a minimum internal temperature of 165°F (74°C) for fifteen seconds.

They decided that the critical limit could be met by placing the chicken breasts in the broiler and cooking them for sixteen minutes.

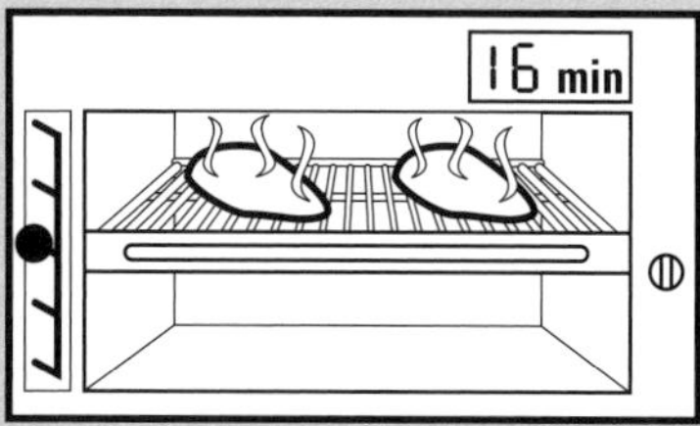

Principle Four: Establish Monitoring Procedures

Once critical limits have been established, determine the best way for your operation to check them to make sure they are consistently met. Identify who will monitor them and how often.

Since each charbroiled chicken breast is cooked to order, *Enrico's* chose to check the critical limit by inserting a clean and sanitized thermocouple probe into the thickest part of each breast.

The grill cook is required to check the temperature of each chicken breast after cooking to ensure that it has reached the minimum internal temperature of 165°F (74°C).

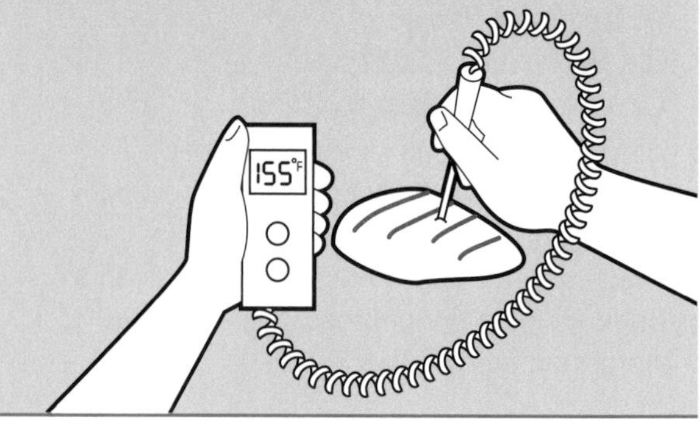

Introduction to the Seven HACCP Principles

Principle Five: Identify Corrective Actions

Identify steps that must be taken when a critical limit is *not* met. These steps should be determined in advance.

If the chicken breast has not reached its critical limit within the sixteen-minute cook time, the grill cook at *Enrico's* must keep cooking the breast until it has.

This and all other corrective actions are noted in the temperature log.

Principle Six: Verify That the System Works

Determine if the plan is working as intended. Plan to evaluate on a regular basis your monitoring charts, records, how you performed your hazard analysis, etc., and determine if your plan adequately prevents, reduces, or eliminates identified hazards.

Enrico's management team performs HACCP checks once per shift to ensure that critical limits were met and appropriate corrective actions were taken when necessary.

Additionally, they check the temperature logs on a weekly basis to identify patterns, or to determine if processes or procedures need to be changed. For example, over several weeks they noticed that toward the end of each week, the chicken breast often failed to meet its critical limit. The appropriate corrective action was being taken; however, management discovered that *Enrico's* received chicken shipments from a different vendor on Thursdays. This vendor provided a six-ounce chicken breast instead of the four-ounce chicken breast listed in *Enrico's* chicken specifications. Management worked with the vendor to ensure they received four-ounce breasts and changed their receiving procedure to include a weight check.

DATE: 1/31/07

Charbroiled Chicken Temperature Log

Time	Temp	Action	Initial
1:00	160° F	Cook to 165° F	CH
2:45	161° F	Cook to 165° F	CH
3:15	159° F	Cook to 165° F	CH
4:45	160° F	Cook to 165° F	CH

Continued on the next page...

Introduction to the Seven HACCP Principles *continued*

Principle Seven: Establish Procedures for Record Keeping and Documentation

Maintain your HACCP plan and keep all documentation created when developing it. In addition, keep records obtained when:

- Monitoring activities are performed
- Corrective action is taken
- Equipment is validated (checked to ensure it is in good working condition)
- Working with suppliers (i.e., shelf-life studies, invoices, specifications, challenge studies, etc.)

Enrico's management team determined that time-temperature logs should be kept for three months and receiving invoices for sixty days.

The team uses this documentation to support and revise their HACCP plan when necessary.

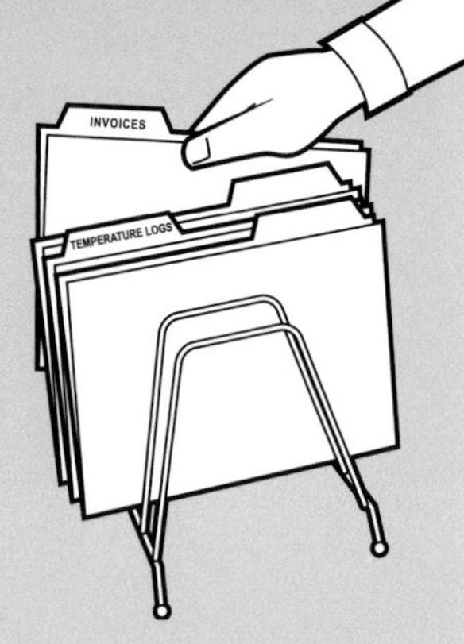

Another HACCP Example

The *Enrico's* example offers insight into the development of one type of HACCP plan. A plan may look much different when it addresses products that are processed more simply, such as those prepared and served without cooking. A good example is the HACCP plan developed by *Fresh Fruit Express,* an innovative fruit-only concept known for their signature item—the Melon Medley salad.

The HACCP team at *Fresh Fruit Express* conducted a hazard analysis for the Melon Medley Salad, which contains fresh watermelon, honeydew, and cantaloupe (Principle 1). They determined that bacteria pose a risk to the fresh-cut melons.

Since the melons are prepared, held, and served without cooking, the HACCP team determined that preparation and holding are CCPs for the salad (Principle 2). They decided that thoroughly cleaning and drying the melon's surface during preparation would reduce bacteria, and holding the melon at the proper temperature could prevent its growth. They determined that receiving is *not* a CCP since *Fresh Fruit Express* requires all melons be purchased from an approved source.

Next, the team established critical limits for each CCP (Principle 3). Since preparation was identified as a CCP, they decided that whole melons must be washed, scrubbed, and thoroughly dried. They developed a detailed SOP with specific techniques for washing the melons. It was also determined that the salad must be held at the critical limit of 41°F (5°C) or lower since it contained cut melons.

Since the melons are prepared the day they are received, the HACCP team decided that the critical limits would be monitored by the team leader (Principle 4). During preparation, he or she is required to visually inspect the washed melons to ensure that all surface dirt and debris have been properly removed. After the melons are cut, mixed, and portioned into individual containers, they are placed in the display cooler. To monitor the critical limit during holding, the team leader is required to check the holding temperature three times per day. A clean, sanitized, and calibrated probe must be inserted into the salad to ensure it is being held at 41°F (5°C) or lower.

If melons still contained surface dirt after preparation, the HACCP team decided that the appropriate corrective action would be to rewash them and get the team leader's approval before they are sliced (Principle 5). If the holding temperature is higher than 41°F (5°C), the team leader must check the temperature of every Melon Medley Salad container in the cooler and throw away all salads that are above 41°F (5°C). The team leader must also record all corrective actions in the Manager Daily HACCP Check Sheet.

The HACCP team wants to verify that the system is working properly (Principle 6). Therefore, the team leader is required to review the Manager Daily HACCP Check Sheet at the end of the shift. The team leader ensures that each item was checked and initialed, and that all corrective actions have been taken and recorded. In addition, *Fresh Fruit Express* formally evaluates the HACCP system once each quarter to determine if it is working properly.

Since a foodborne illness associated with fresh produce can take as long as sixteen weeks to emerge, the team determined that all HACCP records must be maintained for sixteen weeks and kept on file. They also decide that produce suppliers must maintain their records for a minimum of one year (Principle 7).

When a HACCP Plan Is Required

The National Restaurant Association and the Food and Drug Administration (FDA) recommend that all restaurants and foodservice establishments develop and implement a food safety management system. Establishments are required to have a HACCP plan in place, however, if they:

- Smoke or cure food as a method of food preservation
- Use food additives as a method of food preservation
- Package food using a reduced-oxygen packaging method
- Offer live, molluscan shellfish from a display tank
- Custom-process animals for personal use
- Package unpasteurized juice for sale to the consumer without a warning label
- Sprout beans or seeds

CRISIS MANAGEMENT

Implementing the food safety principles you have learned in the ServSafe program can help ensure that the food you serve your customers is safe. However, despite your best efforts, a foodborne-illness outbreak can occur in your establishment at any time. How you respond when that happens can determine whether or not you end up in the middle of a crisis.

The basis of a successful crisis-management program is a written plan that identifies the resources required and procedures that must be followed to handle crises. The time to prepare for a crisis is before one occurs. There is no off-the-shelf disaster plan that works for every establishment. Each plan must meet the operation's individual needs.

Developing a Plan

It is critical to get management's support when developing your crisis-management plan. Once you have it, you should start by identifying the basic objectives of the plan. These usually include meeting the immediate needs of the operation and keeping the business viable. Then, include the plan details. This may consist of a checklist with brief step-by-step procedures to follow or a full-scale plan covering specific tasks, roles, and resources. You should also prepare specific procedures for developing, updating, and distributing the plan.

There are a number of steps you can take to prepare for the possibility of a crisis:

Exhibit 10d

Develop a Crisis-Management Team

The size of the team will vary depending on the size of the establishment.

- **Develop a crisis-management team.** In large, multi-unit operations, the team may include the president and senior managers from the finance, operations, risk management, marketing, franchising, human resources, public relations, and training departments. Crisis-management teams are sometimes much smaller and may vary in makeup depending on the establishment and the situation. (See *Exhibit 10d.*) For example, an independent restaurant's team might include the owner, general manager, and chef.
- **Identify potential crises.** While the greatest threat to customers may be from foodborne illness, do not forget that other crises can include food security threats, robberies, severe weather, power outage, water interruption, fire, flood, or some other trauma.
- **Develop simple instructions for responding to each type of crisis.** In a foodborne-illness outbreak, steps include isolating the suspect food, obtaining samples, and preventing its further sale. Also, exclude suspect employees from handling food, and contact and cooperate with the local health department.
- **Assemble a contact list with names and numbers, and post it by phones.** The list should include all crisis-management team members and outside resources, such as police, fire and health departments, testing labs, issues experts, and management or headquarters personnel.
- **Develop a crisis-communication plan.** It should include:
  - A list of media responses or a question-and-answer sheet suggesting what to say for each crisis.
  - Sample press releases that can be tailored quickly to each incident.

Exhibit 10e

Crisis-Communication Plan

There are several guides available to help you develop a communication plan.

- A list of media contacts to call for press conferences or news briefings. Include a media-relations plan with "do's and don'ts" for dealing with the media.
- A plan for communicating with employees during the crisis. Possibilities include shift meetings, email, a telephone tree, etc. There are several guides available to help you develop a crisis communication plan. (See *Exhibit 10e.*)

- **Assign and train a spokesperson to handle media relations.** Appoint a single spokesperson to handle all media queries and communications. Designating a point person usually results in more consistent messages and allows you to control media access to your staff. The spokesperson should be familiar with interview skills so that he or she knows what to expect and how to respond. Crisis situations can be very stressful, and training will enable your spokesperson to handle it better. Make sure all of your staff know who the spokesperson is, and instruct them to direct questions to that person.
- **Assemble a crisis kit for the establishment.** The kit can take the form of a three-ring notebook or binder enclosing the plan's materials. Keep the kit in an accessible place, such as the manager's or chef's office.
- **Test the plan by running a simulation to make sure it works as intended.** You can test the plan yourself or hire a public relations or consulting firm with crisis-management experience to enact a crisis and test your team's readiness. In either case, the test should be designed to simulate a crisis that will be as close as possible to what could happen.

Exhibit 10f

Responding to a Foodborne-Illness Complaint

Express concern and be sincere, but do not admit responsibility.

Crisis Response

A crisis-management plan should include response procedures to help manage an emergency more effectively. In a foodborne-illness outbreak, you may be able to avert a crisis by responding quickly if you receive customer complaints. Take all customer complaints seriously. Express your concern and be sincere, but do not admit responsibility or accept liability. (See *Exhibit 10f.*) Listen carefully and promise to investigate and respond. With legal guidance, consider developing an incident report to help you through the process. Questions to ask include:

- **What did you eat and drink at our establishment and when?**
- **When did you become ill?** What were the symptoms, and how long did you experience them?
- **Did you eat anything else before or after eating at our establishment?** What and where? Did anyone else eat the same food? Did they become ill?
- **Did you seek medical attention?** Where and how soon after becoming ill? What diagnosis and treatment did you receive?

Evaluate the complaint. If several customers have similar complaints, you may have a potential crisis on your hands. Take steps to control the situation and reassure customers that you are doing everything you can to identify and fix the problem.

At this point, you may consider calling your crisis team together to implement the plan.

Exhibit 10g

When a Foodborne-Illness Outbreak is Confirmed

Cooperate with health officials and accept responsibility.

- **Direct the team to gather information, plan courses of action, and manage events as they unfold.**
- **Work with, not against, the media.** Make sure the spokesperson is fully informed before arranging a press conference. Contacting the media before they contact you helps you to control what they report. Stick to the facts, and be as honest as possible. If you do not have all the facts, say so, and let the media know that you will communicate with them as soon as you do know. Keep a cool head, and do not be defensive. The easiest way to magnify or prolong a crisis is to deny, lie, give wrong information, or change your story.
- **Show concern and be sincere.** If health officials have confirmed that your establishment is the source of the illness, accept responsibility and cooperate with the investigation. (See *Exhibit 10g.*) Accepting responsibility is not the same as admitting liability. While customers may have become ill from eating food in your operation, the cause may have been beyond your control and not your fault. If you do not express your concern and mean it, you will lose credibility with the public, not just customers.

- **Communicate the information directly to all of your key audiences.** Do not depend on the media to relay all the facts. Tell your side of the story to employees, customers, stockholders, and the community. Use newsletters, a Web site, flyers, and newspaper or radio advertising.
- **Fix the problem, and communicate to both the media and to your customers what you have done.** Each time you take a step to resolve the problem, let the media know. Hold briefings when you have news, and go into each briefing or press conference with an agenda. Take control rather than simply respond to questions.

Crisis Recovery and Assessment

During the planning process, you should develop procedures to help you recover from a crisis. Think about what steps you need to take to ensure that the establishment is clean and sanitized and food is safe so that your operation can get back up and running. Once a crisis is over, it is important to determine the causes and effects of the crisis so that your establishment can implement changes to take advantage of the lessons learned. Some things to evaluate include:

- Obstacles faced in returning to normal operations
- How communication with employees and customers was handled
- Assessment of the damages from the crisis
- Overall assessment of the crisis and your response

A foodborne-illness outbreak has the potential to damage your business beyond repair. Investing time and resources in a crisis management plan can help minimize it.

Planning for Possible Crises

In addition to a foodborne-illness outbreak, other possible crises can affect the safety of the food you serve. Among these are power outages, water interruption (including water contamination), fire, and flood. These crises require immediate action to prevent endangering people's health and therefore should be addressed in your crisis-management plan.

Preparedness involves having contingencies in place to deal with a crisis when it occurs. Your crisis-management plan should also include response and recovery steps. Sample contingency plans for each of these crises have been provided in *Exhibit 10h.*

Exhibit 10h

Sample Contingency Plans

<table>
<tr><th>Planning and Preparedness</th><th colspan="2">Response</th><th>Recovery</th></tr>
<tr><td colspan="4">Foodborne-Illness Outbreak</td></tr>
<tr><td></td><td>If</td><td>Then</td><td></td></tr>
<tr><td rowspan="4">■ Assemble an emergency contact list including contact information for the local regulatory agency, testing labs, and management headquarters personnel.
■ Train designated employees to correctly fill out a foodborne-illness incident report.
■ Ensure a proper crisis communication plan is in place.</td><td>The suspected food is still in the establishment</td><td>■ Isolate the suspected food and properly identify it, to prevent further sale.
■ If possible, obtain samples of the suspect food from the customers.</td><td rowspan="4">■ Work with the regulatory agency to create solutions and resolve issues.
■ Clean and sanitize all areas of the establishment to ensure the incident does not reoccur.
■ Throw out all suspect products.
■ Conduct an investigation to determine the cause of the outbreak.
■ Establish new procedures or revise existing ones based on the investigation results to prevent the incident from reoccurring.
■ Develop a plan to reassure customers that the food served in your establishment is safe.</td></tr>
<tr><td>The suspected outbreak is caused by an ill employee</td><td>■ Exclude the suspect employee from the establishment.</td></tr>
<tr><td>The regulatory agency confirms your establishment is the source of the outbreak</td><td>■ Cooperate with the regulatory agency to resolve the crisis.</td></tr>
<tr><td>The media contacts your establishment</td><td>■ Follow your crisis communication plan. Let your spokesperson handle all communications.</td></tr>
</table>

Continued on the next page...

Exhibit 10h

Sample Contingency Plans *continued*

Planning and Preparedness	Response		Recovery
Power Outage			
	If	**Then**	
■ Arrange for access to an electrical generator and a refrigerated truck to use in the event of an emergency. ■ Prepare a menu that includes items that do not require cooking that could be used in the event of an emergency. ■ Develop a policy that addresses when refrigerator doors should be opened. ■ Make a list of electrical equipment that could be negatively affected when the power is restored. ■ Have emergency contact information for the utility company, garbage service, ice supplier, etc.	Refrigeration equipment stops working	■ Write down the time of the power outage. ■ Check and record food temperatures periodically. ■ Keep refrigerator and freezer doors closed. ■ Pack potentially hazardous food in ice bought from an approved supplier.	■ Check refrigeration units frequently after the power has been restored to ensure the equipment can maintain product temperature. ■ Discard potentially hazardous food that has been in the temperature danger zone for more than four hours.
	Ventilation hoods or fans stop working	■ Stop all cooking.	
	Hot-holding equipment stops working	■ Write down the time of the power outage. ■ Throw out all potentially hazardous food held below 135°F (57°C) for more than four hours. ■ Food can be reheated if the power outage was less than four hours.	

Exhibit 10h

Sample Contingency Plans

<table>
<tr><th>Planning and Preparedness</th><th colspan="2">Response</th><th>Recovery</th></tr>
<tr><td colspan="4">Water-Service Interruption</td></tr>
<tr><td></td><td>If</td><td>Then</td><td></td></tr>
<tr><td rowspan="4">■ Water outage or contamination
■ Prepare a menu that includes items that require little or no water that could be used in the event of an emergency.
■ Keep a supply of single-use items.
■ Keep a supply of bottled water, and have a supplier who can provide bottled water in an emergency.
■ Have a supplier who can provide ice in an emergency.
■ Assemble an emergency contact list including contact information for the local regulatory agency, plumber, and water department.</td><td>Hands cannot be washed</td><td>■ Implement an emergency handwashing procedure.
■ Do not touch ready-to-eat food with bare hands.</td><td rowspan="4">Take the following actions when water is restored or the local regulatory agency has lifted any boiled-water advisory:
■ Clean and sanitize equipment with waterline connections such as spray misters, coffee or tea urns, ice machines, and others. Follow manufacturers' instructions.
■ Flush water lines as required by the local regulatory agency.
■ Work with your regulatory agency to resume normal operations.</td></tr>
<tr><td>Toilets do not flush</td><td>■ Find other toilet facilities for employee use during operating hours.
■ Stop operations if toilet facilities are not available.</td></tr>
<tr><td>Drinking water is not available or it is contaminated</td><td>■ Use bottled water.
■ Use water from an approved supplier. Keep water in a covered, sanitized container during hauling or storage.</td></tr>
<tr><td>Food items that require water during preparation cannot be made</td><td>■ Throw out any ready-to-eat food made with water before the contamination was discovered.
■ Use bottled or boiled water for ready-to-eat food.</td></tr>
</table>

Continued on the next page...

Exhibit 10h

Sample Contingency Plans *continued*

Planning and Preparedness	Response		Recovery
Water-Service Interruption *continued*			
	If	**Then**	
■ Develop procedures that minimize water use during the emergency. (i.e., using single-use items for service) ■ Work with your regulatory agency to develop an emergency handwashing procedure that can be used during water service interruptions.	Water is not available for food preparation and cooking	■ Use water from an approved supplier. ■ Use the emergency menu. ■ Use prewashed, packaged produce or frozen or canned fruits and vegetables. ■ Only thaw food in the refrigerator, microwave, or as a part of the cooking process.	
	Ice cannot be made	■ Throw out existing ice. ■ Stop making ice. ■ Use ice from an approved source.	
	Equipment, utensils, facility cannot be cleaned or sanitized	■ Use single-use items. ■ Use bottled water or water from an approved source to clean and sanitize.	
	Beverages made with water cannot be prepared	■ Stop using the drink machines that require water, such as the auto-fill coffee maker, instant hot-water heater, etc.	

Exhibit 10h

Sample Contingency Plans

Planning and Preparedness	Response		Recovery
Fire			
	If	**Then**	
■ Assemble an emergency contact list including contact information for the fire and police departments, local regulatory agency, and management or headquarters personnel. ■ Post the fire department phone number by each phone so it is easy to see.	A fire occurs	■ Stop operations if food can no longer be safely prepared. ■ Block off areas, equipment, utensils, and other items affected by the fire.	■ Throw out all food affected by the fire. ■ Throw out all damaged utensils, linens, or items that cannot be cleaned and sanitized. ■ Clean and sanitize the establishment. ■ If necessary, hire a janitorial service specializing in areas exposed to fires. ■ Check water lines. Using fire hoses may lower water pressure in the area and could cause backflow and water contamination.
Flood			
	If	**Then**	
■ Have a plan to monitor and maintain flood control equipment: plumbing, storm drains, sump pumps, etc.	A water line leaks or water builds up on the floor but food, utensils, etc., are not affected	■ Keep people away from the wet floor. ■ Repair the leak. ■ Block off areas, equipment, utensils, and other items affected by the flood.	■ Throw out all damaged utensils, linens, or items that cannot be cleaned and sanitized. ■ Throw out any food or food packaging that made contact with the water.

Continued on the next page...

Exhibit 10h

Sample Contingency Plans *continued*

Planning and Preparedness	Response		Recovery
Flood *continued*			
	If	**Then**	
■ Assemble an emergency contact list including contact information for the local regulatory agency, plumber, utility companies, etc. ■ Keep a supply of bottled water.	A flood affects or damages food, utensils, etc.	■ Stop all operations.	■ Clean and sanitize the facility, utensils, equipment surfaces, floors, or other affected areas. ■ Hire a janitorial service, if needed, specializing in cleaning areas exposed to floods.
	The flood is the result of a sewage backup in the food-preparation area	■ Stop all operations.	

SUMMARY

A food safety management system is a group of programs, procedures, and measures for preventing foodborne illness by actively controlling risks and hazards throughout the flow of food. HACCP and active managerial control offer two systematic and proactive approaches. In order for the food safety system to be effective, you must first have the necessary food safety programs in place. These include programs for personal hygiene, facility design, supplier selection and specifications, sanitation and pest control, equipment maintenance, and food safety training.

One way to control risks associated with foodborne illness is to implement active managerial control. This approach focuses on controlling the five most common risk factors responsible for foodborne illness as identified by the CDC. These include purchasing food from unsafe sources, failing to cook food adequately, holding food at improper temperatures, using contaminated equipment, and practicing poor personal hygiene.

There are specific steps that should be taken when using active managerial control to manage these risks in your establishment. First, you must consider the five risk factors as they apply throughout the flow of food and then identify any issues that could impact food safety. Next, you will need to develop policies and procedures that address the issues that were identified. These will require regular monitoring to determine if they are being followed. Finally, you must verify that the policies and procedures you have established are actually controlling the risk factors.

A HACCP system focuses on identifying specific points within a product's flow through the operation that are essential to prevent, eliminate, or reduce biological, chemical, or physical hazards to safe levels. To be effective, a HACCP system must be based on a plan specific to a facility's menu, customers, equipment, processes, and operation. The HACCP plan is developed following seven sequential principles—essential steps for building a food safety system.

First, the establishment must identify and assess potential hazards in the food they serve by taking a look at how it is processed. Once common processes have been identified, they can determine where food safety hazards are likely to occur for each one. The establishment must then identify points where they can be prevented, eliminated, or reduced to safe levels. These are the CCPs. Next, the establishment must determine minimum or maximum limits that must be met for each CCP to prevent, eliminate, or reduce the hazard. The establishment must determine how they will monitor the CCPs they have identified and what actions will be taken when critical limits have not been met. Finally, the establishment must find ways to verify that the HACCP system is working, and establish procedures for record keeping and documentation.

A food safety management system can help ensure that the food you serve is safe. Despite your best efforts, however, a foodborne-illness outbreak—and other crises—can occur in your establishment. The time to prepare for a crisis is before one occurs. The key is to start with a written plan that identifies the resources required and procedures that must be followed to handle the crisis. To prepare for a crisis: develop a crisis-management team, spell out specific instructions for handling the crisis, develop a crisis-communication plan, and assign and train a media spokesperson. When receiving customer complaints, listen carefully, express concern, and be sincere. Do not admit responsibility, but promise to investigate and respond. With legal advice, consider developing an incident report to guide you through the process. Call your crisis-management team together, and implement your plan.

Apply Your Knowledge

Use these questions to review the concepts presented in this chapter.

Discussion Questions

1. What food safety programs must be in place for a food safety management system to be effective?
2. What are the five most common risk factors responsible for foodborne illness as identified by the CDC?
3. List the specific steps that should be taken when using active managerial control to manage food safety risks.
4. List the seven HACCP principles in order.
5. When is an establishment required to have a HACCP plan?

For answers, please turn to the Answer Key.

Apply Your Knowledge

Use these questions to test your knowledge of the concepts presented in this chapter.

Multiple-Choice Study Questions

1. The temperature of a roast is checked to see if it has met its critical limit of 145°F (63°C). This is an example of which HACCP principle?
   A. Verification
   B. Monitoring
   C. Record keeping
   D. Hazard analysis

2. The temperature of a pot of beef stew is checked during holding. The stew has *not* met the critical limit of 135°F (57°C) and is discarded according to house policy. Discarding the stew is an example of which HACCP principle?
   A. Monitoring
   B. Corrective action
   C. Hazard analysis
   D. Verification

3. Which is *not* one of the CDC's most common risk factors for foodborne illness?
   A. Failing to cook food adequately
   B. Failing to have a pest control program
   C. Failing to purchase food from safe sources
   D. Failing to hold food at the proper temperature

4. Which is *not* a corrective action?
   A. Continuing to cook a hamburger until it reaches a minimum internal temperature of 155°F (68°C) for fifteen seconds
   B. Discarding cooked chicken that has been held at 120°F (49°C) for five hours
   C. Sanitizing a prep counter before starting a new task
   D. Rejecting a shipment of oysters received at 55°F (13°C) that will be served raw

5. Which program should be in place before you begin developing your food safety system?
   A. Personal hygiene program
   B. Incentive program
   C. Workplace accident prevention program
   D. None of the above

Apply Your Knowledge

Multiple-Choice Study Questions

6. The purpose of a food safety management system is to
   A. identify the proper methods for receiving food.
   B. identify and control possible hazards and risks throughout the flow of food.
   C. keep the establishment pest free.
   D. identify faulty equipment within the establishment.

7. A chef sanitized his thermometer probe and checked the temperature of minestrone soup being held in a hot-holding unit. The temperature was 120°F (49°C), which did not meet the establishment's critical limit of 135°F (57°C). He recorded the temperature in the log and reheated the soup to 165°F (74°C) for fifteen seconds. Which was the corrective action?
   A. Sanitizing the thermometer probe
   B. Taking the temperature of the soup
   C. Reheating the soup
   D. Recording the temperature of the soup in the temperature log

8. A HACCP plan is required when an establishment
   A. serves raw shellfish.
   B. serves undercooked ground beef.
   C. uses mushrooms that have been picked in the wild.
   D. packages unpasteurized juice for sale to consumers without a warning label.

For answers, please turn to the Answer Key.

ADDITIONAL RESOURCES

Web Sites

Association of Food and Drug Officials

www.afdo.org

The Association of Food and Drug Officials (AFDO) fosters uniformity in the adoption and enforcement of food, drug, medical devices, cosmetics and product safety laws, rules, and regulations. AFDO also provides the mechanism and the forum in which regional, national, and international issues are deliberated and resolved uniformly to provide the best public health and consumer protection. Visit this Web site for information on the association's annual conference, its position statements, and food safety information.

Center for Infectious Disease Research & Policy

www.cidrap.umn.edu

The Center for Infectious Disease Research & Policy (CIDRAP) at the University of Minnesota functions to prevent illness and death from infectious diseases through epidemiologic research and the rapid translation of scientific information into real-world practical applications and solutions. The Center's Web site provides access to information on food safety and foodborne illness and news stories on food security and foodborne-illness surveillance.

Conference for Food Protection

www.foodprotect.org

The Conference for Food Protection (CFP) is a nonprofit organization that provides a forum for regulators, industry professionals, academia, professional organizations, and consumers to identify problems, formulate recommendations, and develop and implement practices that ensure food safety. Though the conference has no formal regulatory authority, it is an organization that influences model laws and regulations among all government agencies. Visit this Web site for information regarding previously recommended changes to the *FDA Food Code,* standards for permanent outdoor cooking facilities, and other guidance documents, along with how to participate in the conference.

Food and Drug Administration

www.fda.gov

The Food and Drug Administration (FDA) is responsible for promoting and protecting public health by helping safe and effective products reach consumers, monitoring products for continued safety, and helping the public obtain accurate, science-based information needed to improve health. Visit this Web site for information about all programs regulated by the FDA, including food-product labeling, bottled water, and food products except meat and poultry.

FDA Center for Food Safety and Applied Nutrition

www.cfsan.fda.gov/list.html

As the center within the Food and Drug Administration (FDA) responsible for food safety, the Center for Food Safety and Applied Nutrition (CFSAN) promotes and protects the public health by researching and implementing guidelines, policies, and standards to ensure that food is safe, nutritious, wholesome, and properly labeled. This Web site provides information relevant to all aspects of food safety and security, including corresponding guidelines, policies, and standards.

Food Safety and Inspection Service

www.fsis.usda.gov

The Food Safety and Inspection Service (FSIS) is the public health agency within the U.S. Department of Agriculture responsible for ensuring that the nation's commercial supply of meat, poultry, and egg products is safe, wholesome, and correctly labeled and packaged. Visit this Web site for food safety information and regulations.

Gateway to Government Food Safety Information

www.foodsafety.gov

This Web site provides links to selected government food safety-related information.

National Frozen & Refrigerated Foods Association

www.nfraweb.org

The mission of the National Frozen & Refrigerated Foods Association is to promote the sales and consumption of frozen and refrigerated food through education, training, research, sales planning, and menu development. Visit this Web site for information on the proper receiving, storage, and preparation of frozen and refrigerated food, including shelf-life charts, tips, and recommended food- and equipment-storage temperatures.

Documents and Other Resources

Crisis Management: Imminent Health Hazards

www.michigan.gov/mda/0,1607,7-125--105442--,00.html

The Michigan Department of Agriculture developed emergency action procedures that outline how to plan, prepare, and respond for emergencies with the potential to become imminent health hazards. Visit this site for procedures that were developed using input from the restaurant and foodservice industry and the regulatory community.

FDA Enforcement Report Index

www.fda.gov/opacom/Enforce.html

The Food and Drug Administration's Center for Food Safety and Applied Nutrition (FDA CFSAN) provides the *FDA Enforcement Report,* published weekly on this Web site. It contains information on actions such as recalls, injunctions, and seizures taken against food and drug products that have not met FDA regulatory requirements.

2005 *FDA Food Code*

www.cfsan.fda.gov/~dms/fc05-toc.html

The Food and Drug Administration (FDA) publishes the *FDA Food Code*, a scientifically sound technical and legal document that serves as a model for regulating the retail and foodservice industry at the federal, state, and local level. Updates to the code are issued in the odd-numbered years. The *FDA Food Code* provides a system of safeguards designed to minimize foodborne illness and ensure employee health, food protection manager knowledge, safe food, nontoxic and cleanable equipment, and appropriate sanitation of the food establishment. It is used as the basis for information in this textbook.

FDA Foodborne Illness

www.cfsan.fda.gov/~mow/foodborn.html

The Food and Drug Administration's Center for Food Safety and Applied Nutrition (FDA CFSAN) provides the FDA Foodborne Illness gateway page, which contains resources related to foodborne illness. Visit this Web site to access information about HACCP, foodborne pathogens, specific government food safety initiatives, and other government sources of foodborne-illness information.

Food Establishment Plan Review Guide

www.cfsan.fda.gov/~dms/prev-toc.html

The *Food Establishment Plan Review Guide,* provided by the Food and Drug Administration (FDA) and the Conference for Food Protection (CFP), is available at this Web site. This guide has been developed to provide guidance and assistance in complying with nationally recognized food safety standards. It includes design, installation, and construction recommendations regarding food equipment and facilities.

Hazard Analysis and Critical Control Point Principles and Application Guidelines

www.fsis.usda.gov/OPHS/NACMCF/past/JFP0998.pdf

The National Advisory Committee on Microbiological Criteria for Foods provides HACCP guidelines on this Web site. The guidelines are intended to facilitate the development and implementation of an effective HACCP plan.

Managing Food Safety: A Manual for the Voluntary Use of HACCP Principles for Operators of Foodservice and Retail Establishments

www.cfsan.fda.gov/~dms/hret2toc.html

The Food and Drug Administration's Center for Food Safety and Applied Nutrition (FDA CFSAN) provides this manual at this Web site. It provides the restaurant and foodservice industry with a roadmap for developing and voluntarily implementing a food safety management system based on HACCP principles. Developed by the FDA, the manual has been reviewed and endorsed by the Conference for Food Protection (CFP).

Unit 3

Sanitary Facilities and Pest Management

Traulsen
Chris

11 Sanitary Facilities and Equipment

Inside this chapter:

- Designing a Sanitary Establishment
- Considerations for Other Areas of the Facility
- Sanitation Standards for Equipment
- Installing and Maintaining Kitchen Equipment
- Utilities

After completing this chapter, you should be able to:

- Identify when a plan review is required.
- Identify organizations that certify equipment that meets sanitation standards.
- Identify characteristics of an appropriate food-contact and nonfood-contact surface.
- Identify the requirements for installing stationary and mobile equipment.
- Recognize the importance of maintaining equipment.
- Identify and prevent cross-connection and backflow.
- Identify requirements for handwashing facilities including appropriate locations and numbers.
- Identify the proper response to a wastewater overflow.
- Recognize the importance of properly installing and maintaining grease traps.
- Identify lighting-intensity requirements for different areas of the establishment.
- Identify potable water sources and testing requirements.
- Identify methods for preventing lighting sources from contaminating food.
- Identify methods for preventing ventilation systems from contaminating food and food-contact surfaces.
- Identify requirements for storing indoor and outdoor waste.
- Identify proper methods for cleaning waste receptacles.
- Recognize the need for frequent waste removal to prevent odor and pest problems.
- Identify characteristics of appropriate flooring.
- Recognize the importance of complying with ADA requirements for facility design.
- Recognize the importance of keeping physical facilities in proper repair.
- Identify requirements for dishwashing facilities.

Key Terms

- Porosity
- Resiliency
- Coving
- NSF International
- Underwriters Laboratories (UL)
- Cantilever-mounted equipment
- Potable water
- Booster heater
- Cross-connection
- Backflow
- Air gap

Apply Your Knowledge

Check to see how much you know about the concepts in this chapter. Use the page reference provided with each question to explore the topic.

Test Your Food Safety Knowledge

1 **True or False:** A hose attached to a utility-sink faucet and left sitting in a bucket of dirty water could contaminate the water supply. *(See page 11-23.)*

2 **True or False:** There must be a minimum of twenty foot-candles of light (215 lux) in a food-preparation area. *(See page 11-25.)*

3 **True or False:** Handwashing stations are required in dishwashing and service areas. *(See page 11-10.)*

4 **True or False:** When mounted on legs, stationary equipment must be at least two inches (five centimeters) off the floor. *(See page 11-18.)*

5 **True or False:** Grease on an establishment's ceiling can be a sign of inadequate ventilation. *(See page 11-26.)*

For answers, please turn to the Answer Key.

INTRODUCTION

Many breakdowns in sanitation are caused by facilities and equipment that are simply too difficult to keep clean. Sanitary facilities and equipment are basic parts of a well-designed food safety system. In this chapter, you will find a wide range of information on various equipment and facility-related issues that are key to keeping an establishment safe.

DESIGNING A SANITARY ESTABLISHMENT

When designing or remodeling a facility, consider how the building and the equipment in each area will be kept clean. Poorly designed areas are generally harder to clean and can become a breeding ground for microorganisms. As a result, food passing through these areas runs a much higher risk of contamination. Facilities should be arranged so contact with contaminated sources such as garbage or dirty tableware, utensils, and equipment is unlikely to occur.

This chapter focuses on four topics related to the sanitary layout and design of equipment and facilities:

- Design and arrangement of equipment and fixtures to comply with sanitation standards
- Material selection for walls, floors, and ceilings that will make cleaning these surfaces easier
- Design of utilities to prevent contamination and to make cleaning easier
- Proper waste management to avoid contaminating food and attracting pests

Exhibit 11a

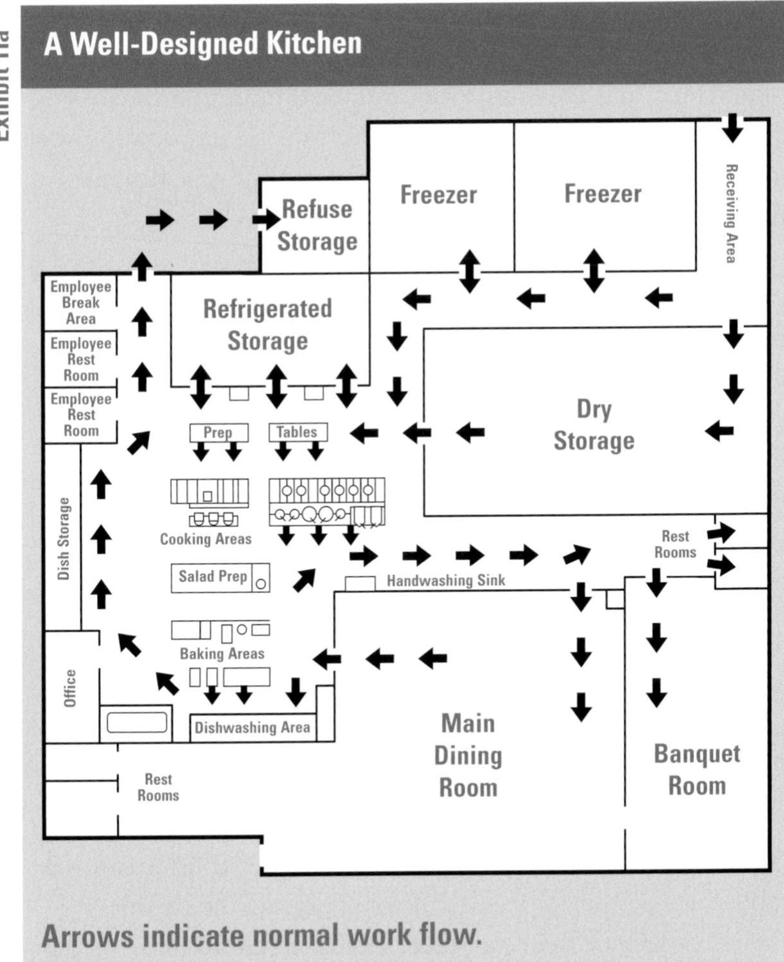

Locating storage areas near receiving areas and prep tables near refrigerators and freezers can minimize time-temperature abuse.

Layout

Well-designed kitchens make it easier to keep food safe. Generally, an efficient kitchen design will result in a more sanitary kitchen. A well-designed kitchen will address the following factors:

- **Work flow.** A work flow must be established that will minimize the amount of time food spends in the temperature danger zone. It must also minimize the number of times food is handled. For example, storage areas should be located near the receiving area to prevent delays in storing food. Prep tables should be located near refrigerators and freezers for the same reason. (See *Exhibit 11a.*)

- **Contamination.** A good layout will minimize the risk of cross-contamination. Dirty equipment should not be placed where it will touch clean equipment or food. For example, it is not a good practice to place the soiled-utensil table next to the salad-preparation sink.
- **Equipment accessibility.** Since hard-to-reach areas are less likely to be cleaned, a well-planned layout will ensure that equipment is accessible for cleaning.

Key Point

Even if local laws do not require it, layout and design plans should be reviewed by the local regulatory agency.

The Plan Review

Consult local regulations prior to new construction or extensive remodeling. The health department or local regulatory agency may require approval of layout and design plans. Plans should be reviewed—even if it is not required. Such reviews ensure compliance, and they often save time and money in the long run. Additionally, these agencies often provide information they consider necessary for good sanitation. Ask for specification guidelines and plan review procedures.

Layout and design plans should include:

- Proposed layout
- Mechanical plans
- Type of construction materials to be used
- Types or models of proposed equipment
- Specifications for utilities, plumbing, and ventilation

Some local jurisdictions will require the approval of design plans by building and zoning departments. In addition, the Americans with Disabilities Act (ADA) requires reasonable accommodation for access to the building by both patrons and employees with disabilities. These guidelines can be found in the *ADA Accessibility Guide* (*ADAAG*).

Once construction has been completed, the establishment must apply for a permit to operate. Before granting the permit, the regulatory agency might conduct an inspection to make certain all design and installation requirements have been met. After the establishment has passed the inspection and obtains a certificate of operation, it can open for business.

Exhibit 11b

Construction Materials

The most important consideration when selecting construction materials is how easy the establishment will be to clean and maintain.

Materials for Interior Construction

Materials used during construction must be selected with several factors in mind. Sound-absorbent surfaces that resist absorption of grease and moisture and reflect light will probably create an environment acceptable to your regulatory agency. The most important consideration when selecting construction materials is how easy the establishment will be to clean and maintain. (See *Exhibit 11b.*)

Flooring

Each area of the establishment has specific flooring needs. While cost and appearance are important considerations, the selection of flooring materials must also be based upon health and safety requirements. Flooring should be strong, durable, and easy to clean. It should also resist wear and help prevent slips. Once installed, flooring should be kept in good condition and be replaced if damaged or worn.

The porosity of flooring material is an especially important consideration. **Porosity** is the extent to which a material will absorb liquids. You should avoid high-porosity flooring for a number of reasons. Its absorbency often makes it an ideal place for microorganisms to grow. High-porosity flooring can also cause people to slip or fall, and it often deteriorates more quickly.

To prevent such problems, nonabsorbent flooring is recommended for specific areas of the establishment, including:

- Walk-in refrigerators
- Food-preparation areas
- Dishwashing areas
- Restrooms
- Other areas subject to moisture, flushing, or spray cleaning

Exhibit 11c

Vinyl Tile

Vinyl tile is nonabsorbent and is easy to clean and maintain.

Nonporous, Resilient Flooring

In most areas of the establishment, nonporous, resilient flooring is the best choice. **Resiliency** means a material has the ability to react to a shock without breaking or cracking.

Nonporous, resilient materials such as vinyl or rubber tiles are relatively inexpensive and are easy to clean and maintain. If individual tiles break, they can be easily replaced. (See *Exhibit 11c.*) They are capable of handling heavy traffic and are resistant to grease and alkalis.

However, this type of flooring does have its disadvantages. Cigarette burns or sharp objects can easily damage it. It also tends to be slippery when wet. Vinyl tile in particular requires a high level of maintenance, including waxing and frequent machine buffing. For this reason, it is usually a poor choice for dining rooms or public areas. However, it is practical for employee dressing rooms, break rooms, and foodservice offices. See *Exhibit 11d* for characteristics and recommended uses of nonporous, resilient flooring.

Exhibit 11d

Characteristics of Nonporous Resilient Flooring

Where to Use	Durability	Advantages	Disadvantages
Rubber tile			
Kitchens; restrooms	Less durable and less resistant to grease and alkalis	Nonslip; resilient; easy to clean	Can only be used in moderate traffic areas
Vinyl sheet			
Offices; kitchens; corridors	Less resistant to grease and alkalis	Very resilient; easy to clean	Can only be used in light or moderate traffic areas
Vinyl tile			
Offices; employee restrooms	Wears out quickly with high traffic	Very resilient; easy to clean	Requires waxing and machine buffing

Key Point

Hard-surface flooring is also commonly used in establishments since it is nonabsorbent and very durable.

Hard-Surface Flooring

Hard-surface flooring is also commonly used since it is nonabsorbent and very durable. These types of flooring are an excellent choice for public restrooms or high-soil areas, especially quarry and ceramic tile.

Still, there are several disadvantages to these flooring materials. They may crack or chip if heavy objects are dropped on them. They may also break objects dropped on them. In addition, hard-surface flooring does not absorb sound. They are more expensive to install and maintain and are somewhat difficult to clean.

While most hard-surface floors, especially marble, can be slippery, unglazed tiles can provide a hard, slip-resistant surface. See *Exhibit 11e* for characteristics and recommended uses of hard-surface flooring.

Exhibit 11e

Characteristics of Hard-Surface Flooring

Where to Use	Durability	Advantages	Disadvantages
Marble; terrazzo			
Public corridors; dining rooms; public restrooms	Wear resistant	Nonporous; good appearance	Nonresilient; heavy; expensive; requires special car; difficult to install
Quarry tile			
Kitchen; service dishwashing, and receiving areas; offices; restrooms; dining rooms	Wear resistant	Nonporous	Nonresilient; heavy; expensive; slippery when wet unless an abrasive is added
Wood			
Offices; dining rooms	Durable in lower traffic areas	Good appearance and sound absorption	Requires frequent polishing and periodic refinishing to maintain surface qualities

Carpeting

Carpeting is a popular choice for some areas of the establishment, such as dining rooms, because it absorbs sound. However, it is not recommended in high-soil areas, such as waitstaff service areas, tray and dish drop-off areas, beverage stations, and major traffic aisles. Carpet can be maintained by simple vacuuming. Areas prone to heavy traffic and moisture will require routine cleaning. Special carpet can be purchased for areas where sanitation, soiling, moisture, and fire safety are concerns.

Exhibit 11f

Coving

Coving is a curved, sealed edge placed between the floor and the wall to eliminate sharp corners or gaps.

Special Flooring Needs

Nonslip surfaces should be used in traffic areas. In fact, nonslip surfaces are best for the entire kitchen, since slips and falls are a potential hazard. Rubber mats are allowed for safety reasons in areas where standing water may occur, such as the dish room. Rubber mats should be picked up and cleaned separately when scrubbing floors.

Coving is required in establishments using resilient or hard-surface flooring materials. **Coving** is a curved, sealed edge placed between the floor and the wall to eliminate sharp corners or gaps that would be impossible to clean. (See *Exhibit 11f.*) The coving tile or strip must adhere tightly to the wall to eliminate hiding places for insects. This will also prevent moisture from deteriorating the wall.

Finishes for Interior Walls and Ceilings

Interior finishes are the materials used on the surface of walls, partitions, or ceilings of an establishment. As with flooring, the most important criteria when choosing interior finishes are ease of cleaning and porosity.

When selecting finishes for walls and ceilings, consider location. A material that might be suitable in one area may be a poor choice for another. Walls and ceilings in food-preparation areas must be light in color to distribute light and to make it easier to spot soil when cleaning. They should be kept in good repair, free of cracks, holes, or peeling paint. The best wall finish in cooking areas is ceramic tile; however, it must be monitored for grout loss and regrouted whenever necessary. Stainless steel is used occasionally because of its durability and resistance to moisture. The most common ceiling materials are acoustic tile, painted drywall, painted plaster, or exposed concrete.

The support structures for walls and ceilings (studs, joists, and rafters) and pipes should not be exposed unless they are finished and sealed for cleaning. Flexible materials such as paper, vinyl, and thin wood veneers are often used for walls and ceilings. Vinyl wall coverings are used in many areas of an establishment because they are attractive, relatively inexpensive, easy to clean, and durable. Vinyl wall coverings are rated for flammability by testing agencies. Plaster or cinder-block walls that have been sealed and painted with oil-resistant, easy-to-wash, glossy paints are appropriate for dry areas of the facility.

CONSIDERATIONS FOR OTHER AREAS OF THE FACILITY

Dry Storage

Dry storerooms should be constructed of easy-to-clean materials that allow good air circulation. (See *Exhibit 11g.*) Shelving, table tops, and bins for dry ingredients should be made of corrosion-resistant metal or food-grade plastic.

Any windows in the storeroom should have frosted glass or shades. Direct sunlight can increase the temperature of the room and affect food quality.

Steam pipes, water lines, and other conduits have no place in a well-designed storeroom. Dripping condensation or leaks in overhead pipes can promote microbial growth in such normally stable items as crackers, flour, and baking

Exhibit 11g

Acceptable Dry-Storage Facility

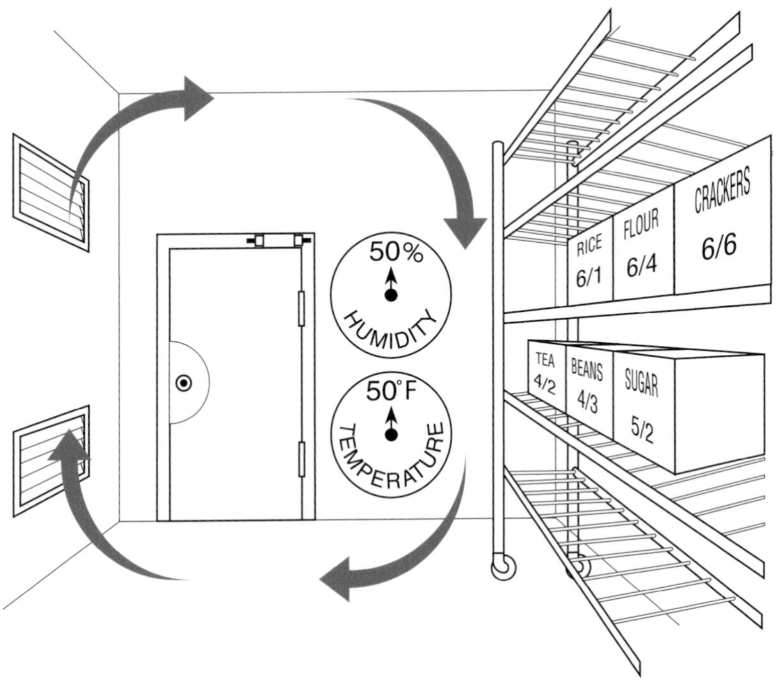

Dry storerooms should be constructed of easy-to-clean materials that allow good air circulation.

powder. Leaking overhead sewer lines can be a source of contamination for any food. Hot water heaters or steam pipes can increase the temperature of the storeroom to levels that will allow foodborne pathogens to grow.

Dry food is especially susceptible to attack by insects and rodents; therefore, cracks and crevices in floors or walls should be filled. Doors leading to the exterior of the building should be self-closing. Screens for windows and doors should be sixteen mesh to the inch, without holes or tears.

Restrooms

Local building and health codes usually specify how many sinks, stalls, toilets, and urinals are required in an establishment. It is best if separate restrooms are provided for employees and customers. If this is not possible, the establishment must be designed so patrons do not pass through food-preparation areas to reach the restroom, since they could contaminate food or food-contact surfaces.

Restrooms should be convenient, sanitary, and have a fully equipped handwashing station, as well as self-closing doors. They must be adequately stocked with toilet paper, and trash receptacles must be provided if disposable paper towels are used. Covered waste containers must be provided in women's restrooms for the disposal of sanitary supplies.

Key Point

Handwashing stations must be conveniently located in food-preparation areas, service areas, dishwashing areas, and restrooms.

Handwashing Stations

Handwashing stations must be conveniently located so employees will be encouraged to wash their hands often. They are required in restrooms and areas used for food preparation, service, and dishwashing. These stations must be operable, stocked, and maintained. A handwashing station must be equipped with the following items (see *Exhibit 11h*):

- **Hot and cold running water.** Hot and cold water should be supplied through a mixing valve or combination faucet at a temperature of at least 100°F (38°C).
- **Soap.** The soap can be liquid, bar, or powder. Liquid soap is generally preferred, and some local codes require it.

- **A means to dry hands.** Most local codes require establishments to supply disposable paper towels in handwashing stations. Installing at least one warm-air dryer in a handwashing station will provide an alternate method for drying hands if paper towels run out. Continuous-cloth towel systems, if allowed, should be used only if the unit is working properly and the towel rolls are checked and changed regularly. The use of common cloth towels is not permitted because they can transmit contaminants from one person's hands to another.
- **Waste container.** Waste containers are required if disposable paper towels are provided.
- **Signage indicating employees are required to wash hands before returning to work.** The sign should reflect all languages used in the establishment.

Exhibit 11h

Acceptable Handwashing Station

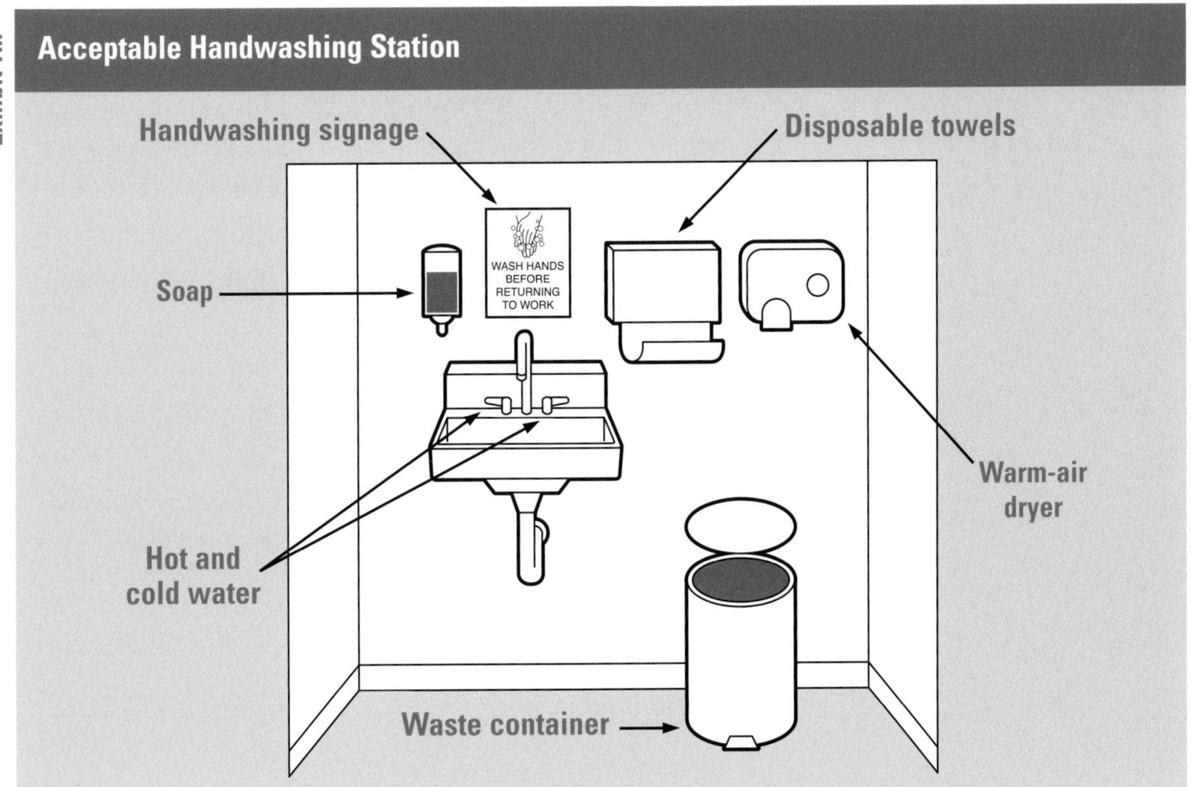

A handwashing station must be equipped with hot and cold running water, soap, a means to dry hands, a waste container (if disposable towels are used), and signage reminding employees to wash hands.

Exhibit 11i

Sink Use

To prevent cross-contamination, employees must use each sink in an establishment for its intended purpose.

Sinks

To prevent cross-contamination, employees must use each sink in an establishment for its intended purpose. (See *Exhibit 11i.*) Handwashing sinks are used for handwashing. Food-preparation sinks are used for food preparation. Service sinks are used for cleaning mops and disposing of wastewater. At least one service sink or curbed drain area is required in an establishment to dispose of soiled water.

Dressing Rooms and Lockers

Dressing rooms are not required. If available, they must not be used for food preparation, storage, or utensil washing. Lockers should be located in a separate room or a room where contamination of food, equipment, utensils, linens, and single-service items will not occur.

Premises

The parking lot and walkways should be kept free of litter and graded so that standing pools of water do not form. In addition, they must be surfaced to minimize dirt and blowing dust. It is recommended that concrete and asphalt be used for walkways and parking lots. Gravel, while acceptable, is not recommended.

Patron traffic through food-preparation areas is prohibited, although guided tours are allowed. The premises may not be used for living or sleeping quarters.

SANITATION STANDARDS FOR EQUIPMENT

It is important to purchase equipment that has been designed with sanitation in mind. Food contact surfaces must be:

- Safe
- Durable
- Corrosion resistant
- Nonabsorbent
- Sufficient in weight and thickness to withstand repeated cleaning

- Smooth and easy to clean
- Resistant to pitting, chipping, crazing (spider cracks), scratching, scoring, distortion, and decomposition

Equipment surfaces that are not designed to come in contact with food but are exposed to splash, spillage, or other food soiling or require frequent cleaning must be:

- Constructed of smooth, nonabsorbent, corrosion-resistant material
- Free of unnecessary ledges, projections, and crevices
- Designed and constructed to allow easy cleaning and maintenance

Exhibit 11j

NSF and UL EPH Marks

Look for the NSF International mark or UL EPH product marks on equipment acceptable for use in a restaurant or foodservice establishment.

The task of choosing equipment designed for sanitation has been simplified by organizations such as **NSF International** and **Underwriters Laboratories (UL).** NSF International develops and publishes standards for sanitary equipment design. The presence of the NSF mark on foodservice equipment means it has been evaluated, tested, and certified by NSF International as meeting international commercial food equipment standards. UL similarly provides sanitation classification listings for equipment found in compliance with NSF International standards. UL also lists products complying with their own published environmental and public health (EPH) standards. Equipment that meets these standards is acceptable for use in a foodservice establishment. Restaurant and foodservice managers should look for the NSF International mark or the UL EPH product mark on foodservice equipment. *Exhibit 11j* shows examples of the NSF International mark and the UL EPH product marks.

Only commercial foodservice equipment should be used in establishments, because household equipment is not built to withstand heavy use.

Although all equipment used in an establishment must meet sanitation standards, certain equipment requires particular attention.

Key Point

The size and type of machine you choose depends on the nature and volume of the items to be cleaned.

Dishwashing Machines

Dishwashing machines vary widely by size, style, and method of sanitizing. High-temperature machines sanitize with extremely hot water; while chemical-sanitizing machines use a chemical solution. Some states require the local regulatory agency's approval before installing a chemical dishwashing system.

The size and type of machine you choose depends upon the nature and volume of the items to be cleaned and the required turnaround time for clean tableware and utensils. Because a dishwashing machine is a big investment, the manager must carefully match the machine to the establishment's needs.

The following types of dishwashing machines are common in restaurant and foodservice establishments:

- **Single-tank, stationary-rack machine, with doors.** This machine holds a stationary rack of tableware and utensils. Items are washed by detergent and water from below and, sometimes, from above the rack. The wash cycle is followed by a hot-water or chemical-sanitizer final rinse.
- **Conveyor machine.** With this machine, a conveyor moves racks of items through the various cycles of washing, rinsing, and sanitizing. The machine may have a single tank or multiple tanks.
- **Carousel or circular conveyor machine.** This multiple-tank machine moves tableware and utensils on a peg-type conveyor or in racks. Some models have an automatic stop after items go through the final rinse cycle. In other models, items must be removed after the final rinse, or they will continue to travel through the machine.
- **Flight-type.** This is a high-capacity, multiple-tank machine with a peg-type conveyor. It may also have a built-in dryer. It is commonly used in institutions and very large establishments.
- **Batch-type, dump.** This stationary-rack machine combines the wash and rinse cycles in a single tank. Each cycle is timed, and the machine automatically dispenses both the detergent and the sanitizing chemical or hot water. Wash and rinse water are drained after each cycle.

- **Recirculating, door-type, nondump machine.** This stationary-rack machine is not completely drained of water between cycles. The wash water is diluted with fresh water and reused from cycle to cycle.

Consider these general guidelines regarding the installation and use of dishwashing machines:

- Water pipes to the dishwashing machine should be as short as possible to prevent the loss of heat.
- The machine must be raised at least six inches (fifteen centimeters) off the floor to permit easy cleaning underneath.
- Materials used in dishwashing machines should be able to withstand wear, including the action of detergents and sanitizers.
- Information should be posted on or near the machine regarding proper water temperature, conveyor speed, water pressure, and chemical concentration.
- The machine's thermometer should be located so it is readable, with a scale in increments no greater than 2°F (1°C).

Clean-in-Place Equipment

Some equipment is designed to be cleaned and sanitized by having detergent solution, a hot-water rinse, and sanitizing solution pass through it. Certain soft-serve ice cream and frozen yogurt dispensers are cleaned and sanitized this way. This process should be performed daily unless otherwise indicated by the manufacturer. Instructions should be followed carefully.

Food-contact cleaning and sanitizing solutions must remain within the tubes and pipes for a predetermined amount of time, reach all food-contact surfaces, and must be completely drained after use.

Refrigerators and Freezers

There are several types of refrigerator and freezer units. The two most common are walk-in and reach-in refrigerators and freezers. These units should be made of stainless steel or a combination of stainless steel and aluminum. The doors should be constructed to withstand heavy use and should close with a slight nudge. Door gaskets can be fixed in place or removable for easy cleaning. A drain must be provided and maintained for disposal of condensation and defrost water. A properly plumbed, indirect drain can be used in the walk-in refrigerator. Excess condensation can be minimized by maintaining a flush-fitting floor sweep (gasket) under the door.

Walk-in refrigerators and freezers with windows in the door may reduce unnecessary opening. Forced-air circulating fans are essential to help provide a quick recovery time so refrigerators and freezers maintain the appropriate temperature.

In addition to the sanitation standards mentioned earlier in this chapter, consider these factors when purchasing a refrigerator or freezer unit:

Exhibit 11k

Commercial-Type Walk-In Refrigerator

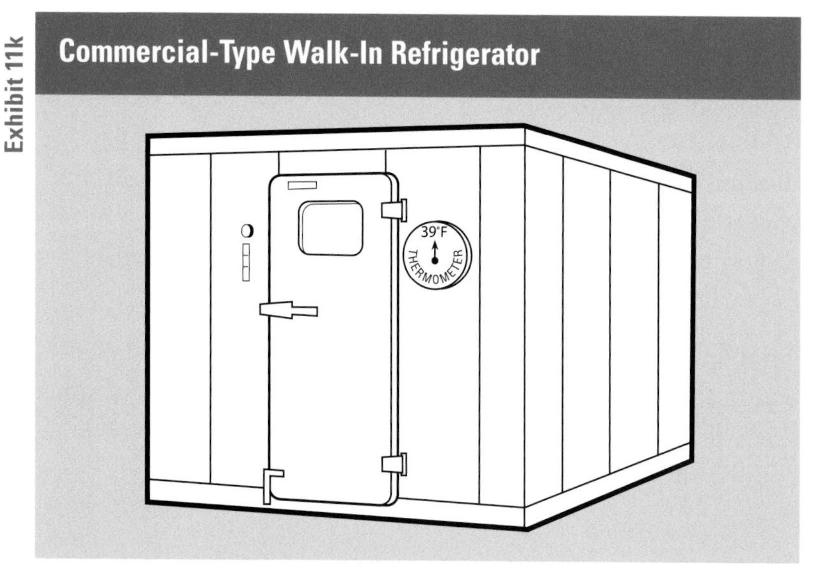

Make sure walk-in units can be sealed to the floor and wall.

- **Choose a unit with adequate storage space.** An uncrowded unit can maintain required holding temperatures, is easier to clean, will prevent moisture buildup, and can minimize breakdowns.
- **Make sure walk-in units can be sealed to the floor and wall.** (See *Exhibit 11k.*) They should offer no access to moisture or rodents. Flooring materials must be able to withstand heavy impact.

Exhibit 11l

Commercial-Type Reach-In Freezer

Note the caster wheels, which make it easier to move the unit for cleaning.

Courtesy of Hobart Corporation

Exhibit 11m

Blast Chiller

Many blast chillers can cool food from 135°F to 37°F (57°C to 3°C) within ninety minutes.

Courtesy of Hobart Corporation

- **Purchase reach-in refrigerator or freezer units with legs that elevate them six inches (fifteen centimeters) off the floor.** Otherwise, mount and seal them on a masonry base. Casters, which make it easier to move the unit for cleaning, are often preferred or required by local regulatory agencies. (See *Exhibit 11l.*)
- **Make sure the unit meets the temperature requirements of the food you store.** Built-in thermometers should be easy to locate and read and be accurate to within 3°F (2°C).

Blast Chillers and Tumble Chillers

Blast chillers are designed to cool food quickly. (See *Exhibit 11m.*) Many are able to cool food from 135°F to 37°F (57°C to 3°C) within ninety minutes. Most units allow the operator to set target chill temperatures and monitor the temperature of food throughout the chill cycle. Once chilled to safe temperatures, the food then can be stored in conventional refrigerators or freezers.

Tumble chillers are also designed to cool food quickly. Prepackaged, hot food is placed into a drum, which rotates inside a reservoir of chilled water. The tumbling action increases the effectiveness of the chilled water in cooling the food.

Cook-Chill Equipment

Some operations prepare food using a cook-chill system. By this method, food is partially cooked, rapidly chilled, and then held in refrigerated storage. When needed, the food simply is reheated. A cook-chill unit is an integrated piece of equipment capable of cooking, cooling, and reheating food.

Cutting Boards

Many jurisdictions allow the use of either wooden or synthetic cutting boards; however, some experts prefer synthetic boards, which can be cleaned and sanitized in a dishwashing machine or by immersion in a three-compartment sink.

If wooden cutting boards and baker's tables are allowed by local codes, they must be made from a nonabsorbent hardwood, such as maple or oak. They must also be free of seams and cracks, nontoxic, and must not transfer any odor or taste to food.

Separate cutting boards should be used for raw and ready-to-eat food to prevent cross-contamination. Cutting boards must be washed, rinsed, and sanitized between uses. Due to the high risk of cross-contamination, procedures for cleaning and sanitizing cutting boards should be included in your standard operating procedures.

INSTALLING AND MAINTAINING KITCHEN EQUIPMENT

When installing equipment, keep in mind that it should be easy for employees to clean the equipment and the surrounding floors, walls, and tabletops. Portable equipment makes it much simpler to do this.

Exhibit 11n

Installing Stationary Equipment

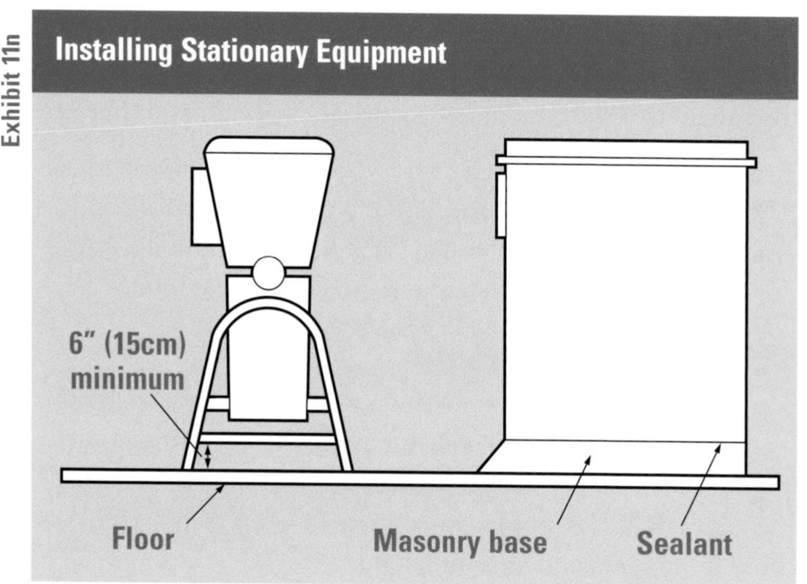

Stationary equipment must be mounted on legs at least six inches (fifteen centimeters) off the floor or sealed to a masonry base.

Installing Kitchen Equipment

When installing stationary equipment, it must be mounted on legs at least six inches (fifteen centimeters) off the floor, or it must be sealed to a masonry base that provides a toe space of four inches (ten centimeters). (See *Exhibit 11n.*) How far the equipment is installed from the wall or from other equipment depends upon its size and the amount of surface to be cleaned. Follow specifications supplied by the manufacturer.

Stationary tabletop equipment should be mounted on legs, providing a minimum clearance of four inches (ten centimeters) between the base of the equipment and the tabletop. Alternatively, the equipment should be tiltable, or it should be sealed to the countertop with a nontoxic, food-grade sealant.

Any crack or seam greater than $\frac{1}{32}$ inch (.8 millimeters) that results when equipment is attached to the floor, wall, or counter

Exhibit 11o

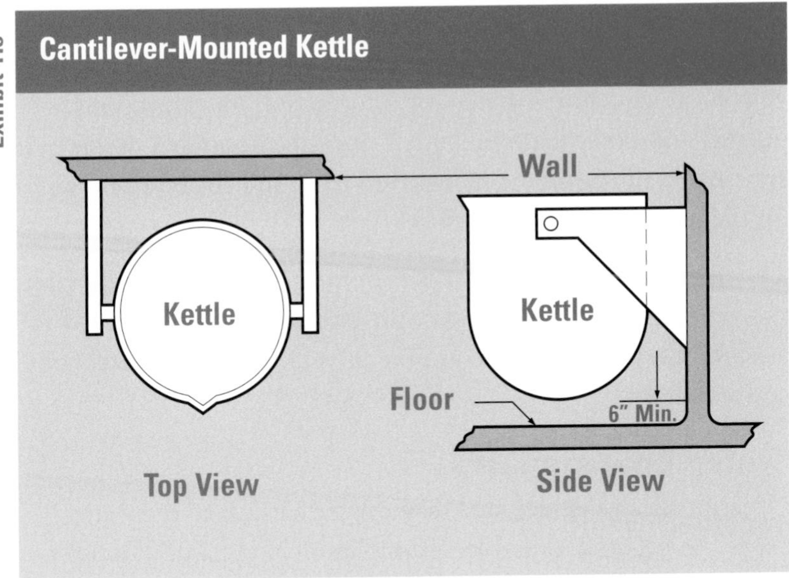

Cantilever mounting allows easy access when cleaning underneath and behind equipment.

must be filled with a nontoxic sealant to prevent food buildup or pests. However, sealant should not be used to cover wide gaps resulting from faulty construction or repairs. Instead, gaps should be properly repaired before equipment is installed.

Cantilever-Mounted Equipment

Attached to a mount or the wall with a bracket, **cantilever-mounted equipment** allows for easier cleaning of surfaces behind and underneath it. (See *Exhibit 11o.*)

Key Point

Equipment must receive regular maintenance from qualified personnel.

Maintaining Equipment

Once equipment has been properly installed, it must receive regular maintenance from qualified personnel. Follow the manufacturers' recommended maintenance schedule.

UTILITIES

Utilities used by an establishment include water and plumbing, electricity, gas, lighting, ventilation, sewage, and wastehandling. There must be enough utilities to meet the cleaning needs of the establishment, and the utility must not contribute to contamination.

Water Supply

Safe water is vital in every establishment, as contaminated water can carry foodborne pathogens. Water that is safe to drink is called **potable water.** Sources of potable water include approved public water mains, private water sources that are regularly maintained and tested, and bottled drinking water. Other sources include closed, portable water containers filled with potable water, on-premises water storage tanks, and properly maintained water-transport vehicles.

If your establishment uses a private water supply, such as a well, rather than an approved public source, you should check with your local regulatory agency for information on inspections, testing, and other requirements. Generally, nonpublic water systems should be tested at least annually and the report kept on file in the establishment.

The use of nonpotable water is extremely limited. If nonpotable water is allowed by local codes, the uses are generally limited to air conditioning, cooling equipment (nonfood), fire protection, and irrigation.

Water Emergencies

Occasionally, an emergency may occur that causes the water supply to become unusable. Examples include natural disasters like floods, problems with a city water supply, or local problems with broken pipes. If there is a problem with the potable water supply, the establishment might have to be closed or an alternate source of water found. Anytime a water emergency occurs, water systems and certain equipment (e.g., ice machines) should be flushed and disinfected.

Establishments might not want to close during a water emergency due to the loss of revenue. Some establishments might need to stay open during the emergency, such as hospitals or nursing homes. In these cases, the regulatory agency might allow the establishment to continue to operate if certain precautions are followed. The issues an establishment must consider in order to continue serving food safely during a water emergency follow on the next page. Contact your local regulatory agency if you have any questions about the safety of a particular practice.

Water Used as a Beverage or Ingredient

If potable water must be obtained from an alternate source for use as a beverage or ingredient, there are several options, including buying bottled water and boiling water (your local regulatory authority should be contacted to determine the proper length of time for boiling). If you have been notified of an upcoming water interruption (such as water being shut off for repairs), you can store potable water in advance using closed, food-grade water containers.

Ice

Most ice machines make new ice and drop it onto previously made ice. Ice made from contaminated water will, therefore, contaminate any previously made ice already in the bin. If the establishment knows in advance that the water supply might become unsafe, previously made ice can be stored before the interruption occurs. During an unexpected water emergency, ice can be purchased. If purchased, ice must be contained in single-use, food-grade plastic bags or wet-strength paper bags filled and sealed at the point of manufacture.

Cleaning

If a water emergency occurs, nonessential cleaning in the establishment should be minimized. When performing essential tasks, such as cleaning and sanitizing pots and pans, the water used should first be boiled according to local regulations. Establishments might consider using single-use items to eliminate the need for using a dishwasher, which requires large amounts of water.

Handwashing

Warm, potable water is required for handwashing. Boiled water can be placed in large plastic containers that dispense water through a spout. Replenish the supply frequently to keep it warm.

Restrooms, Showers, and Laundries

Restrooms, showers, and laundries require large amounts of water. Managing them would be difficult without a potable water system. Portable toilets can be used, but it might not be possible to supply water to showers and laundry facilities. The local regulatory agency should be contacted to determine possible alternatives. Some codes allow the use of nonpotable water for flushing toilets and operating washing machines.

Hot Water

Providing a continuous supply of hot water can be a problem for many establishments serving the public. They must have enough hot water to meet peak demand. Water heaters should be evaluated regularly to make sure they can meet this demand. Consider how quickly the heater produces hot water, the size of the holding tanks, and the location of the heater in relation to sinks or dishwashing machines.

Since most general-purpose water heaters will not heat water to temperatures required for hot-water sanitizing, a **booster heater** may be needed to maintain water temperature. Many dishwashing machines now come with booster heaters.

Key Point

Only licensed plumbers should install and maintain plumbing systems in an establishment.

Plumbing

In almost every community in the United States, plumbing design is regulated by law, and with good reason. Improper plumbing design can have serious consequences. Improperly installed or poorly maintained plumbing that allows the mixing of potable and nonpotable water has been implicated in outbreaks of typhoid fever, dysentery, hepatitis A, Norovirus, and other gastrointestinal illnesses. Improperly installed water pipes can also lead to contamination from metals, such as toxic-metal poisoning from beverage dispensers or contamination from chemicals used in the system, such as detergents, sanitizers, or drain cleaners.

Local plumbing codes vary widely regarding the types of connections permitted and the kind of protection required. If in doubt about the plumbing regulations governing your establishment, contact your local regulatory agency. Only licensed plumbers should install and maintain plumbing systems in an establishment.

Cross-Connections

The greatest challenge to water safety comes from cross-connections. A **cross-connection** is a physical link through which contaminants from drains, sewers, or other wastewater sources can enter a potable water supply. A cross-connection is dangerous because it allows the possibility of **backflow**. Backflow is the unwanted, reverse flow of contaminants through a cross-

Exhibit 11p

Common Cross-Connection

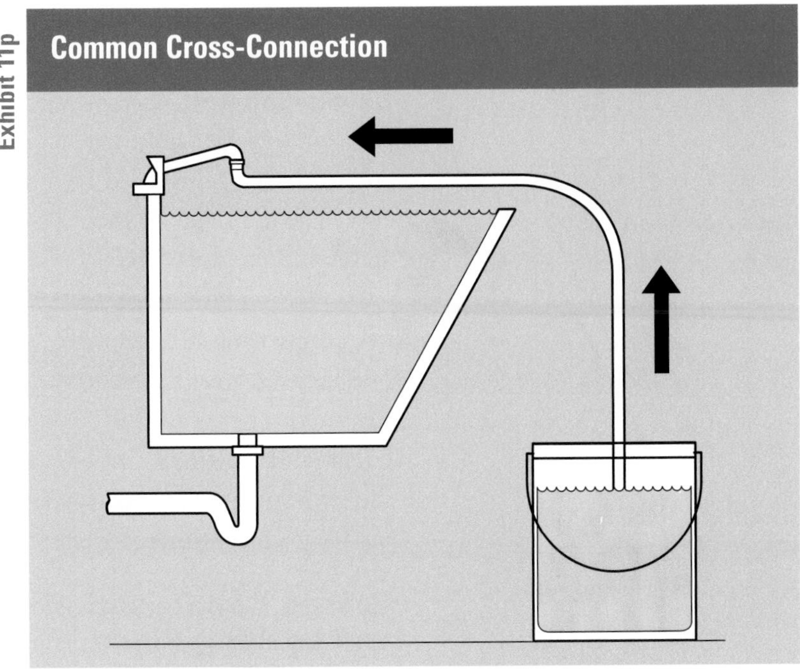

A hose connected to a faucet and left submerged in a mop bucket creates a dangerous cross-connection.

connection into a potable water system. It can occur whenever the pressure in the potable water supply drops below the pressure of the contaminated supply. A running faucet located below the flood rim of a sink or a running hose in a mop bucket are examples of a cross-connection.

Exhibit 11p illustrates a situation in which a hose has been attached to the faucet of a utility sink to add hot water to a partially filled mop bucket. The nozzle of the hose is left submerged in the bucket of dirty water, accidentally creating a cross-connection. Because of heavy water usage somewhere else in the facility, the water pressure could drop low enough that contaminated water from the mop bucket would be drawn back through the hose and into the potable water supply.

Exhibit 11q

Vacuum Breaker

A vacuum breaker should be installed on threaded faucets to prevent backflow.

Backflow Prevention

To prevent cross-connections like this, do not attach a hose to a faucet unless a backflow-prevention device, such as a vacuum breaker, is attached. (See *Exhibit 11q.*) Threaded faucets and connections between two piping systems must have a vacuum breaker or other approved backflow-prevention device.

The only completely reliable method for preventing backflow is creating an **air gap**. An air gap is an air space used to separate a water supply outlet from any potentially contaminated source. A properly designed and installed sink typically has two air gaps to prevent backflow. One is the air space between the faucet and the flood rim of the sink. The other is located between the

Exhibit 11r

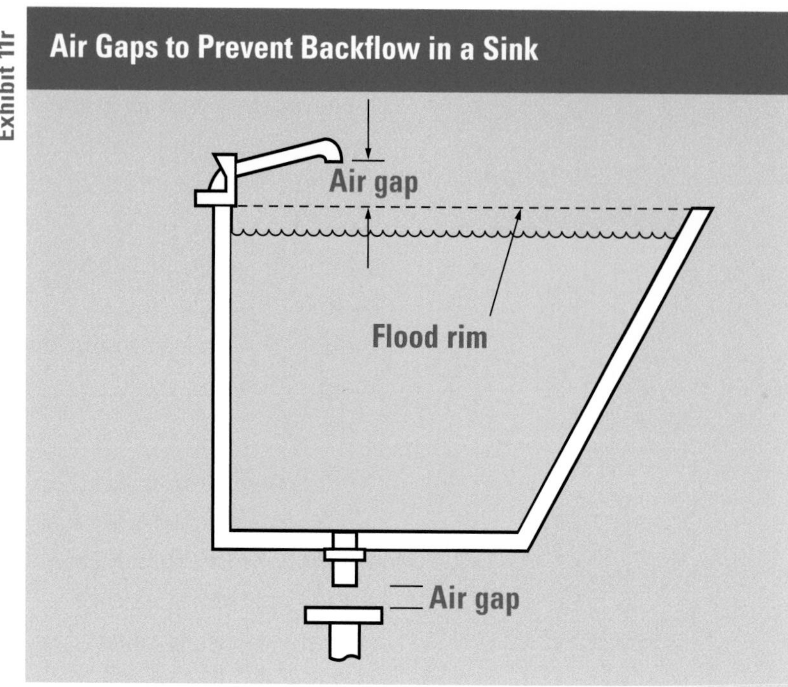

Air gaps between the faucet and the flood rim and between the drainpipe and floor drain of a sink prevent backflow.

drain pipe of the sink and the floor drain of the establishment. (See *Exhibit 11r.*) Typically, the size of the air gap should be twice the diameter of the water supply outlet. For example, if a faucet has an opening with a diameter of one inch (2.5 centimeters), the air gap between the faucet and the flood rim of the sink must be at least two inches (five centimeters).

Grease Condensation and Leaking Pipes

Grease condensation in pipes is another common problem in plumbing systems. Grease traps are often installed to prevent buildup from creating a drain blockage. If used, grease traps must be easily accessible, installed by a licensed plumber, and cleaned periodically according to manufacturers' recommendations. If the traps are not cleaned or are not cleaned properly, a backup of wastewater could lead to odor and contamination.

Overhead wastewater pipes or fire-safety sprinkler systems can leak and become a source of contamination. Even overhead lines carrying potable water can be a problem, since water can condense on the pipes and drip onto food. All piping should be serviced immediately when leaks occur.

Key Point

A backup of raw sewage on the floor is cause for immediate closure of the establishment, correction of the problem, and thorough cleaning.

Sewage

Sewage and wastewater are reservoirs of pathogens, soils, and chemicals. It is absolutely essential to prevent them from contaminating food or food-contact surfaces.

If there is a backup of wastewater, prompt action must be taken. The action taken depends on the type of backup. A backup of raw sewage on the floor is cause for immediate closure of the establishment, correction of the problem, and thorough cleaning. Other backups might not be as serious, but they still require an immediate correction of the problem, followed by a thorough clean up.

The facility must have adequate drainage to handle all wastewater. Any area subjected to heavy water exposure should have its own floor drain. Wastewater from equipment and from the potable supply should be channeled into an open, accessible waste sink or floor drain. The drainage system should be designed to prevent floors from flooding.

Lighting

Good lighting generally results in improved employee work habits, easier and more effective cleaning, and a safer work environment. Building and health codes usually set lighting intensity requirements for specific areas in the establishment. (See *Exhibit 11s.*) These are typically based on a unit of measurement called a foot-candle. Other units of measurement include lumens, luxes, and luminaires.

Overhead or ceiling lights above workstations should be positioned so employees do not cast shadows on the work surface. Use shatter-resistant light bulbs and protective covers made of metal mesh or plastic to prevent broken glass from contaminating food or food-contact surfaces. Shields should also be provided for heat lamps.

Exhibit 11s

Minimum Lighting Intensity Requirements for Different Areas of the Establishment

Minimum Lighting Intensity	Area
50 foot-candles (540 lux)	■ Food-preparation areas
20 foot-candles (215 lux)	■ Handwashing or dishwashing areas ■ Buffets and salad bars ■ Displays for produce or packaged food ■ Utensil-storage areas ■ Wait stations ■ Restrooms ■ Inside some pieces of equipment (e.g., reach-in refrigerators)
10 foot-candles (108 lux)	■ Inside walk-in refrigerators and freezer units ■ Dry-storage areas ■ Dining rooms (for cleaning)

Key Point

If ventilation in the establishment is adequate, there will be little or no buildup of grease and condensation on walls and ceilings.

Ventilation

Proper ventilation helps maintain an establishment's indoor air quality by removing steam, smoke, grease, and heat from the establishment. Adequate ventilation is particularly important in the food-preparation area because it reduces the level of odors, gases, dirt, mold, humidity, grease, and fumes present in the air, all of which can contribute to contamination. Excess humidity can cause condensation on walls and ceilings, which may drip onto food. An accumulation of grease can cause fires. If ventilation is adequate, there will be little or no buildup of grease and condensation on walls and ceilings.

Mechanical ventilation must be used in areas for cooking, frying, and grilling. Ventilation must be designed so hoods, fans, guards, and ductwork do not drip onto food or equipment. Hood filters or grease extractors must be tight fitting and easy to remove, and they should be cleaned on a regular basis. Thorough cleaning of the hood and ductwork should also be done periodically by professionals.

Since so much air is moved through exhaust hoods, clean air must be taken in to replace it. This replacement air is called make-up air, which must be replaced without creating drafts. All outside air intakes must be screened to keep pests out.

In many areas of the United States, clean-air ordinances restrict the use of exhaust fans. Exhaust air containing food odors, smoke, and grease might have to be purified by filters or other devices. It is the establishment's responsibility to see that the ventilation system meets local regulations.

Solid Waste Management

Waste management is an important issue in every establishment. There are many things foodservice managers can do to improve their waste management practices. The Environmental Protection Agency (EPA) has recommended three approaches to managing waste:

1. **Reduce the amount of waste produced.** Eliminate unnecessary packaging.

2 **Reuse when possible.** Before using containers again, they must be cleaned and sanitized. Never reuse chemical containers as food containers.

3 **Recycle materials.** Store recyclables so they cannot contaminate food or equipment or attract pests.

When these practices are followed, the amount of waste can be greatly reduced.

Garbage Disposal

Garbage is wet waste matter, usually containing food, which cannot be recycled. It attracts pests and has the potential to contaminate food, equipment, and utensils.

Garbage containers must be leak proof, waterproof, pest proof, easy to clean, and durable. They can be made of galvanized metal or an approved plastic. They must have tight-fitting lids and must be kept covered when not in use. Plastic bags and wet-strength paper bags may be used to line these containers.

Garbage should be removed from food-preparation areas as soon as possible. Frequent disposal prevents odor and pest problems. Garbage-storage areas, inside or outside, should be large enough to contain all garbage and must be located away from food-preparation and storage areas. When removing garbage, employees should not carry it above or across a food-preparation area.

All garbage containers should be cleaned frequently and thoroughly. Both the inside and outside of containers must be cleaned. A cleaning area equipped with hot and cold water and a floor drain is recommended for indoor cleaning. It must be located so food being prepared or in storage will not be contaminated when garbage containers are being cleaned.

Food waste can also be disposed of through the use of in-drain garbage disposals, which reduce the amount of waste that goes into garbage containers. However, local regulations often limit their use in establishments, since these disposals create large amounts of food and water waste that might overload local wastewater systems.

Exhibit 11t

Outdoor Trash Receptacles

Receptacles and compactor systems should be located on or above a smooth surface of nonabsorbent material, such as concrete or machine-laid asphalt.

Pulpers, or grinders, are another garbage-disposal alternative. Pulpers grind food and some other types of waste (such as paper) into small parts that are flushed with water. The water is then removed so the processed solid waste weighs less and is more compact for easier disposal.

Outdoor trash receptacles should be kept covered with their drain plugs in place at all times except during cleaning. Receptacles and compactor systems should be located on or above a smooth surface of nonabsorbent material such as sealed concrete or machine-laid asphalt. The area must be kept clean. (See *Exhibit 11t.*)

SUMMARY

An establishment that is difficult to clean will not be cleaned well. Sanitation efforts will be more effective if the establishment is designed and equipped with ease of cleaning in mind. In most communities, plans for new construction or extensive remodeling are subject to review and approval by local regulatory agencies.

Sanitation can be built into the facility through the proper design and construction of floors, walls, and ceilings. The selection of materials should be based on ease of cleaning and durability, as well as appearance.

Separate restrooms should be provided for employees and patrons. If this is not possible, restrooms should be positioned so that patrons do not pass through food-preparation areas to reach them. Restrooms should be cleaned regularly and have a fully equipped handwashing station and self-closing doors. They should also be stocked adequately with toilet paper and have trash receptacles.

Handwashing stations must be conveniently located, operable, and fully stocked and maintained. They are required in food-preparation areas, service areas, dishwashing areas, and restrooms. They must be equipped with hot and cold running water, soap, a means to dry hands, a waste container, and signage reminding employees to wash hands.

It is important to purchase equipment that has been designed with sanitation in mind. Food-contact surfaces must be corrosion resistant, nonabsorbent, and smooth, and they must resist pitting and scratching. Equipment should be installed so both the equipment and the area surrounding it can be cleaned easily. Stationary equipment must be mounted on legs at least six inches (fifteen centimeters) off the floor, or it must be sealed to a masonry base. Stationary tabletop equipment should be mounted on legs with a clearance of four inches (ten centimeters) between the equipment and the tabletop, or it should be sealed to the tabletop. Any gap between a piece of equipment and the floor, wall, or tabletop greater than $\frac{1}{32}$ inch (.8 millimeters) should be filled with a nontoxic sealant to prevent food buildup and pests.

Utilities must be designed to meet the establishment's cleaning needs, and they must not contribute to contamination. Potable water—water that is safe to drink—is vital in an establishment. Sources include public water mains, private water sources that are regularly maintained and tested, and bottled drinking water. In a water emergency, an establishment might be allowed to remain open if certain precautions are followed. These could include boiling or purchasing water, and boiling water for handwashing and essential tasks. The local regulatory agency should be contacted any time there is a question about the safety of a particular practice.

Improperly installed or maintained plumbing can have serious consequences. Only licensed plumbers should install and maintain plumbing systems. The greatest challenge to water safety comes from cross-connections—a physical link through which contaminants from drains, sewers, and other wastewater sources can flow back into the potable-water supply. Vacuum breakers and air gaps can be used to prevent backflow. It is essential to prevent wastewater from contaminating food and food-contact surfaces.

A backup of raw sewage is cause for immediate closure of the establishment, correction of the problem, and thorough cleaning.

Good lighting in an establishment generally results in improved employee work habits, easier and more effective cleaning, and a safer work environment. Follow the lighting intensity requirements for each area of the establishment. Use shatter-resistant bulbs and protective covers to prevent broken glass from contaminating food or food-contact surfaces. Proper ventilation improves the indoor air quality of the establishment by removing smoke, grease, steam, and heat. If ventilation is adequate, there will be little or no buildup of grease and condensation on walls and ceilings. Ventilation must be designed so hoods, fans, guards, and ductwork do not drip onto food or equipment. Hood filters and grease extractors must be cleaned regularly.

Garbage containers must be leak proof, waterproof, pest proof, easy to clean, and durable. They must have tight-fitting lids and must be kept covered when not in use. All garbage containers should be cleaned frequently and thoroughly both inside and out. Garbage should be removed from food-preparation areas as soon as possible, and must not be carried above or across a food-preparation area.

Apply Your Knowledge

1. Why did the people become ill?
2. What should Carlos do to correct the problem?

A Case in Point

Several people became ill shortly after drinking beverages at the bar in a local restaurant. They all complained that their iced drinks had an odd taste. At the time of the incident, the glasswasher in the bar had been out of service.

When interviewed, Carlos, the manager, explained that the glasswasher was functional but that the unit could not be used because the large volume of water discharged after each wash load was worsening a recent drain-blockage problem. Carlos also mentioned there had been intermittent backups in the plumbing during the previous week and a large pool of water had been found under the glasswasher. Drain cleaners had been used repeatedly with no change in the blockage.

The icemaker shared piping with the glasswasher in the bar and the grease trap on the sink in the restaurant. Carlos revealed he had installed the plumbing himself.

When ice cubes were removed from the icemaker, congealed grease and food debris were found on them. Grease and debris also covered the bottom of the ice bin.

For answers, please turn to the Answer Key.

Apply Your Knowledge

Use these questions to review the concepts presented in this chapter.

Discussion Questions

1. What is one of the most important considerations when choosing flooring for food-preparation areas?
2. What action must be taken in the event of a backup of raw sewage in an establishment?
3. What can be done to prevent backflow in an establishment?
4. What are some potable-water sources for an establishment? What are the testing requirements for nonpublic water systems?
5. What are the requirements of a handwashing station? In what areas of an establishment are handwashing stations required?
6. What are the requirements for installing stationary equipment?

For answers, please turn to the Answer Key.

Apply Your Knowledge

Use these questions to test your knowledge of the concepts presented in this chapter.

Multiple-Choice Study Questions

1. Generally, establishments that use a private water source such as a well must have it tested at least
   - A. once a year.
   - B. every two years.
   - C. every three years.
   - D. every five years.
2. An establishment should respond to a backup of raw sewage by
   - A. closing.
   - B. correcting the problem that caused the backup.
   - C. thoroughly cleaning affected areas of the establishment.
   - D. All of the above
3. Which will *not* prevent backflow?
   - A. An air gap between the sink drain pipe and the floor drain
   - B. The air space between the faucet and the flood rim of a sink
   - C. A vacuum breaker
   - D. A cross-connection
4. When designing the layout of a foodservice establishment, what is the most important consideration for keeping food safe?
   - A. Where the establishment will be located
   - B. How easy it will be to clean and maintain the facility
   - C. The number of employees that will be required to run the establishment
   - D. The number of customers that will patronize the establishment each night
5. When mounting tabletop equipment on legs, the clearance between the base of the equipment and the tabletop must be at least
   - A. one inch (2.5 centimeters).
   - B. two inches (five centimeters).
   - C. four inches (ten centimeters).
   - D. six inches (fifteen centimeters).

Continued on the next page...

Apply Your Knowledge

Multiple-Choice Study Questions *continued*

6. Equipment food-contact surfaces must meet all of the following conditions *except*
   A. they must be corrosion resistant.
   B. they must be absorbent.
   C. they must be smooth.
   D. they must resist pitting.

7. Which statement about dishwashing areas is true?
   A. Water pipes to the dishwashing machine should be as long as possible.
   B. The machine should be raised two inches (five centimeters) from the floor to permit easy cleaning.
   C. Information should be posted on the machine regarding proper water temperature.
   D. The machine's thermometer should have a scale in increments no greater than 5°F (–15°C).

8. What is the purpose of a grease trap?
   A. To prevent backflow
   B. To replace an air gap
   C. To store used fryer shortening
   D. To prevent grease buildup from blocking a drain

9. Which practice will prevent overhead lights from contaminating food?
   A. Using shatter-resistant bulbs
   B. Using fluorescent bulbs
   C. Using extended-life bulbs
   D. Using glass covers over bulbs

10. Which is *not* true regarding garbage containers?
    A. They must be leak proof.
    B. They must remain uncovered.
    C. They must be pest proof.
    D. They must be waterproof.

Apply Your Knowledge

Multiple-Choice Study Questions

11. The lighting intensity in a dry-storage area should be at least
    A. fifty foot-candles (540 lux).
    B. twenty foot-candles (215 lux).
    C. ten foot-candles (108 lux).
    D. five foot-candles (54 lux).
12. Which organization certifies equipment that meets sanitation standards?
    A. NSF International
    B. FDA
    C. EPA
    D. OSHA

For answers, please turn to the Answer Key.

ADDITIONAL RESOURCES

Articles and Texts

Barbaran, Regina S. and Joseph F. Durocher. *Successful Restaurant Design.* Hoboken, NJ: John Wiley & Sons, Inc. 2001.

Katsigris, Costas and Chris Thomas. *Design and Equipment for Restaurants and Foodservice: A Management View.* Hoboken, NJ: John Wiley & Sons, Inc. 2005.

Web Sites

Environmental Protection Agency

www.epa.gov

Since 1970, the U.S. Environmental Protection Agency (EPA) has been instrumental in creating a cleaner environment and protecting the health of all Americans. The EPA has a staff who research and set national standards for a variety of environmental programs. Visit this Web site to find out the latest news and information relating to food safety and the environment.

FDA Center for Food Safety and Applied Nutrition

www.cfsan.fda.gov/list.html

As the center within the Food and Drug Administration (FDA) responsible for food safety, the Center for Food Safety and Applied Nutrition (CFSAN) promotes and protects the public health by researching and implementing guidelines, policies, and standards to ensure that food is safe, nutritious, wholesome, and properly labeled. This Web site provides information relevant to all aspects of food safety and security, including corresponding guidelines, policies, and standards.

NSF International

www.nsf.org

NSF International is a nonprofit, nongovernmental organization focused on the development of standards in the areas of food, water, indoor air, and the environment. The organization also certifies products against these standards. Visit this Web site for NSF International's food equipment standards and a listing of certified food equipment.

Underwriters Laboratories, Inc.

www.ul.com

Underwriters Laboratories, Inc. (UL) is an independent, nonprofit testing and certification organization for product safety. UL both develops and tests products against standards essential to helping ensure public safety and confidence, reducing costs, and improving quality. This Web site provides access to UL's Standards for Safety and listings of UL certified food equipment.

Documents and Other Resources

Food Establishment Plan Review Guide

www.cfsan.fda.gov/~dms/prev-toc.html

The Conference for Food Protection (CFP) and the FDA's *Food Establishment Plan Review Guide* is available at this Web site. This guide has been developed to provide guidance and assistance in complying with nationally recognized food safety standards. It includes design, installation, and construction recommendations regarding food equipment and facilities.

2005 *FDA Food Code*

www.cfsan.fda.gov/~dms/fc05-toc.html

The U.S. Food and Drug Administration (FDA) publishes the *FDA Food Code,* a scientifically sound technical and legal document that serves as a model for regulating the retail and foodservice industry at the federal, state, and local level. Updates to the code are issued on the odd-numbered years. The *FDA Food Code* provides a system of safeguards designed to minimize foodborne illness and ensure employee health, food protection manager knowledge, safe food, nontoxic and cleanable equipment, and appropriate sanitation of the food establishment. It is used as the basis for information in this textbook.

POT AND PAN CLEANING AND SANITIZING PROCEDURES
PROCEDIMIENTOS DE LIMPIEZA Y SANEAMIENTO DE OLLAS Y SARTENES
PRE-SCRAPE & WASH
CÓMO TALLAR PREVIAMENTE Y LAVAR
RINSE & SANITIZE
CÓMO ENJUAGAR Y SANEAR
ECOLAB
SPARTA

Cleaning and Sanitizing

Inside this chapter:

- Cleaning versus Sanitizing
- Cleaning
- Sanitizing
- Machine Dishwashing
- Manual Dishwashing
- Cleaning and Sanitizing Equipment
- Cleaning the Kitchen
- Cleaning the Premises
- Tools for Cleaning
- Storing Utensils, Tableware, and Equipment
- Using Hazardous Materials
- Developing a Cleaning Program

After completing this chapter, you should be able to:

- Explain the difference between cleaning and sanitizing.
- Identify approved sanitizers.
- Identify factors affecting the efficiency of sanitizers (i.e., time, temperature, concentration, water hardness, and pH).
- Follow the requirements for frequency of cleaning and sanitizing food-contact surfaces.
- Follow the legal requirements for the use of poisonous or toxic material in a food establishment.
- Properly clean and sanitize items in a three-compartment sink.
- Properly clean and sanitize food-contact surfaces.
- Properly clean nonfood-contact surfaces.
- Identify proper machine-dishwashing techniques.
- Identify storage requirements for poisonous or toxic materials.
- Dispose of poisonous or toxic materials according to legal requirements.
- Properly store tools, equipment, and utensils that have been sanitized.
- Use the appropriate test kit for each sanitizer.

Key Terms

- Cleaning
- Sanitizing
- Detergent
- Solvent cleaners
- Acid cleaners
- Abrasive cleaners
- Heat sanitizing
- Chemical sanitizing
- Sanitizer
- Master cleaning schedule

Apply Your Knowledge

Check to see how much you know about the concepts in this chapter. Use the page references provided with each question to explore the topic.

Test Your Food Safety Knowledge

1. **True or False:** Chemicals can be stored in food-preparation areas if they are properly labeled. *(See page 12-19.)*

2. **True or False:** The temperature of the final sanitizing rinse in a high-temperature dishwashing machine should be 140°F (60°C). *(See page 12-9.)*

3. **True or False:** Cleaning reduces the number of microorganisms on a surface to safe levels. *(See page 12-3.)*

4. **True or False:** Utensils cleaned and sanitized in a three-compartment sink should be dried with a clean towel. *(See page 12-11.)*

5. **True or False:** Tableware and utensils that have been cleaned and sanitized should be stored at least two inches off of the floor. *(See page 12-18.)*

For answers, please turn to the Answer Key.

INTRODUCTION

In Chapter 10, you learned that a good food safety management system depends on food safety programs. A cleaning and sanitation program is one of the most important of these.

If you do not keep your facility and equipment clean and sanitary, food can easily become contaminated. No matter how carefully you prepare and cook food, without a clean and sanitary environment, bacteria and viruses—such as those that cause salmonellosis and hepatitis A—can quickly spread to both cooked and uncooked food. Cleaning and sanitizing must be done carefully and correctly. If not used properly, cleaning and sanitizing chemicals can be just as harmful to customers and employees as the illnesses they help prevent.

Cleaning & Sanitizing

Surfaces must *first* be cleaned and rinsed *before* being sanitized.

CLEANING VERSUS SANITIZING

It is important to understand the difference between cleaning and sanitizing. **Cleaning** is the process of removing food and other types of soil from a surface, such as a countertop or plate. **Sanitizing** is the process of reducing the number of microorganisms on that surface to safe levels. To be effective, cleaning and sanitizing must be a two-step process. Surfaces must *first* be cleaned and rinsed *before* being sanitized.

Everything in your operation must be kept clean; however, any surface that comes in contact with food, such as knives, utensils, and cutting boards, must be cleaned *and* sanitized.

All food-contact surfaces must be washed, rinsed, and sanitized:

- After each use
- Any time you begin working with another type of food
- Any time you are interrupted during a task and the tools or items you have been working with may have been contaminated
- At four-hour intervals, if the items are in constant use

CLEANING

Several factors affect the cleaning process. These include:

- **Type of soil:** Certain types of soil require special cleaning methods.
- **Condition of the soil:** The condition of the soil or stain affects how easily it can be removed. Dried or baked-on stains will be more difficult to remove than soft, fresh stains.
- **Water hardness:** Cleaning is more difficult in hard water because minerals react with the detergent, decreasing its effectiveness. Hard water can cause scale or lime deposits to build up on equipment, requiring the use of lime-removal cleaners.

- **Water temperature:** In general, the higher the water temperature, the better a detergent will dissolve and the more effective it will be in loosening dirt.
- **Surface being cleaned:** Different surfaces require different cleaning agents. Some cleaners work well in one situation but might not work well or might even damage equipment when used in another.
- **Agitation or pressure:** Scouring or scrubbing a surface helps remove the outer layer of soil, allowing a cleaning agent to penetrate deeper.
- **Length of treatment:** The longer soil on a surface is exposed to a cleaning agent, the easier it is to remove.

Cleaning Agents

Cleaning agents are chemical compounds that remove food, soil, rust, stains, minerals, or other deposits. They must be stable, noncorrosive, and safe for employee use. Since cleaning agents have different cleaning properties, ask your supplier to help you select those that will best meet your needs.

For best results, use cleaning agents as directed. They can be ineffective and even dangerous if misused. Follow manufacturers' instructions carefully. Employees should never combine compounds or attempt to make up their own cleaning agents. Also, do not substitute one type of detergent for another unless the intended use is stated clearly on the label. Detergents used for dishwashing machines, for example, can cause severe burns to the skin if used for manual dishwashing.

Cleaning agents are divided into four categories: **detergents, solvent cleaners, acid cleaners,** and **abrasive cleaners.** Some categories may overlap. For example, most abrasive cleaners and some acid cleaners contain detergents. Some detergents also may contain solvents.

Detergents

There are different types of detergents for different cleaning tasks. All detergents, however, contain surfactants (surface-acting agents) that reduce surface tension between the soil and the surface being cleaned. These allow the detergent to quickly penetrate and soften the soil. General-purpose detergents are mildly alkaline cleaners that remove fresh soil from floors, walls, ceilings, prep surfaces, and most equipment and utensils. Heavy-duty detergents are highly alkaline cleaners that remove wax, aged or dried soil, and baked-on grease. Dishwashing detergents, for example, are highly alkaline.

Solvent Cleaners

Solvent cleaners, often called degreasers, are alkaline detergents containing a grease-dissolving agent. These cleaners work well in areas where grease has been burned on, such as grill backsplashes, oven doors, and range hoods. Solvents are usually effective only at full strength, making them costly to use on large areas.

Exhibit 12a

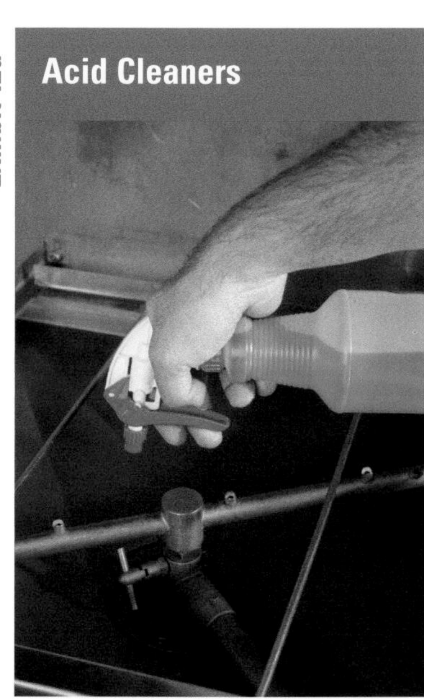

Acid cleaners are often used to remove scale in dishwashing machines.

Acid Cleaners

Acid cleaners are used on mineral deposits and other soils that alkaline cleaners cannot remove. These cleaners are often used to remove scale in dishwashing machines and steam tables, as well as rust stains and tarnish on copper and brass. (See *Exhibit 12a.*) The type and strength of the acid varies with the cleaner's purpose. Follow the instructions carefully and use acid cleaners with caution.

Abrasive Cleaners

Abrasive cleaners contain a scouring agent that helps scrub hard-to-remove soil. These cleaners are often used on floors or to remove baked-on food in pots and pans. Use abrasives with caution since they can scratch surfaces.

SANITIZING

There are two methods used to sanitize surfaces: **heat sanitizing** and **chemical sanitizing.**

Heat Sanitizing

One way to heat-sanitize tableware, utensils, or equipment is to immerse it in hot water. To be effective, the water must be at least 171°F (77°C) and the items must be immersed for thirty seconds. It may be necessary to install a heating device to maintain this temperature. Use a thermometer to check water temperature when heat sanitizing by immersion.

High-temperature dishwashing machines use hot water to sanitize tableware, utensils, and other items. To check the water temperature in these machines, attach temperature-sensitive labels, tape, or a high-temperature probe to items that will be run through the machine.

Chemical Sanitizing

Chemical **sanitizers** are regulated by state and federal environmental protection agencies (EPAs). The three most common types are chlorine, iodine, and quaternary ammonium compounds (quats). The advantages and disadvantages of each type are listed in *Exhibit 12b*. Refer to your local or state regulatory agency for recommendations on selecting a sanitizer.

Exhibit 12b

Advantages and Disadvantages of Different Sanitizers

Advantages	Disadvantages
Chlorine	
■ Most commonly used sanitizer ■ Kills a wide range of vegetative microorganisms ■ Least expensive of the three ■ Effective in hard water	■ Inactivated by presence of soil ■ Corrosive to some metals, such as stainless steel and aluminum, when used improperly ■ Can be irritating to skin ■ Does not remain active after it has dried
Iodine	
■ Remains active for a short period of time after it has dried ■ Not as quickly inactivated by soil as chlorine ■ Nonirritating to skin	■ Less effective in reducing microorganisms than chlorine ■ Somewhat corrosive to surfaces ■ Most expensive of the three ■ Slightly affected by the presence of soil
Quats	
■ Not as quickly inactivated by soil as chlorine ■ Remains active for a short period of time after it has dried ■ Noncorrosive to surfaces ■ Nonirritating to skin	■ Easily affected by presence of detergent residue ■ Less effective against certain types of microorganisms ■ Hard water reduces effectiveness

For a list of approved sanitizers, check the Code of Federal Regulations 40CFR180.940—"Food-Contact Surface Sanitizing Solutions."

Chemical sanitizing is done in two ways: either by immersing a clean object in a specific concentration of sanitizing solution for a specific amount of time or by rinsing, swabbing, or spraying the object with a specific concentration of sanitizing solution.

In some instances, detergent sanitizer blends may be used to sanitize surfaces, but items still must be cleaned and rinsed first. Scented or oxygen bleaches are not acceptable as sanitizers for food-contact surfaces. Household bleaches are acceptable only if the labels indicate they are EPA registered.

Exhibit 12c

Sanitizer Test Kit

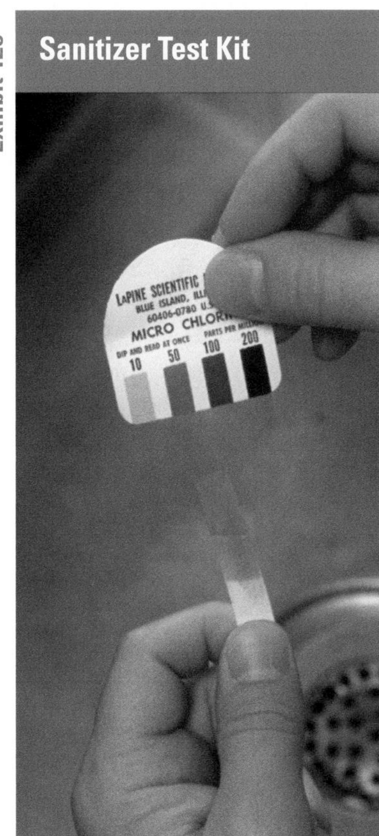

Use a test kit to check the concentration of a sanitizing solution.

Factors Influencing the Effectiveness of Sanitizers

Several factors influence the effectiveness of chemical sanitizers. The most critical include contact time, temperature, and concentration.

Contact Time

In order for a sanitizing solution to kill microorganisms, it must make contact with the object for a specific amount of time. Since minimum times may differ for each sanitizer, check with your supplier.

Temperature

Follow manufacturers' recommendations for the proper temperature.

Concentration

Chemical sanitizers are mixed with water until the proper concentration, or ratio of sanitizer to water, is reached. Mixing the sanitizer to the proper concentration is critical, since concentrations below those required in your jurisdiction or recommended by the manufacturer could fail to sanitize objects. Concentrations higher than recommended can be unsafe, leave an odor or bad taste on objects, and corrode metals. Concentration is measured using a sanitizer test kit and is expressed as ppm—parts per million. (See *Exhibit 12c.*)

The test kit should be designed for the sanitizer you are using and is usually available from the manufacturer or your supplier. The concentration of a sanitizing solution must be checked frequently since the sanitizer is depleted during use. Hard water, food particles, and detergent inadequately rinsed from a surface can quickly reduce the sanitizer's effectiveness. A sanitizing solution must be changed when it is visibly dirty or when its concentration has dropped below the required level. *Exhibit 12d* provides some general guidelines for using chlorine, iodine, and quats effectively.

Exhibit 12d

General Guidelines for Using Chlorine, Iodine, and Quats

Chlorine				Iodine	Quats
Temperature					
120°F (49°C)	100°F (38°C)	75°F (24°C)	55°F (13°C)	75°F (24°C)	75°F (24°C)
Concentration					
25 ppm	50 ppm	50 ppm	100 ppm	12.5–25 ppm	As recommended
pH					
<8–10	<10	<8	<8–10	≤5	As recommended
Contact Time					
10 sec	7 sec	7 sec	10 sec	30 sec	30 sec

MACHINE DISHWASHING

Most tableware, utensils, and pots and pans can be cleaned and sanitized in a dishwashing machine. Dishwashing machines sanitize by using either hot water or a chemical-sanitizing solution.

High-Temperature Machines

High-temperature machines rely on hot water to clean and sanitize. Water temperature is critical. If the water is *not* hot enough, items will not be properly sanitized. If the water is too hot, it may vaporize before tableware and utensils have been sanitized. Extremely hot water can also bake food onto these items.

The temperature of the final sanitizing rinse must be at least 180°F (82°C). For stationary rack single-temperature machines, it must be at least 165°F (74°C). The dishwasher must be equipped with a built-in thermometer that measures water temperature at the manifold—the point where the water sprays into the tank. Establishments that clean and sanitize high volumes of tableware may need to install a heating device to keep up with the demand for hot water.

Chemical-Sanitizing Machines

Chemical-sanitizing machines can clean and sanitize at much lower temperatures, but not lower than 120°F (49°C). Since different sanitizers require different rinse-water temperatures, it is important to follow the dishwashing temperature guidelines provided by the manufacturer. Items washed and rinsed at these lower temperatures may take longer to air-dry, so you may need more room at the clean end of the machine and more tableware during peak periods.

The effectiveness of your dishwashing program will depend on a number of factors:

- A well-planned layout in the dishwashing area, with a scraping and soaking area and adequate space for both soiled and clean items
- Sufficient water supply, especially hot water
- Separate area for cleaning pots and pans
- Devices that indicate water pressure and temperature of the wash and rinse cycles
- Protected storage areas for clean tableware and utensils
- Employees who are trained to operate and maintain the equipment and use the proper chemicals

All dishwashing machines should be operated according to manufacturers' instructions. These instructions will typically be located on the machine.

Cleaning & Sanitizing

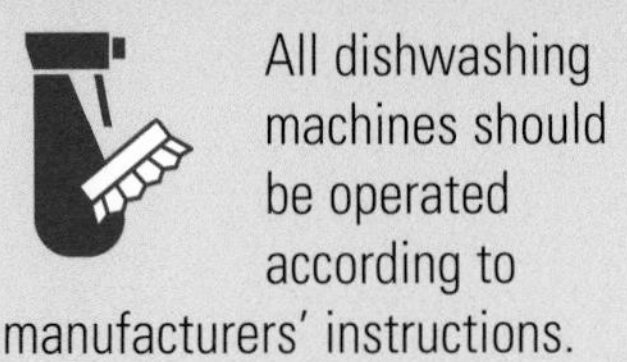

Exhibit 12e

Loading Dish Racks

Never overload dish racks, and make sure all surfaces are exposed to the spray action of the dishwasher.

Dishwashing Machine Operation

There are general procedures to follow to clean and sanitize tableware, utensils, and related items in a dishwashing machine:

- **Check the machine for cleanliness at least once a day, cleaning it as often as needed.** Fill tanks with clean water. Clear detergent trays and spray nozzles of food and foreign objects. Use an acid cleaner on the machine whenever necessary to remove mineral deposits caused by hard water. Make sure detergent and sanitizer dispensers are properly filled.
- **Scrape, rinse, or soak items before washing.** Presoak items with dried-on food.
- **Load dish racks correctly.** Make sure all surfaces are exposed to the spray action. Use racks designed for the items being washed, and never overload them. (See *Exhibit 12e.*)
- **Check temperatures and pressure.** Follow manufacturer's recommendations.
- **Check each rack for soiled items as it comes out of the machine.** Run dirty items through again until they are clean. Most items will need only one pass if the water temperature is correct and proper procedures are followed.
- **Air-dry all items.** Towels can recontaminate items.
- **Keep your dishwashing machine in good repair.**

MANUAL DISHWASHING

Establishments that do not have a dishwashing machine may use a three-compartment sink to wash items (some local regulatory agencies allow the use of two-compartment sinks; others require four-compartment sinks). These sinks are often used to wash larger items. A properly set up station includes:

- Area for rinsing away food or for scraping food into garbage containers
- Drain boards to hold both soiled and clean items
- Thermometer to measure water temperature
- Clock with a second hand, allowing employees to time how long items have been immersed in the sanitizing solution

Cleaning and Sanitizing in a Three-Compartment Sink

Before cleaning and sanitizing items in a three-compartment sink, each sink and all work surfaces must be cleaned and sanitized. Follow the steps below and as shown in *Exhibit 12f* on the next page when manually cleaning and sanitizing tableware, utensils, and equipment.

1. **Rinse, scrape, or soak all items before washing.**
2. **Wash items in the first sink in a detergent solution at least 110°F (43°C).** Use a brush, cloth, or nylon scrub pad to loosen the remaining soil. Replace the detergent solution when the suds are gone or the water is dirty.
3. **Immerse or spray-rinse items in the second sink.** Remove all traces of food and detergent. If using the immersion method, replace the rinse water when it becomes cloudy or dirty.

Key Point

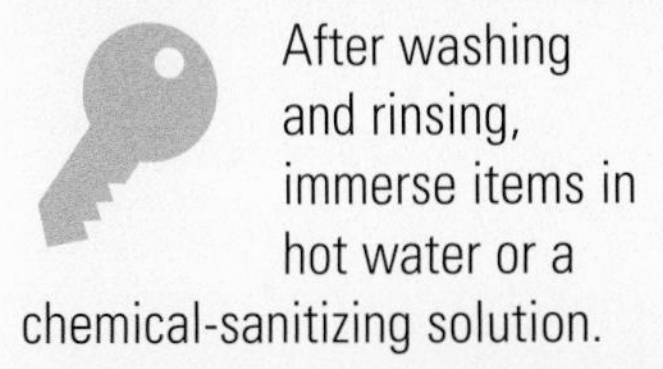

After washing and rinsing, immerse items in hot water or a chemical-sanitizing solution.

4. **Immerse items in the third sink in hot water or a chemical-sanitizing solution.** If hot-water immersion is used, the water must be at least 171°F (77°C) and the items must be immersed for thirty seconds. A heating device may be needed to maintain this temperature. If chemical sanitizing is used, the sanitizer must be mixed at the proper concentration and the water temperature must be correct. Check the concentration of the sanitizing solution at regular intervals with a test kit.
5. **Air-dry all items.**

Wood surfaces, such as cutting boards, handles, and bakers' tables, need special care. After each use, scour them in a solution with a stiff-bristle nylon brush, rinse them in clean water, and sanitize them. Do not soak wood surfaces in detergent or sanitizing solutions.

Exhibit 12f

Steps for Cleaning and Sanitizing Items in a Three-Compartment Sink

CLEANING AND SANITIZING EQUIPMENT

To prevent foodborne illness, it is important to teach employees how to clean and sanitize equipment properly.

Cleaning & Sanitizing

Clean-in-place equipment used to hold and dispense potentially hazardous food must be cleaned and sanitized everyday unless otherwise indicated by the manufacturer.

Clean-in-Place Equipment

Some pieces of equipment, such as soft-serve yogurt machines, are designed to have cleaning and sanitizing solutions pumped through them. Since many of them hold and dispense potentially hazardous food, they must be cleaned and sanitized everyday unless otherwise indicated by the manufacturer.

Stationary Equipment

Equipment manufacturers will usually provide instruction for cleaning stationary equipment, such as a slicer. In general, follow these steps:

- Turn off and unplug equipment before cleaning.
- Remove food and soil underneath and around the equipment.
- Remove detachable parts and manually wash, rinse, and sanitize them, or run them through a dishwasher if permitted. Allow them to air-dry.

Cleaning & Sanitizing

Keep cloths used for wiping surfaces that have come in contact with raw meat, fish, or poultry separate from other cleaning cloths.

- Wash and rinse fixed, food-contact surfaces, then wipe or spray them with a chemical-sanitizing solution.
- Keep cloths used for wiping surfaces that have come in contact with raw meat, fish, or poultry separate from other cleaning cloths.
- Air-dry all parts, then reassemble according to directions. Tighten all parts and guards. Test equipment at recommended settings, then turn it off.
- Resanitize the food-contact surfaces handled when putting the unit back together by wiping with a cloth that has been submerged in sanitizing solution.

In some cases, you may be able to spray-clean fixed equipment. Check with the manufacturer. If allowed, spray each part with solution in the right concentration and let it sit for the required amount of time.

Key Point

Refrigerators and freezers should be cleaned and sanitized regularly.

Refrigerated Units

Clean up spills in refrigerators and freezers immediately. These units should regularly undergo thorough cleaning and sanitizing to remove soil, mold, and odors. When cleaning and sanitizing units, follow these suggestions:

- Clean before storing deliveries so less food has to be moved.
- Move food to another unit before starting to clean.
- Clean shelves regularly. Thoroughly clean walls, floors, door edges, and gaskets.

CLEANING THE KITCHEN

Kitchen floors, walls, shelves, equipment exteriors, ceilings, light fixtures, drains, and restrooms are nonfood-contact surfaces. However, they still require regular cleaning to prevent accumulation of dust, dirt, food residue, and other debris. Sanitizing these surfaces is not required. How often these surfaces are cleaned will depend on several factors, including the type of food served, surfaces to be cleaned, and rate of ventilation in the kitchen. Spills should be cleaned up immediately. Floors and walls around food-preparation and cooking areas should be cleaned at least daily—before or after each shift is preferable. Other nonfood-contact surfaces may need to be cleaned less often.

Key Point

Floors require regular cleaning to prevent soils and spills from contaminating the entire operation.

Floors

Floors are a safety hazard if they are not cleaned regularly or rinsed properly. They can also be a source of cross-contamination since soil and spills can be tracked through the entire operation. To clean floors, follow these steps:

- Mark the area being cleaned with signs or safety cones to prevent slips and falls.
- Sweep the floor.
- Use a deck or scrub brush and full-strength detergent on heavily soiled areas to remove grease and dirt.
- Mop or pressure-spray the area, working from the walls toward the floor drain. Soak the mop in a bucket of detergent solution and wring it out. Clean a ten-foot by ten-foot area with both sides of the mop, using a figure-eight motion. Soak and wring out the mop, then clean the same area again.
- Remove excess water with a damp mop or squeegee, working away from the walls and toward the floor drain.
- Rinse the floor thoroughly with clean water, using the same mopping procedure.

Walls and Shelves

Clean tile and stainless-steel surfaces by spraying or sponging with a detergent solution. Use a nylon scrub brush to clean dried-on soil or grease. Rinse with clean water. When spray-cleaning, take care not to damage walls. Protect food, equipment, and nearby supplies. Use a wet cloth to clean other wall surfaces, such as painted drywall.

Ceilings and Light Fixtures

Ceilings do not need to be cleaned as often as floors or walls. Check ceilings and light fixtures daily to ensure that cobwebs, dust, dirt, or condensation will not fall and contaminate food or food-contact surfaces below. Wipe and rinse ceilings and light fixtures with a cloth.

CLEANING THE PREMISES

Although the kitchen requires the most attention, all areas of your operation must be kept clean. Areas such as restrooms, bussing or serving stations, and tables and booths need to be kept clean and sanitary as well. Pay attention to the exterior of the facility, too. Dirty receiving docks and garbage areas can attract pests. A dirty parking lot or exterior is unsightly and can hurt business.

Tables

Tables, booths, and counters are nonfood-contact surfaces that should always be kept clean. If you use table linens or butcher's paper, change it before seating new customers.

- Use a dry wiping cloth to clean crumbs and dry-food spills from tables. Replace the cloth whenever it becomes soiled or sticky, so that it does not transfer soil and microorganisms to other surfaces.
- Use a moist cloth to clean up other types of food spills. Keep moist cloths in a bucket of chemical-sanitizing solution. Replace both the sanitizing solution and wiping cloths as necessary. Check the concentration of the sanitizing solution often.

Serving Stations

Serving stations might be little more than a small space to store tableware and linens. In many operations, however, they are also used to dispense water, coffee, and other beverages; prepare and serve bread and condiment baskets; and serve desserts. Serving stations often contain a sink, a coffee brewer, beverage dispensers, and ice makers or bins. Some stations even have small coolers to hold butter, cream, condiments, and other temperature-sensitive items. To keep serving stations clean, follow these procedures:

- **Clean up spills immediately.** Wipe or sweep up dry food with a dry cloth or broom. Clean wet spills with a damp cloth kept in sanitizing solution.

Cleaning & Sanitizing

Service-station areas used to prepare potentially hazardous food should be cleaned at least every four hours.

- **Wash, rinse, and sanitize sinks and countertops either daily or after each shift.** Any work area used to prepare potentially hazardous food, such as cheesecake, should be cleaned at least every four hours.
- **Clean equipment daily or as often as recommended by the manufacturer.** Items such as ice bins, beverage-dispensing nozzles and lines, and coffee grinders should be cleaned as often as needed to prevent the accumulation of dirt or mold.
- **Clean and sanitize bus tubs manually or in the dishwashing machine.** This should be performed daily or after each shift.

Cleaning & Sanitizing

Many customers associate the cleanliness of the restroom with the establishment's commitment to food safety.

Public Restrooms

Unsanitary restrooms pose a danger to your customers and your business. Never underestimate cleaning needs in this area. Restrooms can quickly become visibly dirty and may harbor unpleasant odors and disease-causing microorganisms. Clean restrooms are important to customers. Many customers associate the cleanliness of the restroom with the establishment's commitment to food safety. When maintaining restrooms, consider the following suggestions:

- Check the condition of public and employee restrooms regularly.
- Restock soap, toilet paper, and towels before they run out.
- Clean sinks, mirrors, walls, floors, counters, dispensers, toilets, urinals, and waste receptacles at least daily. Sanitize toilets and urinals at least once daily. Clean up spills as often as necessary.
- Remove trash at least once daily, or as often as necessary.

Exterior Premises

The exterior of the premises should be kept clean. Windows, walls, and fixtures should be cleaned on a regular basis and should be included on the master cleaning schedule. Check the grounds at least daily, pick up trash, and sweep walkways. Clean garbage areas as often as necessary to prevent odors or trash from attracting pests.

Key Point

Color-coding can help ensure that the right tools are used for the right tasks.

TOOLS FOR CLEANING

Cleaning is easier when you have the right cleaning tools. For example, worn-out tools will not give you the pressure or friction needed for cleaning, and tools that are the wrong size will be ineffective. However, even the correct tools can contaminate surfaces if they are not handled carefully. Cleaning all tools before putting them away can help prevent this, as does designating tools for specific tasks. For example, some establishments designate one set of tools to clean food-contact surfaces and another set for nonfood-contact surfaces. Similarly, one set of tools can be designated for cleaning and another for sanitizing.

Color-coding each set of tools often helps reinforce these different uses. Whatever you do, always use a separate set of tools for the restroom.

Something to Think About... **The Contamination Culprit**

A local health department noted that a routine sample of soft-serve ice cream from an establishment tested extremely high for *E. coli.* The health inspector and manager examined all aspects of the ice-cream mix, the soft-serve machine, and the maintenance and cleaning of the machine. Everything appeared to be in order. The mix was fine, the machine was in perfect working order, and the employees cleaned and sanitized the machine every night according to the manufacturer's specifications.

The source of the *E. coli* was found, eventually, in the utility closet. A new employee responsible for cleaning and sanitizing the machine had been using a brush that was also used for heavy cleaning of the restaurant itself, including the restrooms. Each night as he cleaned the machine with the dirty brush, the employee was unknowingly contaminating the machine with *E. coli.*

What could have been done to prevent this situation?

Brushes

Brushes apply more effective pressure than wiping cloths, and the bristles loosen soil more easily. Worn brushes will not clean effectively and can be a source of contamination.

Brushes come in different shapes and sizes for each task. Lacquered wood or plastic brushes with synthetic bristles are preferred. They do not absorb moisture, are nonabrasive, and last longer. Use the right brush for the job.

Scouring Pads

Steel wool and other abrasives are sometimes used to clean heavily soiled pots and pans, equipment, or floors. However, metal scouring pads can break apart and leave residue on surfaces, which can later contaminate food. Nylon scouring pads provide an alternative.

Mops and Brooms

Keep both light- and heavy-duty mops and brooms on hand. Mop heads can be all-cotton or synthetic blends. It makes sense to have a bucket and wringer for both the front and back of the house. Both vertical and push-type brooms will also be useful.

Store flatware and utensils with handles up so employees can pick them up without touching food-contact surfaces.

STORING UTENSILS, TABLEWARE, AND EQUIPMENT

Once tableware, utensils, and equipment are clean and sanitary, store them so they stay that way. It is equally important to ensure that cleaning tools and supplies are stored properly.

Tableware and Equipment

- Store tableware and utensils at least six inches (fifteen centimeters) off the floor. Keep them covered or otherwise protected from dirt and condensation.
- Clean and sanitize drawers and shelves before clean items are stored.
- Clean and sanitize trays and carts used to carry clean tableware and utensils. Do this daily or as often as necessary.
- Store glasses and cups upside down. Store flatware and utensils with handles up so employees can pick them up without touching food-contact surfaces.
- Keep the food-contact surfaces of clean-in-place equipment covered until ready for use.

Cleaning Tools and Supplies

Cleaning tools and chemicals should be placed in a storage area away from food and food-preparation sites. The area should be well lighted so employees can identify chemicals easily. It should also be equipped with hooks for hanging mops, brooms, and other cleaning tools. A utility sink should be provided for filling buckets and cleaning tools, as well as a floor drain for dumping dirty water. (See *Exhibit 12g.*) Never clean mops, brushes, or other tools in sinks designated for handwashing, food preparation, or dishwashing.

When storing tools and supplies, consider the following:

- Air-dry wiping cloths overnight.
- Hang mops, brooms, and brushes on hooks to air-dry. Do not leave brooms or brushes standing on their bristles.
- Clean, rinse, and sanitize buckets. Let them air-dry, and store them with other tools.

Exhibit 12g

Storage Area for Cleaning Tools and Supplies

Tools and chemicals should be placed in a storage area away from food and food-preparation sites.

USING HAZARDOUS MATERIALS

Chemicals are both useful and necessary to keep an establishment clean, sanitary, and pest free. Used properly, they pose little threat to an employee's safety. Used improperly, they can become a health hazard that can cause injury. To reduce this risk, you should only purchase chemicals that are approved for use in a restaurant or foodservice establishment.

> **Key Point**
>
> 
>
> OSHA requires employers to tell employees about the chemical hazards they might be exposed to and to train employees to safely use the chemicals they work with.

Because of the potential dangers of chemicals used in the workplace, the Occupational Safety and Health Administration (OSHA) requires employers to comply with its Hazard Communication Standard (HCS). This standard, also known as Right-to-Know or HAZCOM, requires employers to tell their employees about chemical hazards to which they might be exposed at the establishment. It also requires employers to train employees on how to safely use the chemicals they work with. Employers must comply with OSHA's HCS by developing a hazard communication program for their establishment.

A hazard communication program must include the following components:

- Inventory of hazardous chemicals used at the establishment
- Chemical labeling procedures
- Material Safety Data Sheets (MSDS)
- Employee training
- Written plan addressing the HCS

Inventory of Hazardous Chemicals

A hazardous chemical is any chemical that poses a physical or health hazard to humans. Chemicals known to have acute or chronic health effects or that are explosive, flammable, or unstable are considered hazardous. Virtually any chemical with toxic properties should be included in an establishment's HAZCOM program.

Take an inventory of the hazardous chemicals stored in your establishment. List the name of the chemical and where it is stored. Update the list when chemicals are added or no longer used.

Key Point

If chemicals are transferred to a new container, it must be labeled with the name of the chemical and its potential hazards, as well as the name and address of the manufacturer.

Labeling Procedures

OSHA requires chemical manufacturers to clearly label the outside of containers with the chemical name, manufacturer's name and address, and possible hazards. When receiving chemicals, only accept containers with proper labels, and make sure they remain readable and attached to the container.

If chemicals are transferred to a new container, the label on that container must include the following information:

- Chemical name
- Manufacturer's name and address
- Potential hazards of the chemical

Material Safety Data Sheets (MSDS)

OSHA requires chemical suppliers and manufacturers to provide Material Safety Data Sheets (MSDS) (see *Exhibit 12h* on the next page) for each hazardous chemical at your establishment. These sheets are sent periodically with shipments of the chemical or can be requested by the establishment. MSDS are part of employees' right to know about the hazardous chemicals they work with and, therefore, must be kept in a location accessible to all employees while on the job. MSDS contain the following information about the chemical:

- Information about safe use and handling
- Physical, health, fire, and reactivity hazards
- Precautions
- Appropriate personal protective equipment (PPE) to wear when using the chemical
- First-aid information and steps to take in an emergency
- Manufacturer's name, address, and phone number
- Preparation date of MSDS
- Hazardous ingredients and identity information

Exhibit 12h

Sample Material Safety Data Sheet

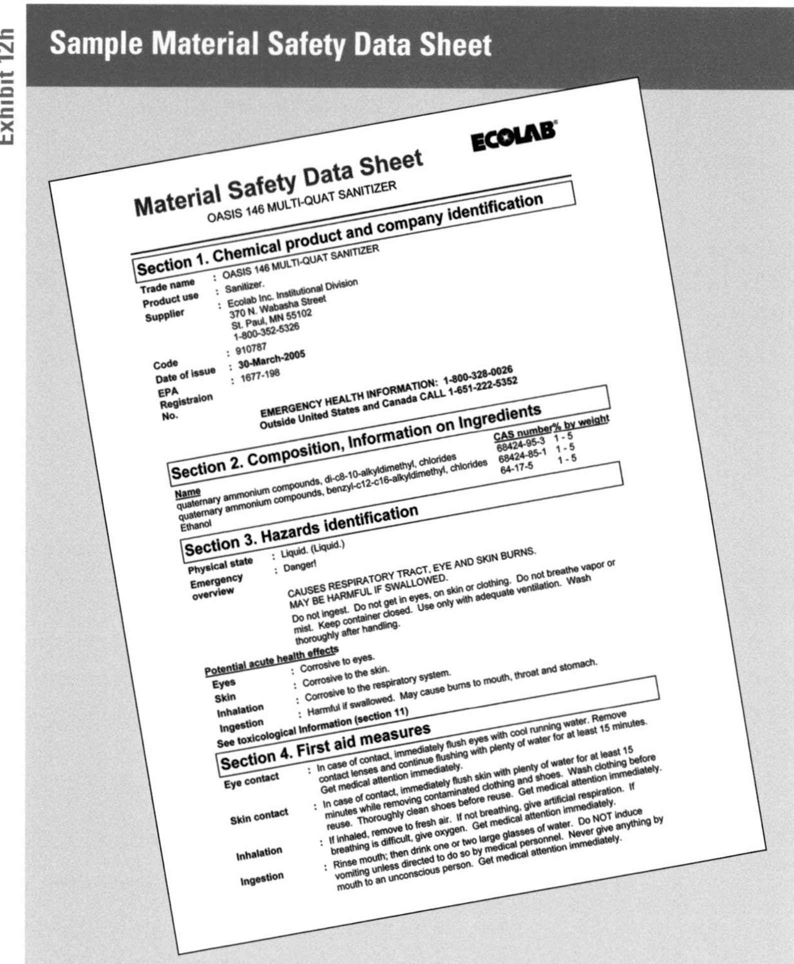

ECOLAB

Material Safety Data Sheet

OASIS 146 MULTI-QUAT SANITIZER

Section 1. Chemical product and company identification

Trade name : OASIS 146 MULTI-QUAT SANITIZER
Product use : Sanitizer.
Supplier : Ecolab Inc. Institutional Division
370 N. Wabasha Street
St. Paul, MN 55102
1-800-352-5326

Code : 910787
Date of issue : **30-March-2005**
EPA Registraion No. : 1677-198

EMERGENCY HEALTH INFORMATION: 1-800-328-0026
Outside United States and Canada CALL 1-651-222-5352

Section 2. Composition, Information on Ingredients

Name	CAS number	% by weight
quaternary ammonium compounds, di-c8-10-alkyldimethyl, chlorides	68424-95-3	1 - 5
quaternary ammonium compounds, benzyl-c12-c16-alkyldimethyl, chlorides	68424-85-1	1 - 5
Ethanol	64-17-5	1 - 5

Section 3. Hazards identification

Physical state : Liquid. (Liquid.)
Emergency overview : Danger!

CAUSES RESPIRATORY TRACT, EYE AND SKIN BURNS.
MAY BE HARMFUL IF SWALLOWED.
Do not ingest. Do not get in eyes, on skin or clothing. Do not breathe vapor or mist. Keep container closed. Use only with adequate ventilation. Wash thoroughly after handling.

Potential acute health effects
Eyes : Corrosive to eyes.
Skin : Corrosive to the skin.
Inhalation : Corrosive to the respiratory system.
Ingestion : Harmful if swallowed. May cause burns to mouth, throat and stomach.
See toxicological Information (section 11)

Section 4. First aid measures

Eye contact : In case of contact, immediately flush eyes with cool running water. Remove contact lenses and continue flushing with plenty of water for at least 15 minutes. Get medical attention immediately.
Skin contact : In case of contact, immediately flush skin with plenty of water for at least 15 minutes while removing contaminated clothing and shoes. Wash clothing before reuse. Thoroughly clean shoes before reuse. Get medical attention immediately.
Inhalation : If inhaled, remove to fresh air. If not breathing, give artificial respiration. If breathing is difficult, give oxygen. Get medical attention immediately.
Ingestion : Rinse mouth; then drink one or two large glasses of water. Do NOT induce vomiting unless directed to do so by medical personnel. Never give anything by mouth to an unconscious person. Get medical attention immediately.

Courtesy of Ecolab, Inc®, St. Paul, MN

Training

OSHA requires that every employee who might be exposed to hazardous chemicals during normal or regular working conditions be informed of the hazards and trained to use the chemicals properly. Employees should receive this training annually, and new employees must receive it when first assigned to a department or area. The following topics should be covered during training:

- Existence and requirements of the HCS
- How the HCS is implemented in the workplace
- Operations and processes in which hazardous chemicals are used
- Inventory of chemicals in your establishment

- Location of MSDS
- How to read MSDS and product labels
- Physical and health hazards of all chemicals used
- Specific procedures adopted to provide protection, such as work practices
- Use of PPE and steps to prevent or reduce exposure to chemicals
- Safety and emergency procedures
- Information on the normal use of chemicals

Key Point

OSHA requires employers to develop a written plan that describes how they will meet the requirements of the Hazard Communication Standard (HCS) in their establishment.

Written Plan

OSHA requires employers to develop a written plan describing how they will meet the requirements of the HCS in their establishment. The following items should be included in your written plan:

- List of hazardous chemicals stored on the premises and their amounts
- Purchasing specifications for chemicals
- Procedures for receiving and storing chemicals
- Labeling requirements in your establishment
- Procedures for accessing MSDS
- List of PPE
- Employee training procedures
- Reporting and record-keeping procedures
- How the employer will inform employees of the hazards of nonroutine tasks

Disposing of Hazardous Materials

Many chemicals used in the establishment pose a hazard to people and the environment if not disposed of properly. When disposing of chemicals, follow the instructions on the label and any local regulations that may apply.

Cleaning & Sanitizing

An effective cleaning program takes commitment from management and the involvement of employees.

DEVELOPING A CLEANING PROGRAM

A clean and sanitary establishment is a prerequisite for a successful food safety management system. Keeping the establishment in this condition requires an effective cleaning program. To develop this program, the needs of the establishment must first be identified and a master cleaning schedule created. Employees must receive training so they know how to properly clean equipment and surfaces. Finally, you need to monitor the program to ensure it is effective.

Identifying Cleaning Needs

- **Identify all surfaces, tools, and equipment in the facility that need cleaning.** Walk through every area of the facility.
- **Look at the way cleaning is done currently.** Get input from employees. Ask them how and why they clean a certain way. Find out which procedures can be improved.
- **Estimate the time and skills needed for each task.** Some jobs may be done more efficiently by two or more people. Others might require an outside contractor. Determine how often things need to be cleaned.

Creating a Master Cleaning Schedule

Use information gathered while identifying your cleaning needs to develop a **master cleaning schedule.** The schedule should include the following:

- **What should be cleaned.** Arrange the schedule in a logical way so nothing is left out. List all cleaning jobs in one area, or list jobs in the order they should be performed. Keep the schedule flexible enough so you can make changes if needed.
- **Who should clean it.** Assign each task to a specific individual. In general, employees should clean their own areas. Rotate other cleaning tasks to distribute them fairly.
- **When it should be cleaned.** Employees should clean as they go *and* clean and sanitize at the end of their shifts. Schedule major cleaning when food will not be contaminated or service interrupted—usually after closing. Schedule work shifts to

allow enough time. Employees rushing to clean before their shifts end may cut corners.

- **How it should be cleaned.** Provide clearly written procedures for cleaning. Lead employees through the process step by step. Always follow manufacturers' instructions when cleaning equipment. Specify cleaning tools and chemicals by name. Post cleaning instructions near the item to be cleaned. A sample master cleaning schedule is provided in *Exhibit 12i* on the next page.

Choosing Cleaning Materials

When selecting cleaning tools for your establishment, consider the following:

- **Select tools and cleaning agents according to the needs identified on the master cleaning schedule.** Talk to suppliers for suggestions on which tools and supplies are appropriate for your operation. Make sure your supplies match the needs listed on the master schedule.
- **Replace worn tools.** Equipment that is worn or soiled may not clean or sanitize surfaces properly.
- **Provide employees with the right protective gear.** Make sure there is an adequate supply of rubber gloves, aprons, goggles, and other supplies.

Implementing the Cleaning Program

Training is critical to the success of the cleaning program. Employees must understand the tasks you want them to perform and the level of quality expected. To ensure the success of the program, follow these guidelines:

- **Schedule a kickoff meeting to introduce the program to employees.** Explain the reason behind it. Stress how important cleanliness is to food safety. If people understand why they are supposed to do something, they are more likely to do it.

Exhibit 12i

Sample Master Cleaning Schedule for a Food-Preparation Area

What should be cleaned?	Who should clean it?	When it should be cleaned?	How it should be cleaned?
Floors			
■ Wipe up spills	Bussers	■ Immediately	■ Check written procedure
■ Damp mop		■ Once per shift, between rushes	■ Cloth mop and bucket, broom and dustpan
■ Scrub		■ Daily, at closing	■ Mop, bucket, safety signs
■ Strip, reseal		■ Every six months	■ Brushes, squeegee, bucket, detergent, safety signs
Walls and Ceilings			
■ Wipe up splashes	Dishwashing staff	■ As soon as possible	■ Clean using cloths and detergent
■ Wash walls		■ Food-prep and cooking areas: daily ■ All other areas: first of month	
Worktables			
■ Clean and sanitize tops	Prep cooks	■ Between uses and at the end of day	■ See cleaning procedure for each table
■ Empty, clean, and sanitize drawers		■ Weekly	■ See cleaning procedure for table

Exhibit 12j

Employee Training

Employees should be shown how to clean equipment and surfaces in each area.

- **Schedule enough time for training.** Work with small groups, or conduct training by area. Show employees how to clean equipment and surfaces in each area. (See *Exhibit 12j.*)
- **Provide plenty of motivation.** Reward employees for any job well done. Create small incentives for individuals or teams, such as "Clean Team of the Month" awards. Tie performance to specific measurements or goals, such as achieving high marks during health department inspections.

Monitoring the Program

Once you have implemented the cleaning program, you must monitor it to make sure it is working. This includes:

- **Supervising daily cleaning routines.**
- **Checking the daily completion of all cleaning tasks against the master cleaning schedule.**
- **Modifying the master schedule to reflect any changes in menu, procedures, or equipment.**
- **Requesting employee input on the program during staff meetings.** Ask employees if they need additional equipment, supplies, staff, time, or training to get cleaning jobs done. Find out if they have suggestions for improving the program.
- **Conducting spot inspections to ensure the program is being followed.**

SUMMARY

All the work that goes into a food safety management system can be undermined if you do not keep your utensils, equipment, and facility clean and sanitary. Cleaning is the process of removing food and other types of soil from a surface. Sanitizing is the process of reducing the number of harmful microorganisms on a clean surface to safe levels. You must clean and rinse a surface before it can be sanitized. Surfaces can be sanitized with hot water or with a chemical-sanitizing solution.

All surfaces should be cleaned on a regular basis. Food-contact surfaces must be cleaned and then sanitized after every use, whenever you begin working with another type of food, any time a task is interrupted, and at four-hour intervals if items are in constant use.

Dishwashing machines can be used to clean, rinse, and sanitize most tableware and utensils. Follow manufacturers' instructions, and make sure your machine is clean and in good working condition. Check the temperature and pressure of wash and rinse cycles daily.

Items that are too large to be placed into a dishwashing machine can be cleaned and sanitized manually. This may be done in a three-compartment sink or, if the items are stationary, by cleaning and then spraying them with a sanitizing solution. Items cleaned in a three-compartment sink should be presoaked or scraped clean, washed in a detergent solution, rinsed in clean water, and sanitized in either hot water or in a chemical sanitizing solution for a predetermined amount of time. All items should then be air-dried.

Clean all nonfood-contact surfaces regularly. Areas such as public restrooms, floors, shelves, and floor drains should be cleaned daily—or more often as needed—and sanitized when appropriate. Ceilings, walls, and fixtures, as well as exterior areas such as docks, garbage containers, driveways, and parking lots, can be cleaned less frequently.

Cleaning tools and chemicals should be placed in a storage area away from food and food-preparation areas. Make sure chemicals are clearly labeled. Keep MSDS for each chemical in a location accessible to all employees while on the job.

Develop and implement a cleaning program. Identify cleaning needs by walking through the operation and talking to employees. Create a master cleaning schedule listing all cleaning tasks, as well as when and how tasks should be completed. Assign responsibility for each task by job title. Enlist employees' support by including their input in the program's design and rewarding good performance. Explain to employees the important relationship between cleaning and sanitizing and food safety.

Monitor the cleaning program to keep it effective. Supervise cleaning procedures. Check completion of each job against the master schedule. Adjust cleaning and sanitizing procedures when there is a change in menu, equipment, or procedures.

Apply Your Knowledge

1. What do you think went wrong?
2. How can Tim prevent this from happening again?
3. How should any cleaning policy changes be introduced?

A Case in Point 1

It was only 9:05 a.m., but Tim, the day shift manager, already could tell it was going to be a bad day. While the executives from American Widget munched unenthusiastically on complimentary doughnuts, they threw annoyed looks at Tim and the busser, who were cleaning the banquet room American Widget had reserved for 9:00 a.m. Tim had opened the room at 8:50 a.m. and found, to his dismay, the remains of the annual banquet of the Pine Valley Martial Arts Club still strewn about the room.

Later, Tim sat down with his cleaning schedule and tried to figure out what had gone wrong. The banquet had been scheduled to end at 11:30 p.m. the previous night. The busser was scheduled to clean the room at midnight. But a note from Norman, the night shift manager, told Tim the banquet had been a wild one and the last guest had left long after the 1:00 a.m. closing time. The busser had punched out at 12:30 a.m., as he always did. Tim sighed.

For answers, please turn to the Answer Key.

Apply Your Knowledge

A Case in Point 2

1 What alternative methods can be used to clean and sanitize the tableware since the machine is not functioning properly?

The kitchen employees in the university cafeteria had scraped and rinsed every piece of tableware, placed them in racks, and fed them into the high-temperature dishwashing machine. When the wash and rinse cycles were complete, one of the employees noticed that the items were spotted. The thermometer registering the final rinse temperature for the sanitizing cycle had a reading of 140°F (60°C), rather than the required 180°F (82°C) indicated on the manufacturer's label. The employee went to inform the manager, who called for service.

For answers, please turn to the Answer Key.

Apply Your Knowledge

Use these questions to review the concepts presented in this chapter.

Discussion Questions

1. When should food-contact surfaces be cleaned and sanitized?
2. What is the difference between cleaning and sanitizing?
3. What are the steps that should be taken (in order) when cleaning and sanitizing items in a three-compartment sink?
4. How should clean and sanitized tableware, utensils, and equipment be stored?
5. What factors affect the efficiency of a sanitizer?
6. What are the requirements for storing cleaning materials?

For answers, please turn to the Answer Key.

Apply Your Knowledge

Use these questions to test your knowledge of the concepts presented in this chapter.

Multiple-Choice Study Questions

1. Which factor influences the effectiveness of a chemical sanitizer?
   A. Amount of time the sanitizer is in contact with the item
   B. Temperature of the sanitizing solution
   C. Concentration of the sanitizer in the solution
   D. All of the above

2. You want to make a spray solution for use in sanitizing food-contact surfaces in the establishment. What should you do to ensure that you have made a proper sanitizing solution?
   A. Compare the color of the solution to another solution of known strength.
   B. Try out the solution on a food-contact surface.
   C. Test the solution with a sanitizer test kit.
   D. Use very hot water when making the solution.

3. What is the proper procedure for sanitizing a table that has been used to prepare food?
   A. Spray it with a strong sanitizing solution, then wipe it dry.
   B. Wash it with a detergent, rinse it, then wipe it with a sanitizing solution.
   C. Wash it with a detergent, then wipe it dry.
   D. Wipe it with a dry cloth, then wipe it with a sanitizing solution.

4. Which item needs to be both cleaned and sanitized?
   A. Floors
   B. Walls
   C. Cutting boards
   D. Ceilings

5. If food-contact surfaces are in constant use, they must be cleaned and sanitized at
   A. four-hour intervals.
   B. five-hour intervals.
   C. six-hour intervals.
   D. eight-hour intervals.

Apply Your Knowledge

Multiple-Choice Study Questions

6. Which step is *incorrect* for cleaning and sanitizing a standing food mixer?
   A. Clean the food and dirt from around the base of the mixer.
   B. Remove the detachable parts, and wash them in the dishwashing machine.
   C. Wash and rinse nondetachable food-contact surfaces. Wipe them with a sanitizing solution.
   D. Dry the detachable parts with a clean cloth, and reassemble the machine.

7. Your dishwashing machine is not working and you must use your three-compartment sink to clean and sanitize tableware. What is the first thing you must do?
   A. Fill the first sink with hot water and detergent.
   B. Clean and sanitize the sinks and drainboards.
   C. Prepare the sanitizing solution in the third sink.
   D. Gather clean towels for drying items.

8. Which is an improper method for storing clean and sanitized tableware and equipment?
   A. Storing glasses and cups upside down
   B. Storing tableware six inches off the floor
   C. Storing flatware in containers with the handles down
   D. Storing utensils in a covered container until needed

9. Managers should only purchase chemicals that
   A. are cost effective and do the job.
   B. will serve two purposes (clean inside and outside).
   C. are approved for use in a restaurant or foodservice establishment.
   D. are approved by OSHA.

10. When disposing of chemicals, managers should
    A. throw them away with other garbage.
    B. place them in a separate bag and then throw them away.
    C. pay someone to remove them from the establishment.
    D. follow the instructions on the label and any local regulations that may apply.

For answers, please turn to the Answer Key.

Take It Back*

The following food safety concepts from this chapter should be taught to your employees:

- The difference between cleaning and sanitizing
- How to make sure sanitizers are effective
- Cleaning and sanitizing in a three-compartment sink
- How to store cleaning supplies

The tools below can be used to teach these concepts in fifteen minutes or less using the directions below. Each tool includes content and language appropriate for employees. Choose the tool or tools that work best.

Tool #1:	Tool #2:	Tool #3:	Tool #4:
ServSafe Video 6: *Facilities, Cleaning and Sanitizing, and Pest Management*	*ServSafe Employee Guide*	ServSafe Posters and Quiz Sheets	ServSafe Fact Sheets and Optional Activities
			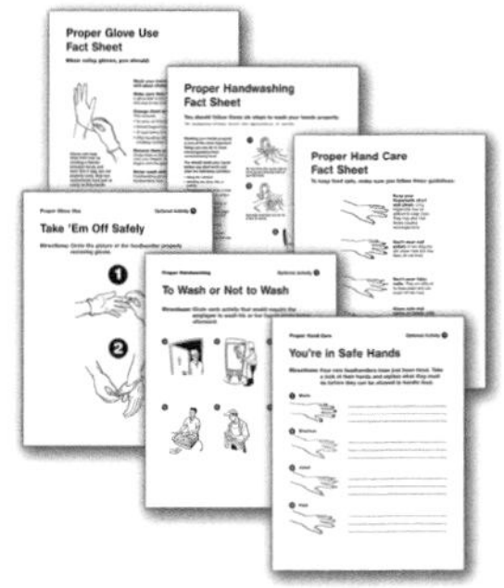

The Difference between Cleaning and Sanitizing

Video segment on cleaning vs. sanitizing

1. Show employees the segment.
2. Ask employees to explain the difference between cleaning and sanitizing and identify when food-contact surfaces must be cleaned and sanitized.

Section 5

Cleaning versus Sanitizing

Discuss with employees the difference between cleaning and sanitizing. Identify when food-contact surfaces must be cleaned and sanitized.

Poster: Cleaning versus Sanitizing

1. Discuss with employees the difference between cleaning and sanitizing. Identify when items must be cleaned and sanitized.
2. Have employees complete the Quiz Sheet: Cleaning versus Sanitizing.

*** Visit the Food Safety Resource Center at *www.ServSafe.com/FoodSafety/resource* to download free posters, quiz sheets, fact sheets, and optional activities and to learn how to obtain *Employee Guides* and videos/DVDs.**

Take It Back*

Tool #1: ServSafe Video 6: *Facilities, Cleaning and Sanitizing, and Pest Management*	Tool #2: *ServSafe Employee Guide*	Tool #3: ServSafe Posters and Quiz Sheets	Tool #4: ServSafe Fact Sheets and Optional Activities
How to Make Sure Sanitizers Are Effective			
Video segment on sanitizing methods 1 Show employees the segment. 2 Ask employees to identify the three factors that determine the effectiveness of a sanitizer.	**Section 5** **How to Make Sure That Sanitizers Are Effective** Discuss with employees the three factors that determine the effectiveness of a sanitizer.		
Cleaning and Sanitizing in a Three-Compartment Sink			
Video segment on dishwashing in a three-compartment sink 1 Show employees the segment. 2 Ask employees to identify the proper steps for cleaning and sanitizing items in a three-compartment sink	**Section 5** **How to Clean and Sanitize in a Three-Compartment Sink** 1 Discuss with employees the proper steps for cleaning and sanitizing in a three-compartment sink. 2 Complete the What's Your Order? activity.	**Poster: How to Clean and Sanitize in a Three-Compartment Sink** 1 Discuss with employees the proper steps for cleaning and sanitizing in a three-compartment sink. 2 Have employees complete the Quiz Sheet: How To Clean And Sanitize In A Three-Compartment Sink.	

* Visit the Food Safety Resource Center at *www.ServSafe.com/FoodSafety/resource* to download free posters, quiz sheets, fact sheets, and optional activities and to learn how to obtain *Employee Guides* and videos/DVDs.

Take It Back*

Tool #1: ServSafe Video 6: *Facilities, Cleaning and Sanitizing, and Pest Management*	Tool #2: *ServSafe Employee Guide*	Tool #3: ServSafe Posters and Quiz Sheets	Tool #4: ServSafe Fact Sheets and Optional Activities
How to Store Cleaning Supplies			
Video segment on storing cleaning tools and supplies 1 Show employees the segment. 2 Ask employees to identify the proper way to store cleaning tools and supplies.	**Section 5** **How to Store Cleaning Supplies** Discuss with employees the proper procedure for storing cleaning supplies.		

*** Visit the Food Safety Resource Center at *www.ServSafe.com/FoodSafety/resource* to download free posters, quiz sheets, fact sheets, and optional activities and to learn how to obtain *Employee Guides* and videos/DVDs.**

Notes

ADDITIONAL RESOURCES

Articles and Texts

Marriott, Norman G. and Robert B. Gravani. *Principles of Food Sanitation.* New York: Kluwer Academic/Plenum Publishers, 2005.

Web Sites

Chlorine Chemistry Council®

www.c3.org

The Chlorine Chemistry Council® is a national trade association representing the manufacturers and users of chlorine and chlorine-related products. Visit this Web site for facts on how chlorine contributes to the safety of food served in restaurants.

Code of Federal Regulations

www.access.gpo.gov/nara/cfr/cfr-table-search.html

The Code of Federal Regulations (CFR) is the codification of the general and permanent rules published in the Federal Register by the executive departments and agencies of the federal government. It is divided into fifty titles that represent broad areas subject to federal regulation. Visit this Web site to search a CFR title. Those titles relevant to the restaurant and foodservice industry include 7, 21, 40, and 42.

Environmental Protection Agency

www.epa.gov

Since 1970, the U.S. Environmental Protection Agency (EPA) has been instrumental in creating a cleaner environment and protecting the health of Americans. EPA staff research and set national standards for a variety of environmental programs. Visit this Web site to find out the latest news and information relating to food safety and the environment.

FDA Center for Food Safety and Applied Nutrition

www.cfsan.fda.gov/list.html

As the center within the Food and Drug Administration (FDA) responsible for food safety, the Center for Food Safety and Applied Nutrition (CFSAN) promotes and protects the public health by researching and implementing guidelines, policies, and standards to ensure that food is safe, nutritious, wholesome, and properly labeled. This Web site provides information relevant to all aspects of food safety and security, including corresponding guidelines, policies, and standards.

National Institute for Occupational Safety and Health

www.cdc.gov/niosh/homepage.html

The National Institute for Occupational Safety and Health (NIOSH) is part of the Centers for Disease Control and Prevention (CDC) and the federal agency responsible for helping assure safe and healthful working conditions for working men and women by providing research, information, education, and training in the field of occupational safety and health. This Web site provides access to information about NIOSH programs and training sessions; guidance documents on illness; accident prevention and safety, including proper handling of chemicals; and preventing latex allergies.

Occupational Safety and Health Administration

www.osha.gov

Occupational Safety and Health Administration's (OSHA) role is to assure the safety and health of American workers by setting and enforcing standards; providing training, outreach, and education; and encouraging continual improvement in workplace safety and health. This Web site provides links to regulation and compliance documents and valuable information on the control of occupational hazards.

Documents and Other Resources

2005 *FDA Food Code*

www.cfsan.fda.gov/~dms/fc05-toc.html

The Food and Drug Administration (FDA) publishes the *FDA Food Code,* a scientifically sound technical and legal document that serves as a model for regulating the retail and foodservice industry at the federal, state, and local level. Updates to the code are issued every other year on the odd-numbered years. The *FDA Food Code* provides a system of safeguards designed to minimize foodborne illness and ensure employee health, food protection manager knowledge, safe food, nontoxic and cleanable equipment, and appropriate sanitation of the food establishment. It is used as the basis for information in this textbook.

Integrated Pest Management

Inside this chapter:

- The Integrated Pest Management (IPM) Program
- Identifying Pests
- Working with a Pest Control Operator (PCO)
- Treatment
- Control Measures
- Using and Storing Pesticides

After completing this chapter, you should be able to:

- Identify requirements of an integrated pest management program.
- Differentiate between pest prevention and pest control.
- Identify ways to prevent pests from entering the facility.
- Identify the signs of pest infestation and/or activity.
- Identify requirements for applying pesticides.
- Identify proper storage requirements for pesticides and pest-application products.

Key Terms

- Infestation
- Integrated pest management
- Pest control operator
- Air curtains
- Pesticide
- Residual sprays
- Contact sprays
- Glue boards

Apply Your Knowledge

Check to see how much you know about the concepts in this chapter. Use the page references provided with each question to explore the topic.

Test Your Food Safety Knowledge

1. **True or False:** A strong oily odor may indicate the presence of roaches. *(See page 13-7.)*
2. **True or False:** The main purpose of an integrated pest management (IPM) program is to control pests once they have entered the establishment. *(See page 13-3.)*
3. **True or False:** Stationary equipment should not be covered before applying pesticides since it gives pests a place to hide. *(See page 13-17.)*
4. **True or False:** Glue traps are used to prevent roaches from entering the establishment. *(See page 13-7.)*
5. **True or False:** Pesticides can be stored in food-storage areas if they are labeled properly and closed tightly. *(See page 13-18.)*

For answers, please turn to the Answer Key.

INTRODUCTION

Pests such as insects and rodents can pose serious problems for restaurants and foodservice establishments. Not only are they unsightly to customers, they also damage food, supplies, and facilities. The greatest danger from pests comes from their ability to spread diseases, including foodborne illnesses.

THE INTEGRATED PEST MANAGEMENT (IPM) PROGRAM

Once pests have come into the facility in large numbers—an **infestation**—they can be very difficult to eliminate. Developing and implementing an **integrated pest management** program is the key. An IPM program uses *prevention* measures to keep pests from entering the establishment and *control* measures to eliminate any pests that do get inside.

For your IPM program to be successful, it is best to work closely with a licensed **pest control operator**. These professionals use safe, up-to-date methods to prevent and control pests.

Prevention is critical in pest control. If you wait until there is evidence of pests in your establishment, they may already be there in large numbers.

An IPM program has three basic rules:

1. Deny pests access to the establishment.
2. Deny pests food, water, and a hiding or nesting place.
3. Work with a licensed PCO to eliminate pests that do enter.

Deny Pests Access to the Establishment

Pests can enter an establishment in one of two ways. They either are brought inside with deliveries, or they enter through openings in the building itself.

Key Point

Check all deliveries before they enter the establishment, and refuse any shipment in which you find pests or signs of infestation.

Deliveries

To prevent pests from entering the establishment with deliveries:

- Use reputable suppliers.
- Check all deliveries before they enter your establishment.
  - Refuse shipments in which you find pests or signs of infestation, such as egg cases and body parts (legs, wings, etc.).

Doors, Windows, and Vents

To prevent pests from entering the establishment through doors, windows and vents:

- **Screen all windows and vents with at least sixteen mesh per square inch screening.** Anything larger might let in mosquitoes or flies. Check screens regularly, and clean and replace them as needed.
- **Install self-closing devices and door sweeps on all doors.** Repair gaps and cracks in door frames and thresholds. Use weather stripping on the bottoms of doors with no threshold.
- **Install** air curtains **(also called air doors or fly fans) above or alongside doors.** These devices blow a steady stream of air across the entryway, creating an air shield around doors left open.
- **Keep all exterior openings closed tightly.** Drive-through windows should be closed when not in use.

Exhibit 13a

Denying Entry to Pests

Concrete

Sheet Metal

Fill openings or holes around pipes with concrete or cover them with sheet metal.

Pipes

Mice, rats, and insects such as cockroaches use pipes as highways through a facility. To prevent this:

- **Use concrete to fill holes or sheet metal to cover openings around pipes.** (See *Exhibit 13a.*)
- **Install screens over ventilation pipes and ducts on the roof.**
- **Cover floor drains with hinged grates to keep rodents out.** Rats are very good swimmers and can enter buildings through drainpipes.

Floors and Walls

Rodents often burrow into buildings through decaying masonry or cracks in building foundations. They move through floors and walls the same way. To prevent this:

- **Seal all cracks in floors and walls.** Use a permanent sealant recommended by your PCO or local health department.
- **Properly seal spaces or cracks where stationary equipment is fitted to the floor.** Use an approved sealant or concrete, depending on the size of the spaces.

Deny Food and Shelter

Pests are usually attracted to damp, dark, and dirty places. A clean and sanitary establishment offers them little in the way of food and shelter. The stray pest that might get in cannot thrive or multiply in a clean kitchen. Besides adhering to your master cleaning schedule, follow these additional guidelines:

- **Dispose of garbage quickly and correctly.** Garbage attracts pests and provides them with a breeding ground. Keep garbage containers clean, in good condition, and tightly covered in all areas (both indoor and outdoor). Clean up spills around garbage containers immediately. Wash and rinse containers regularly.
- **Store recyclables in clean, pest-proof containers as far away from your building as local regulations allow.** Bottles, cans, paper, and packaging material provide shelter and food for pests.

- **Store all food and supplies properly and as quickly as possible.**
  - Keep all food and supplies away from walls and at least six inches (fifteen centimeters) off the floor.
  - When possible, keep humidity at 50 percent or lower. Low humidity helps prevent roach eggs from hatching.
  - Refrigerate food such as powdered milk, cocoa, and nuts after opening. Most insects that might be attracted to this food become inactive at temperatures below 41°F (5°C).
  - Rotate products so pests do not have time to settle into them and breed.
- **Clean the establishment thoroughly.** Careful cleaning eliminates the pests' food supply, destroys insect eggs, and reduces the number of places pests can safely take shelter. (See *Exhibit 13b.*)
  - Clean up food and beverage spills immediately, including crumbs and scraps.
  - Clean toilets and restrooms as often as necessary.
  - Train employees to keep lockers and break areas clean. Food and dirty clothes should *not* be kept in or around lockers. Break rooms should be cleaned properly after use.
  - Keep cleaning tools and supplies clean and dry. Store wet mops on hooks rather than on the floor, since roaches frequently hide in them.
  - Empty water from buckets to keep from attracting rodents.

Exhibit 13b

Thorough Cleaning

Cleaning eliminates pests' food supply, destroys insect eggs, and reduces the number of places pests can take shelter.

Grounds and Outdoor Dining Areas

The popularity of outdoor dining offers a different set of pest concerns. Birds, flies, bees, and wasps can be annoying and dangerous to the health of your customers. As with indoor pests, the key to controlling them lies in denying food and shelter. (See *Exhibit 13c* on the next page.)

- Mow the grass, pull weeds, get rid of standing water, and pick up litter.

Exhibit 13c

Minimize Pests in Outdoor Dining Areas

Cover garbage containers, remove dirty dishes, and clean up spills to deny food to pests.

- Cover all outdoor garbage containers.
- Remove dirty dishes and uneaten food from tables, cleaning them as quickly as possible.
- Do not allow employees or customers to feed birds or wildlife on the grounds.
- Locate electronic insect eliminators, or "zappers," away from food, customers, employees, and serving areas.
- Call your PCO to remove hives and nests.

IDENTIFYING PESTS

Pests may still get into your establishment even if you take measures to prevent them. Pests are good hitchhikers, hiding in delivery boxes and even coming in on employees' clothing or personal belongings. It is important to be able to spot signs when pests are present and to determine the type you are dealing with. Record the time, date, and location when you spot signs of pests, and report this to your PCO. Early detection allows the PCO to start treatment as soon as possible.

Cockroaches

Roaches often carry disease-causing microorganisms such as *Salmonella* spp., fungi, parasite eggs, and viruses. Research shows that many people are allergic to residue left by roaches on food and surfaces. Roaches reproduce quickly and can adapt to some pesticides, making it difficult to control them.

Exhibit 13d

Common Roaches Found in Restaurants and Foodservice Establishments

American

German

Brown-banded

Oriental

Courtesy of Orkin Commercial Services

There are several different types of roaches. (See *Exhibit 13d.*) Most live and breed in dark, warm, moist, hard-to-clean places. You will typically find them in the following areas:

- Behind refrigerators, freezers, and stoves
- In sink and floor drains
- In spaces around hot-water pipes
- Inside equipment, often near motors and other electrical devices
- Under shelf liners and wallpaper
- Underneath rubber mats
- In delivery bags and boxes
- Behind unsealed coving (especially rubber-based)

Roaches generally feed in the dark. If you see a cockroach in daylight, you may have a major infestation, since only the weakest roaches come out in daylight. If you suspect you have a roach problem, check for these signs:

- Strong oily odor
- Droppings (feces) that look like grains of black pepper
- Capsule-shaped egg cases that are brown, dark red, or black and may appear leathery, smooth, or shiny

You may have problems with more than one type of roach. Glue traps—containers with sticky glue on the bottom—should be used to find out what type of roaches might be present. Work with your PCO to place the traps where roaches typically can be found. If possible, place them on the floor in the corner where two walls meet. Check the traps after twenty-four hours, and show them to your PCO. The type of roach present and its stage of development (nymph or adult) will determine the type of treatment needed.

Key Point

Flies transmit foodborne illnesses such as typhoid fever and dysentery.

Flies

The common housefly is also a great threat to human health. Because they feed on garbage and animal waste, flies transmit foodborne illnesses such as typhoid fever and dysentery. They also spread pathogens such as *Shigella* spp. and *Staphylococcus*, which can stick to their feet, hair, and mouths. Their feces and vomitus, swarming with bacteria, contaminate food. Flies have no teeth, so they only can eat liquids or dissolved food. To eat, they inject a long spike and vomit into solid food, let it dissolve, and then eat it.

In general, houseflies have the following characteristics:

- They prefer calm air and the edges of objects, such as rims of garbage cans.
- To find food and lay their eggs, they are drawn to odors of decay, garbage, and animal waste.
- In warm weather, they reproduce rapidly—eggs can hatch in as few as thirty hours. For their eggs to hatch, they need warm, moist, decayed material located out of the sun. Eggs hatch into maggots, which can grow into adult flies in six days.

Other types of flies can be just as bothersome. Fruit flies are somewhat less harmful because they are primarily attracted to spoiled fruit, not animal waste. They can still transmit diseases, however. "Biting" flies, such as deer flies and horse flies, can be an additional nuisance to outdoor diners.

Other Insects

- **Beetles, weevils, and moths.** Flour-moth larvae and beetles are usually found in dry-storage areas. Look for insect bodies, wings, or webs, as well as clumped-together food and holes in food and packaging. To prevent these pests, cover food tightly, keep storage areas clean and sanitary, and practice the first in, first out (FIFO) rule.
- **Ants.** Ants often nest in walls and floors near stoves and hot-water pipes. They are drawn to grease and sweet food. Clean up all food scraps and spills to keep them out.

- **Termites and carpenter ants.** These insects can cause great structural damage, boring into wood and weakening walls, floors, and ceilings. They are rarely visible. Carpenter ants nest in wood but forage for food elsewhere. Look for signs of sawdust that has fallen from the ceiling. Call your PCO immediately if you see signs of these insects.
- **Spiders.** Look for webs in corners and around fixtures. Spiders can be controlled by checking for and removing webs daily. The bites of some spiders, such as the black widow and brown recluse, are poisonous and can cause illness but are rarely fatal. If you see either type, call your PCO.
- **Bees, wasps, and hornets.** Bees, wasps, and hornets can sting outdoor diners. Some people are allergic to their venom and can go into shock or even die from just one or two stings. In general, these insects are drawn to sweet food. Here are some suggestions for handling them:
  - Bees make hives in hollow trees, under eaves, or in other protected places. They usually will not bother people unless their hives are threatened. If you find a hive, call your PCO to remove it.
  - Wasps and hornets often build nests under eaves. Call your PCO to remove them.
  - Yellow jackets usually nest in the ground and are drawn to proteins in early summer and sweets in late summer. They aggressively defend their nests, so call a PCO to remove them.
- **Mosquitoes and gnats.** Mosquitoes carry diseases such as malaria, typhoid, yellow fever, and encephalitis. These members of the fly family feed on the blood of animals and humans. Mosquitoes are drawn to heat and are most active at dawn and dusk. Gnats are annoying to outdoor diners. To minimize their presence, keep all outdoor areas free of stagnant water and move lights away from the building.
- **Deer ticks.** Ticks can carry Lyme disease and cause other serious illnesses. Minimize risk by keeping the grass short around dining areas.

Exhibit 13e

Common Types of Rodents

Roof rat

Common house mouse

Norway rat

Courtesy of Orkin Commercial Services

Rodents

Rodents are a serious health hazard. They eat and ruin food, damage property, and can spread disease. Most rodents have a simple digestive system. They urinate and defecate as they move around a facility. Their waste can fall into food and can contaminate surfaces.

Rats and mice are the most common types of rodent. (See *Exhibit 13e.*) They hide during the day and search for food at night. Like other pests, they reproduce often. Typically, they do not travel far from their nests—rats travel only 100 to 150 feet, while mice travel only ten to thirty feet. Mice can squeeze through a hole the size of a dime to enter a facility, while rats can fit through quarter-sized holes. Rats can stretch to reach an item as high as eighteen inches (forty-six centimeters), can jump three feet (one meter) in the air, and can even climb straight up brick walls. Rats and mice have very good senses of hearing, touch, and smell and are smart enough to avoid poison bait and poorly laid traps. Effective control of these rodents requires the knowledge and experience of professionals.

A building can be infested with both rats and mice at the same time. Look for these signs:

- **Signs of gnawing.** Rats and mice gnaw to reach food and to wear down their teeth, which grow continuously. Rats' teeth are so strong they can gnaw through pipes, concrete, and wood.
- **Droppings.** Fresh droppings are shiny and black. Older droppings are gray.
- **Tracks.** Check dusty surfaces by shining a light across them at a low angle.
- **Nesting materials.** Rats and mice use soft materials such as scraps of paper, cloth, hair, feathers, and grass to build their nests.
- **Holes.** Rats usually nest in holes in quiet places. Nests are often found near food and water and may be found next to buildings.

Birds

Bird droppings carry fungi and bacteria that can make people sick. They also may carry mites and microorganisms that can cause encephalitis and other diseases. Birds can be drawn to crumbs and food scraps in outdoor dining areas. Keep these areas clean, and remove food from tables quickly. Post signs asking customers not to feed birds. If birds become a problem, call a PCO specializing in bird-control measures.

Other Animals

Though less common, other animals such as bats, raccoons, and squirrels can infest your building too. Building damage, bites, and the possible spread of rabies are the biggest concerns. Prevent these types of animals from getting inside the establishment. If there is evidence any of these pests have entered your establishment, work with a PCO or your local animal-control department to remove them.

- **Bats.** Bats will nest in high places that are warm, dark, and dry, often in eaves or attics. Bats are beneficial since they eat insects. In many places, it is against the law to kill them. However, they will bite if cornered and can spread rabies. Established bat colonies can be hard to eliminate. At dusk, after bats have left to feed, block all crevices and entrances. Light the area for several days to keep them away until they find a new place to roost.
- **Raccoons.** Raccoons have adapted to humans and urban growth. They feed at night, but it is not unusual to see them during the day. They prefer wooded areas, but females will nest in dark places, such as attics and chimneys, to give birth. Raccoons can contract rabies, and they will bite if cornered.
- **Squirrels.** Like raccoons, squirrels will nest in attics if given the chance. Clear branches away from buildings, and fill any holes or cracks around eaves and gutters.

Key Point

Most pest control measures should be carried out by professional pest control operators (PCOs).

WORKING WITH A PEST CONTROL OPERATOR (PCO)

Few pest problems are solved simply by spraying **pesticides**—chemical agents used to destroy pests. Although you can take many preventive measures to reduce the risk of infestation, most control measures should be carried out by professionals. Employ a licensed, certified PCO to handle pest control. Working as a team, you and the PCO can prevent and/or eliminate pests and keep them from coming back. You can rely on your PCO to:

- Help you develop an integrated approach to pest management. This may include using a combination of chemical and nonchemical treatments to solve and prevent problems.
- Stay up to date on new equipment and products.
- Provide prompt service to address problems as they occur. Contracts should stipulate regular visits plus immediate service when pests are spotted.
- Keep records that document all steps taken to prevent and control pests.

Key Point

When choosing a PCO, make sure he or she is licensed or certified, belongs to professional organizations, and is insured.

How to Choose a PCO

Hiring a PCO is like choosing any other service provider—you must do your homework. Use these guidelines to help make your decision:

- **Talk to other managers.** Find out what PCO they use and what their experiences have been. When you call a PCO, ask for references and check them thoroughly. Make sure the PCO has experience working with restaurants and other foodservice establishments.
- **Make sure the PCO is licensed or certified by your state, as required by federal law.** Certification means the PCO has passed a test on proper pesticide use and other control methods.
- **Ask the PCO if he or she belongs to any professional organizations.** Membership in the National Pest Management Association (NPMA) or state or local groups usually means the PCO has current information on IPM.

- **Ask for proof of insurance.** Make sure the PCO has adequate coverage to protect you and your employees, customers, and facility.
- **Weigh all factors, not just price.** Make sure the PCO you choose has the expertise and resources you need to provide the service promised. A low bid may be costly in the long run.

Key Point

Always require a contract from your PCO. Service contracts should spell out the work to be performed and what is expected from both you and your PCO.

Service Contract

Always require a written service contract from your PCO. Service contracts outline the work to be performed and what is expected from both you and the PCO. Read your contract carefully and have your lawyer review it, if possible. A contract should include the following:

- Description of services to be provided, including an initial inspection, regular monitoring visits, follow-up visits, and emergency service
- Warranty for work to be done
- Legal liability of the PCO
- Period of service
- Your duties, including preventive measures and facility preparation before and after treatment
- Records to be kept by the PCO, such as:
  - Pests sighted and trapped; species, location, and actions taken
  - All chemicals used and Material Safety Data Sheets (MSDS) for each (copies should be accessible to employees)
  - Building and maintenance problems noted and fixed
  - Maps or photos of the facilities noting location of traps, bait, and problem spots
  - Schedule for checking and cleaning traps, replacing bait, and reapplying chemicals
  - Regular written summary reports from the PCO; copies should be kept on file in the establishment for reference when planning improvements and assessing facility goals

Key Point

After the initial inspection, your PCO should outline a treatment plan in writing.

Courtesy of the National Pest Management Association

TREATMENT

Effective treatment starts with a thorough inspection of your facility and grounds. Give the PCO complete access to the building and cooperate fully during the inspection.

- Prepare employees to answer the PCO's questions.
- Provide building plans and equipment layouts.
- Point out possible trouble spots.

After the initial inspection, your PCO should provide a treatment plan in writing. In addition to price, the plan should:

- **Specify exactly what treatment will be used for each area or problem and the potential risks involved.**
- **Indicate dates and times of each treatment.** The federal government requires a PCO to give you enough advance warning to prepare the facility properly. Employees must not be on site during the treatment.
- **Provide steps you can take to control pests.**
- **Detail building defects that may cause problems for prevention and control measures.**
- **Determine timing of follow-up visits.** The PCO should review how well treatment is working and suggest alternate treatments if pests reappear.

CONTROL MEASURES

PCOs can use a variety of pest-control methods that are environmentally sound and safe for establishments. They are trained to know which techniques will work best to control different types of pests in your area. Since new technologies are being developed all the time, the more you know about each of these methods, the better you can evaluate how well your PCO is doing.

Controlling Insects

There are several methods your PCO can use to control insects, depending on the type of insect and degree of infestation.

- **Repellents.** Repellents are liquids, powders, or mists that keep insects away from an area but do not kill them. Repellents are often used in hard-to-reach places, such as the spaces behind wallboards and plaster.

- **Sprays.** Chemical pesticide sprays are used to control insects. They include residual and contact sprays.
  - **Residual sprays** leave a film of insecticide that insects absorb as they crawl across it. Used in cracks and crevices like those along baseboards, these sprays can be liquid or a dust, such as boric acid.
  - **Contact sprays** kill insects on contact. They are usually used on groups of insects, such as clusters of roaches or a nest of ants.
- **Bait.** Chemical bait sometimes is used to control roaches or ants. The bait contains an attractant. When insects eat it, the chemical kills them. The advantage of using baits is that kitchen areas do not need to be prepped and people can remain on site while they are set.
- **Traps.** There are several types of traps.
  - Light-only units simply entice insects to crawl inside where they find it hard to escape.
  - Electronic insect eliminators, or "zappers," use an electrically charged grid to kill insects attracted to the light.
  - Other units use both light and chemical attractants to lure insects onto a glue board.

Wasp and hornet traps are designed to hold nectar, which attracts these insects. Once inside, they cannot escape. Most fly traps use a light source—usually UV light—to attract insects to the trap. Tests have shown that different insects respond to different types or ranges of UV light. Make sure your PCO is using the most current technology.

The placement of traps is important. Never place them above or near food-preparation or storage areas or food-contact surfaces.

Exhibit 13f

Methods for Controlling Rodents

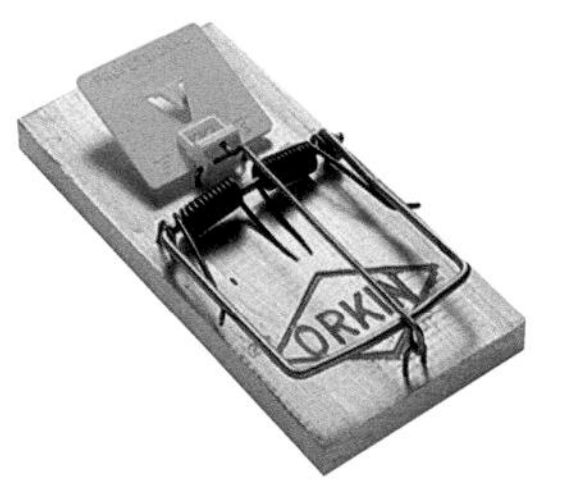

Rodent spring trap

Box trap

Glue board

Several devices can be used to control rodents.

Courtesy of Orkin Commercial Services

Controlling Rodents

Rats and mice tend to use the same routes through an establishment. Your PCO will choose the best method to eliminate these pests, which could include the following (see *Exhibit 13f*):

- **Traps.** Traps are a safe, effective way to kill rats and mice. If the infestation is large, however, traps will take time. Work with your PCO to set traps near or in rodent runways. Check traps often, and remove dead rodents carefully. If a trapped rodent is still alive, have your PCO remove it.
- **Glue boards.** Glue boards kill mice. When these devices are placed in runways, the mice stick to the board and die in several hours from exhaustion or lack of water or air. Check boards often, and throw away any with trapped mice. Glue boards are not effective for controlling rats since these rodents are usually strong enough to escape.
- **Bait.** Chemical bait should only be used by a PCO in areas where it cannot contaminate food or food-contact surfaces. It is usually placed in special covered, locked containers near rodent runways and possible entry points. Your PCO may change the baits and their locations often until they work properly. Rats can easily detect chemical bait and often avoid it.

Controlling Birds

Birds can be a serious problem in outdoor dining areas. While there is no way to eliminate birds, your PCO can use several techniques to keep them from nesting and roosting on your building.

- **Repellents.** Chemical pastes are sometimes used on gutters and ledges to repel birds. Paste must be used carefully so it does not fall into food or onto tables.
- **Netting.** Fine-mesh wire netting is used to keep birds from roosting on statues and bas-relief carvings on buildings.

- **Wires.** A PCO might string wires across the roof to prevent birds from roosting. The wires can be electrified to deliver a mild shock.
- **Sound.** In some instances, birds can be frightened away by the sound of other birds in distress. Prerecorded tapes are played through loudspeakers near roosting sites, though birds might eventually ignore the sounds and return to roost.
- **Balloons.** Birds sometimes can be frightened away with helium-filled mylar balloons. Ask your PCO about this technique.

Exhibit 13g

Pesticide Use

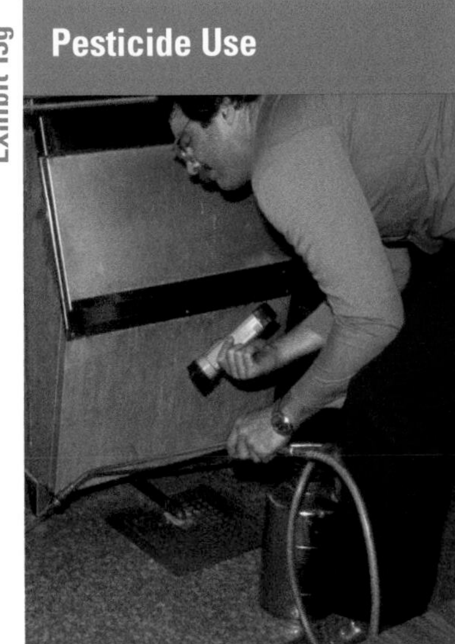

Rely on your PCO to decide if and when pesticides should be used in your establishment.

Courtesy of the National Pest Management Association

USING AND STORING PESTICIDES

While it may seem more cost effective to purchase and apply pesticides on your own, there are many reasons *not* to do so:

- Pesticides can be dangerous to employees, customers, and food.
- Applied improperly, they will be ineffective.
- Pests can develop resistance and immunity to pesticides.
- Each region has its own pest-control problems, and some control measures are more effective than others.
- Pesticides are regulated by federal, state, and local laws; some are not approved for use in restaurants and foodservice establishments.

Rely on your PCO to decide if and when pesticides should be used in your establishment. They are trained to determine the best pesticide for each pest and how and where to apply it. (See *Exhibit 13g.*)

To minimize the hazard to people, have your PCO use pesticides only when you are closed for business and employees are not on site. When pesticides will be applied, prepare the area to be sprayed by removing all food and movable food-contact surfaces. Cover equipment and food-contact surfaces that cannot be moved. Wash, rinse, and sanitize food-contact surfaces after the area has been sprayed.

Anytime pesticides are used or stored on the premises, you should have a corresponding MSDS, since they are hazardous materials. Your PCO should store and dispose of all pesticides used in your facility. If they are stored on the premises, follow these guidelines:

- **Keep pesticides in their original containers.**
- **Store pesticides in locked cabinets away from areas where food is stored or prepared.**
- **Store aerosol or pressurized spray cans in a cool place.** Exposure to temperatures higher than 120°F (49°C) could cause the cans to explode.
- **Check local regulations before disposing of pesticides.** Many are considered hazardous waste. Dispose of empty containers according to manufacturers' directions and local regulations.

SUMMARY

Pests can carry and spread a variety of diseases. Once they have infested a facility, it can be very difficult to eliminate them. Developing and implementing an integrated pest management (IPM) program is the key. An IPM program uses prevention measures to keep pests from entering the establishment and control measures to eliminate any pests that do get inside. To be successful, establishments must deny pests access to the facility, food, water, and shelter. Finally, you must work with a licensed pest control operator (PCO) to eliminate pests that do enter.

Pests can be brought inside the establishment with deliveries, or they can enter through openings in the building itself. To prevent them from getting inside, check deliveries before they enter your facility and refuse any shipment in which you find pests or signs of infestation. Screen all windows and vents, install self-closing doors and air curtains, and keep exterior openings closed when not in use. Fill or cover holes around pipes, and seal cracks in floors and walls.

Pests are usually attracted to damp, dark, and dirty places. A clean and sanitary establishment offers them little food and shelter. Stick to your master cleaning schedule. Dispose of garbage quickly, and keep containers clean and tightly covered in all areas. Store recyclables as far from your building as allowed. Keep food and supplies away from walls and at least six inches (fifteen centimeters) off the floor. Rotate products so pests do not have time to settle into them and breed. Protect outdoor diners by denying pests food and shelter in these areas as well. Mow grass, remove standing water, and pick up litter. Cover outdoor garbage containers. Remove dirty dishes and uneaten food from tables, and clean them as quickly as possible. Do not allow employees or customers to feed birds or wildlife on the grounds.

Understanding pests is the key to controlling them. Roaches live and breed in dark, warm, and moist places. Check for a strong oily odor; droppings, which look like grains of black pepper; and egg cases. Use glue traps to find out what types of roaches are present. Rodents also are a serious health hazard. A building can be infested with both rats and mice at the same time. Look for droppings, signs of gnawing, tracks, nesting materials, and holes.

Although you can take many prevention measures yourself, most control measures should be carried out by professionals. Employ a licensed, certified PCO to handle pest control, and require him or her to provide a written service contract outlining the work to be performed, as well as what is expected from both of you. Working as a team, you can prevent and/or eliminate pests and keep them from coming back.

There are several methods your PCO can use to control insects, depending on the type of insect and degree of infestation. Control methods include repellents, sprays, chemical bait, and traps. Rodents can be controlled through the use of traps set near runways, glue boards (for mice), and chemical bait.

While it might seem cost effective to apply pesticides yourself, there are many reasons for not doing so. Pesticides can be dangerous to your employees, customers, and food. Some pesticides are not approved for use in a restaurant or foodservice establishment, and pests can develop immunity to them.

Rely on your PCO to decide if and when pesticides should be used in your establishment. PCOs are trained to determine the best pesticide for each pest and how and where to apply it.

Pesticides are hazardous materials. Anytime they are used or stored on the premises, you should have corresponding Material Safety Data Sheets (MSDS). To minimize the hazard to people, have your PCO use pesticides only when you are closed for business and your employees are not on site. Your PCO should store and dispose of all pesticides used in your facility. If they are stored on the premises, they should be kept in their original containers and stored in locked cabinets away from food-storage and food-preparation areas.

Apply Your Knowledge

1. What factors might Fred have overlooked in his recent remodeling that could have led to this infestation?
2. What should Fred do to eliminate the roaches?

A Case in Point

Now We're Cooking is a restaurant located in the middle of town on the first floor of a landmark ninety-year-old building. The building is in a shopping area that includes several other restaurants. Fred, the manager, recently remodeled the restaurant. He chose materials that are easy to clean. All of the new foodservice equipment is designed for ease of cleaning as well. The cleaning procedures listed on Fred's master cleaning schedule are written out in detail and completed by the employees as scheduled, with frequent self inspections. Spills are cleaned up immediately. FIFO is followed, and food is stored on metal racks away from walls and six inches (fifteen centimeters) off the floor.

During a self inspection two weeks after he finished remodeling, Fred found live cockroaches behind the sinks, in the vegetable-storage area, in the public restrooms, and near the garbage-storage area.

For answers, please turn to the Answer Key.

Apply Your Knowledge

Use these questions to review the concepts presented in this chapter.

Discussion Questions

1. What is the purpose of an integrated pest management program?
2. How can you prevent pests from entering your establishment?
3. How can you tell if your establishment has been infested with cockroaches or rodents?
4. What are the storage requirements for pesticides?
5. What precautions must be taken both before and after pesticides are applied in your establishment?

For answers, please turn to the Answer Key.

Apply Your Knowledge

Use these questions to test your knowledge of the concepts presented in this chapter.

Multiple-Choice Study Questions

1. Which is a sign you might have a problem with rodents?
   A. You find capsule-shaped egg cases.
   B. You find scraps of paper and cloth gathered in the corner of a drawer.
   C. You see droppings that look like black grains of pepper.
   D. You smell a strong, oily odor.

2. Cockroaches typically are found in places that are
   A. cold, dry, and light.
   B. cold, moist, and dark.
   C. warm, moist, and dark.
   D. warm, dry, and light.

3. Which is a sign you might have a problem with cockroaches?
   A. You find signs of gnawing on the storeroom wall.
   B. You find droppings that look like black grains of pepper underneath a refrigeration unit.
   C. You see small piles of sawdust that appear to have fallen from the ceiling.
   D. You find webs and wings in the dry-storage area.

4. Why does careful cleaning help prevent insect infestations?
   A. It eliminates their food supply.
   B. It destroys insect eggs.
   C. It reduces the number of places pests can take shelter.
   D. All of the above

5. To prevent a pest infestation, you should store food at least _____ off the floor.
   A. one inch (2.5 centimeters)
   B. two inches (five centimeters)
   C. four inches (ten centimeters)
   D. six inches (fifteen centimeters)

Continued on the next page...

Apply Your Knowledge

Multiple-Choice Study Questions *continued*

6. All of these are critical components of an integrated pest management program *except*
   A. denying pests access to the establishment.
   B. denying pests food, water, and a hiding or nesting place.
   C. working with a licensed PCO to eliminate pests that do enter.
   D. notifying the EPA that pesticides are being used in the establishment.

7. Which is *not* used to control roaches?
   A. Repellent
   B. Contact spray
   C. Glue traps
   D. Chemical bait

8. When pesticides are applied in the establishment, you must do all of these *except*
   A. leave stationary equipment uncovered.
   B. remove all movable, food-contact surfaces.
   C. wash, rinse, and sanitize food-contact surfaces that have been sprayed.
   D. make a corresponding MSDS available to employees for the pesticide used.

9. Anytime pesticides are stored on the premises,
   A. they must be transferred from the original container to smaller containers.
   B. they must be stored away from food-preparation areas.
   C. they must be kept in dry-storage areas.
   D. they must be registered with the local regulatory agency.

For answers, please turn to the Answer Key.

ADDITIONAL RESOURCES

Articles and Texts

Marriott, Norman G. and Robert B. Gravani. *Principles of Food Sanitation.* New York: Kluwer Academic/Plenum Publishers, 2005.

Web Sites

Environmental Protection Agency

www.epa.gov

Since 1970, the U.S. Environmental Protection Agency (EPA) has worked to create a cleaner environment and protect the health of Americans. EPA staff research and set national standards for a variety of environmental programs. Visit this Web site to find out the latest news and information relating to food safety and the environment.

FDA Center for Food Safety and Applied Nutrition

www.cfsan.fda.gov/list.html

As the center within the Food and Drug Administration (FDA) responsible for food safety, the Center for Food Safety and Applied Nutrition (CFSAN) promotes and protects the public health by researching and implementing guidelines, policies, and standards to ensure that food is safe, nutritious, wholesome, and properly labeled. This Web site provides information relevant to all aspects of food safety and security, including corresponding guidelines, policies, and standards.

National Center for Infectious Diseases

www.cdc.gov/ncidod/index.htm

The National Center for Infectious Disease (NCID), one of the Centers for Disease Control and Prevention, works in partnership with local and state public-health officials, other federal agencies, medical and public-health professional associations, infectious-disease experts from academic and clinical practice, and international and public-service organizations to fulfill its mission to prevent illness, disability, and death caused by infectious diseases in the United States. This Web site houses information on bacterial, viral, and parasitic diseases.

National Pest Management Association

www.pestworld.org

The National Pest Management Association is a nonprofit organization with the goal of supporting the professional pest control industry's commitment to the protection of public health, food, and property. Consult this Web site to obtain information on habits and threats of specific pests, obtain tips on pest prevention, and ask an expert pest-related questions.

Occupational Safety and Health Administration

www.osha.gov

The Occupational Safety and Health Administration's (OSHA) role is to assure the safety and health of American workers by setting and enforcing standards; providing training, outreach, and education; and encouraging continual improvement in workplace safety and health. This Web site provides links to regulation and compliance documents and valuable information on the control of occupational hazards.

Documents and Other Resources

2005 *FDA Food Code*

www.cfsan.fda.gov/~dms/fc05-toc.html

The Food and Drug Administration (FDA) publishes the *FDA Food Code,* a scientifically sound technical and legal document that serves as a model for regulating the retail and foodservice industry at the federal, state, and local level. Updates to the code are issued on the odd-numbered years. The *FDA Food Code* provides a system of safeguards designed to minimize foodborne illness and ensure employee health, food protection manager knowledge, safe food, nontoxic and cleanable equipment, and appropriate sanitation of the food establishment. It is used as the basis for information in this textbook.

Notes

Food Establishment Inspection Report
FOODBORNE RISK FACTORS AND PUBLIC HEALTH INTERVENTIONS
Compliance Status
Compliance Status
GOOD RETAIL PRACTICES (X = not in compliance)

Unit 4

Sanitation Management

Food Safety Regulation and Standards

Inside this chapter:

- Objectives of a Foodservice Inspection Program
- Government Regulatory System for Food
- The *FDA Food Code*
- The Inspection Process
- Self-Inspections
- Federal Regulatory Agencies
- Voluntary Controls within the Industry

After completing this chapter, you should be able to:

- Identify the principles and procedures needed to comply with food safety regulations.
- Identify state and local regulatory agencies and regulations that require food safety compliance.
- Prepare for a regulatory inspection.
- Identify the proper procedures for guiding a health inspector through the establishment.

Key Terms

- U.S. Department of Agriculture (USDA)
- Food and Drug Administration (FDA)
- Regulations
- Health inspector
- Centers for Disease Control and Prevention (CDC)
- Environmental Protection Agency (EPA)
- National Marine Fisheries Service (NMFS)

Apply Your Knowledge

Check to see how much you know about the concepts in this chapter. Use the page references provided with each question to explore the topic.

Test Your Food Safety Knowledge

1. **True or False:** The Food and Drug Administration (FDA) issues food regulations that must be followed by each establishment. *(See page 14-4.)*
2. **True or False:** Health inspectors are employees of the Centers for Disease Control and Prevention (CDC). *(See page 14-4.)*
3. **True or False:** You should ask to accompany the health inspector during the inspection of your establishment. *(See page 14-11.)*
4. **True or False:** Critical violations noted during a health inspection usually must be corrected within one week of the inspection. *(See page 14-7.)*
5. **True or False:** A HACCP-based inspection focuses on the flow of food in an establishment as opposed to the sanitary appearance of the facility. *(See page 14-10.)*

For answers, please turn to the Answer Key.

INTRODUCTION

There are several reasons why it is important to have a foodservice inspection program. Most important is that failure to ensure food safety can jeopardize the health of your customers and could cost you your business. All establishments, including quick-service and fine-dining restaurants, delicatessens, hospitals, nursing homes, and schools, must follow standard food safety practices critical to the safety and quality of the food served. An inspection system lets the establishment know how well it is following these practices.

OBJECTIVES OF A FOODSERVICE INSPECTION PROGRAM

All establishments serving the public must provide safe food and are subject to inspection. It does not matter whether there is a charge for the food or whether the food is consumed on or off premises.

The purpose of an inspection program is to:

- Evaluate whether the establishment is meeting minimum food safety standards
- Protect the public's health by requiring establishments to provide food that is safe, uncontaminated, and properly presented
- Convey new food safety information to the establishment
- Provide an establishment with a written report, noting deficiencies, so it can be brought into compliance with safe food practices

GOVERNMENT REGULATORY SYSTEM FOR FOOD

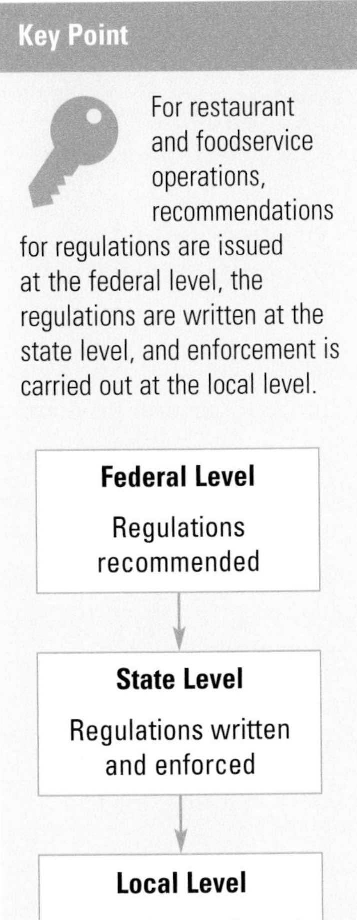

Every country has a history of government involvement in the development of health laws. Today, government control of food in the United States is exercised at three levels: federal, state, and local.

At the federal level, the **U.S. Department of Agriculture (USDA)** and the **Food and Drug Administration (FDA)** are directly involved in the inspection process.

The USDA is responsible for inspection and quality grading of meat, meat products, poultry, dairy products, eggs and egg products, and fruit and vegetables shipped across state lines. The USDA provides these services through the Food Safety and Inspection Service (FSIS) agency.

The FDA issues the *FDA Food Code* jointly with the USDA and Centers for Disease Control and Prevention (CDC). In addition, it inspects foodservice operations that cross state borders because they overlap the jurisdictions of two or more states. This includes interstate foodservice establishments such as those on planes and trains, as well as food manufacturers and processors. The FDA shares responsibility with the USDA for inspecting food-processing plants to ensure standards of purity, wholesomeness, and compliance with labeling requirements.

In the United States, most food **regulations** affecting restaurant and foodservice operations are written at the state level (except regulations for interstate or international establishments, which are determined at the federal level). Each state decides whether to adopt the *FDA Food Code* or some modified form of it.

State regulations may be enforced by state or local (city or county) health departments. In a large city, the city health department will probably be responsible for enforcing health codes. In smaller cities or in rural areas, a county or state health department may be responsible for enforcement. In any case, the manager must be familiar with the local agencies and their enforcement system. City, county, or state **health inspectors** (also called sanitarians, health officials, or environmental health specialists) conduct restaurant and foodservice inspections in most states. They generally are trained in food safety, sanitation, and public health principles.

Key Point

The FDA issues the *FDA Food Code*, which recommends regulations for restaurant and foodservice operations.

THE *FDA FOOD CODE*

The *FDA Food Code* is issued by the FDA based on input from the Conference for Food Protection (CFP). The *FDA Food Code* lists the government's recommendations for foodservice regulations. Currently, these recommendations are updated every two years to reflect developments in the restaurant and foodservice industry and the field of food safety.

The *FDA Food Code* is intended to assist state health departments in developing regulations for a foodservice inspection program. It is not an actual law. Although the FDA recommends adoption by the states, it cannot require it. Rather, the *FDA Food Code* represents the FDA's best advice for a uniform system of regulation to ensure food safety. Some states use the *FDA Food Code* as a basis for their own codes instead of adopting it in its entirety. The *FDA Food Code* covers the following areas:

- **Foodhandling and preparation:** criteria for receiving, storage, display, service, transportation
- **Personnel:** health, personal cleanliness, clothing, hygiene practices
- **Equipment and utensils:** materials, design, installation, storage
- **Cleaning and sanitizing:** facility, equipment

- **Utilities and services:** water, sewage, plumbing, restrooms, waste disposal, integrated pest management (IPM)
- **Construction and maintenance:** floors, walls, ceilings, lighting, ventilation, dressing rooms, locker areas, storage areas
- **Foodservice units:** mobile, temporary
- **Compliance procedures:** restaurant and foodservice inspections, enforcement actions

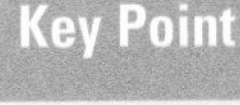

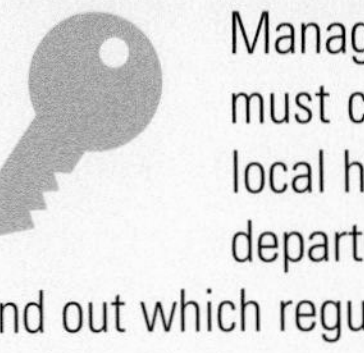

Managers must contact local health departments to find out which regulations apply to their operations.

Adoption and interpretation of food safety standards may vary widely from one state to another or, in some cases, from one locality to another. Therefore, restaurant and foodservice managers must consult with local health departments to find out which specific regulations apply to their operations.

Currently, when a state adopts, develops, or amends its food code, it is required to provide time for public input and comment. This is the time when any interested person or association can comment on how the proposed changes will impact an establishment. If you are interested in participating in this process, make sure you keep in contact with your health department or State Restaurant Association.

The lack of uniformity in local food codes can frustrate efforts by the restaurant industry to establish uniform food safety standards. For example, some jurisdictions require refrigerated food temperatures to be 45°F (7°C) or lower, while others require 41°F (5°C) or lower. States may also differ in the recommended inspection frequency. Some states require inspections for restaurant and foodservice establishments at least every six months, while others schedule inspections more or less frequently. Some states may even differ as to which establishments should be inspected; for example, some states do not inspect convenience food stores. Some states might have separate regulations or agencies for different types of foodservice operations (such as vending machines or delicatessens in grocery stores).

Although all inspectors focus on food safety practices, the areas emphasized during the inspection can vary among jurisdictions or individual inspectors. For example, some inspectors are more concerned with refrigeration temperatures while others may focus on the physical appearance of the facility. One inspector may examine the flow of food while another may focus primarily on the personal hygiene of employees. It is the responsibility of the manager to keep food safe and wholesome throughout the establishment at all times, regardless of the inspector or the inspection process.

THE INSPECTION PROCESS

All establishments serving the public, including quick-service and fine-dining restaurants, delicatessens, hospitals, nursing homes and schools, are subject to inspection. It does not matter whether there is a charge for the food or whether the food is consumed on or off the premises. The inspection lets the establishment know how well it is following practices critical to the safety and quality of the food it serves.

During health department inspections, the local health code serves as the inspector's guide. You should keep a current copy of your local or state sanitation regulations and be familiar with them. Regularly compare the code to procedures at your establishment, but remember that code requirements are only minimum standards to keep food safe.

Keep in mind that changes continue to occur in the inspection process. Some areas use traditional, scored inspection systems with a number or letter grade, while others use HACCP-based inspections or a combination of the two.

The Traditional Inspection

Many health departments use a traditional inspection to rate establishments. By this method, scoring is based on a demerit scale. Usually, the highest possible score is one hundred points. For every violation, one to five points are subtracted from one hundred points to get the final score. (See *Exhibit 14a* on the next page.) By this method:

- Noncritical violations worth one or two points must be corrected by the time of the next routine inspection.
- Critical violations worth four or five points must be corrected within a time frame specified by the inspector (usually forty-eight hours or less).
- If a low score is received upon reinspection, the establishment may be fined or even closed.

Key Point

Establishments can be closed if a low score is received upon reinspection.

Some health departments use a letter to score the establishment. Whatever scoring system is used, make sure you understand it, as well as your options for improving an unsatisfactory score.

One limitation of the traditional inspection is that multiple violations of the same type do not cause any more demerits than a single violation. For example, leaving one potentially hazardous food at room temperature incurs the same number of demerits as leaving all food at room temperature. Clearly, leaving all food at room temperature is a greater health concern and should be addressed appropriately.

As a result, a new type of inspection has been implemented. The *FDA Food Code* now recommends the use of a risk-based inspection form. This new form focuses on risk factors most likely to cause foodborne illness, such as approved food sources, contamination prevention, and good hygiene practices. Rather than deducting points for violations, the new form simply indicates whether or not an operation is in compliance for a risk factor. (See *Exhibit 14b* on page 14-9.)

Exhibit 14a

Traditional Inspection Form

FOOD AND SANITATION INSPECTION REPORT

Establishment Name	Address	Inspection Date
Manager's Name	Inspector	Inspection Time

Based on an inspection this day, the items marked below identify the violations in the operation or facilities that must be corrected by the next routine inspection or such shorter time as may be specified in writing by the regulators. Failure to comply with any time limits for corrections specified in this notice may result in cessation of your foodservice operations.

Item #	Wt	
01	5#	**FOOD** From approved, licensed sources, free of spoilage, no home-processed food, no dented or damaged cans.
02	1	Food containers properly labeled.
03	5*	**FOOD PROTECTION** Potentially hazardous food must be maintained at 41°F or 140°F or above; dairy products, meat, poultry, fish, cooked or broiled potatoes, beans, and rice. Cooling: 1) Shallow pans, product 3 inches in depth, refrigerated. 2) Precooled in ice bath with frequent stirring to 41°F. NO COOLING AT ROOM TEMPERATURE.
04	4*	Adequate equipment to maintain proper food temperature.
05	1	Accurate thermometers; conspicuous.
06	2	Potentially hazardous food properly thawed 1) in a refrigerator, 2) under cold running water, 3) by microwave, 4) as part of cooking.
07	4*	Unwrapped and potentially hazardous food not re-served.
08	2	Food protected from contamination; i.e., covered, off floors, etc.
09	2	Handling of food (ice) minimized, proper utensils provided and used.
10	1	Food-dispensing utensils properly stored when in use.
11	5*	**PERSONNEL** Personnel with infectious disease, cuts or burns restricted from handling food, cleaning utensils/ equipment.
12	5*	Hands washed, good hygienic practices. No smoking, eating, or drinking in kitchen.
13	1	Clothes clean, hair restrained.
14	2	**FOOD EQUIPMENT** Food-contact surfaces nontoxic, smooth, durable, nonabsorbent, and easily cleanable.
15	1	Nonfood surfaces smooth, durable, nonabsorbent, and easily cleanable.

Item #	Wt	
16	2	Dishwashing facilities properly designed, constructed, maintained, installed, located, operated.
17	1	Dishwashing facilities provided with accurate thermometers, pressure gauge, chemical test kit.
18	1	Soiled equipment, dishes, and utensils preflushed, scraped, soaked.
19	2	Wash and rinse water clean, hot.
20	4*	Washing and Sanitizing 3-compartment sink: 1) wash 2) rinse 3) sanitize 4) air-dry Sanitizers: 50 ppm chlorine, 200 ppm quaternary ammonia, or 12.5 ppm iodine for 1 minute. Dishwashing machine: 180°F final rinse temperature or 50 ppm chlorine at dish level.
21	1	Wiping cloths: Stored in 100 ppm chlorine or 25 ppm iodine.
22	2	Food-contact surfaces of equipment: cutting boards, meat slicers, can openers, work counters, etc., shall be washed, rinsed, and sanitized. Surfaces free of abrasives and detergents.
23	1	Nonfood-contact surfaces of equipment and utensils clean.
24	1	Clean equipment and utensils properly stored and handled to prevent contamination.
25	1	Single-service items properly stored
26	2	No reuse of single-service articles.
27	5*	**WATER** Safe water source. Hot and cold water provided at all times.
28	4*	**SEWAGE** Sewage and wastewater (mop) properly disposed.
29	1	**PLUMBING** Plumbing properly installed and maintained.
30	5*	No cross-connection between potable and wastewater. Backflow devices present.
31	4*	**TOILET AND HANDWASHING FACILITIES** Handwashing sinks accessible.

Item #	Wt	
32	2	Handsinks provided with soap and single-service towels. Toilet rooms clean, in good repair, doors self closing, trash receptacles.
33		**GARBAGE AND REFUSE** Garbage containers sanitarily maintained, covered.
34	1	Outside refuse areas maintained, clean.
35	4*	**INSECT AND RODENT CONTROL** No insects, rodents, or other animals present. Outer openings protected. IPM policies in place.
36	1	**FLOORS, WALLS, AND CEILINGS** Floors clean and constructed to be smooth, durable, nonabsorbent, and properly covered.
37	1	Walls, ceilings, and attached equipment clean and constructed to be smooth, durable, and nonabsorbent.
38	1	**LIGHTING** Lighting adequately shielded.
39	1	**VENTILATION** Rooms and equipment vented as required.
40	1	**DRESSING ROOMS** Employees' personal items properly stored in locker or separate area.
41	5*	**OTHER OPERATIONS** Chemicals/cleaning agents stored away from food-preparation and utensil-washing areas; chemicals labeled.
42	1	Premises clean, maintained. No unnecessary articles. Cleaning/ maintenance equipment properly stored. Authorized personnel only.
43	1	**FIRE EXTINGUISHERS** Sufficient number, servicing performed, personnel familiar with use.
44	1	**FOOD AND SANITATION TRAINING PROGRAM**

Exhibit 14b

New Inspection Form Recommended by the FDA

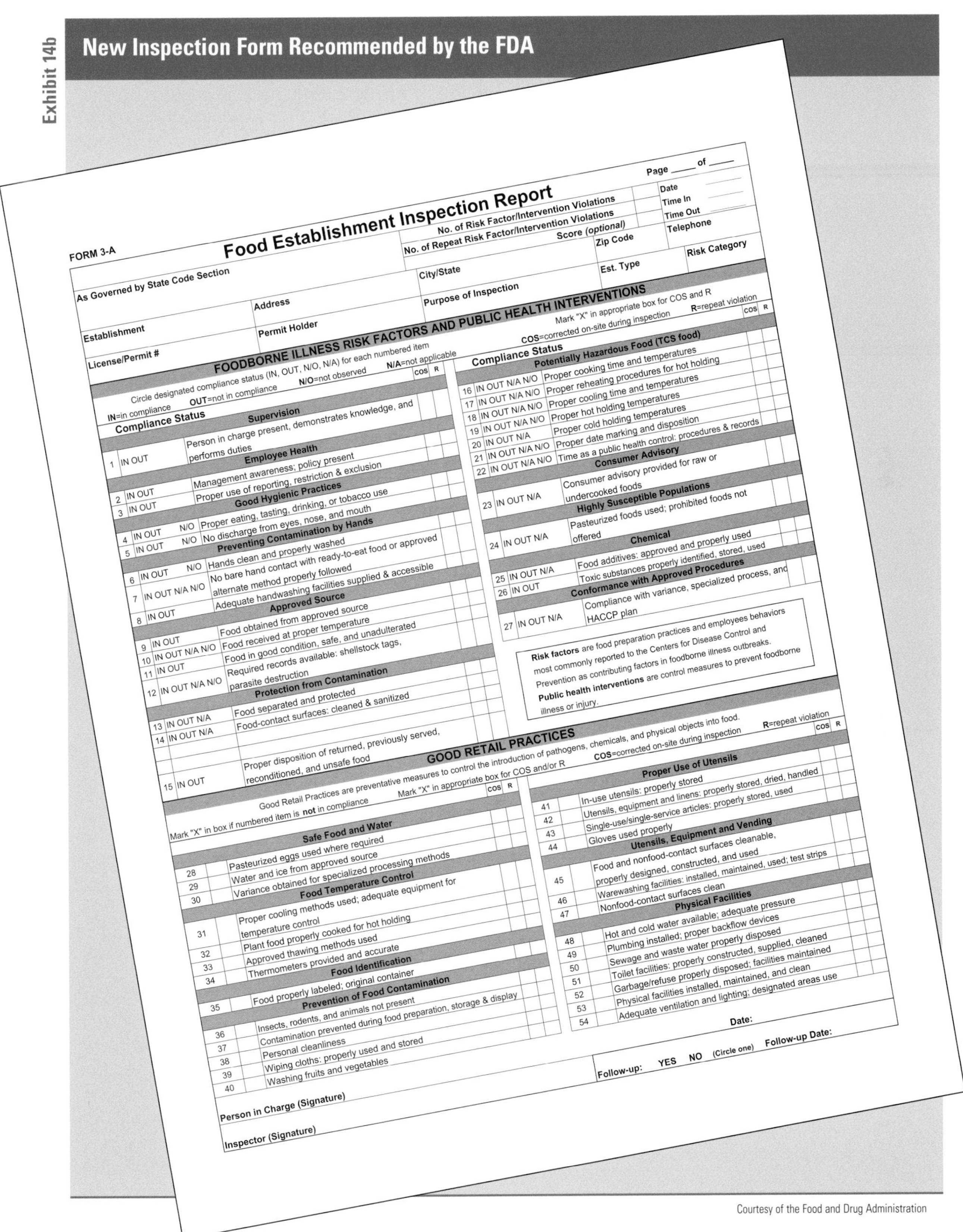

Page ____ of ____

FORM 3-A

Food Establishment Inspection Report

No. of Risk Factor/Intervention Violations | Date
No. of Repeat Risk Factor/Intervention Violations | Time In
Score (optional) | Time Out

Establishment	License/Permit #		Telephone
Address	City/State	Zip Code	
Permit Holder	Purpose of Inspection	Est. Type	Risk Category
As Governed by State Code Section			

FOODBORNE ILLNESS RISK FACTORS AND PUBLIC HEALTH INTERVENTIONS

Circle designated compliance status (IN, OUT, N/O, N/A) for each numbered item
IN=in compliance OUT=not in compliance N/O=not observed N/A=not applicable
Mark "X" in appropriate box for COS and R
COS=corrected on-site during inspection R=repeat violation

	Compliance Status		COS	R
		Supervision		
1	IN OUT	Person in charge present, demonstrates knowledge, and performs duties		
		Employee Health		
2	IN OUT	Management awareness; policy present		
3	IN OUT	Proper use of reporting, restriction & exclusion		
		Good Hygienic Practices		
4	IN OUT N/O	Proper eating, tasting, drinking, or tobacco use		
5	IN OUT N/O	No discharge from eyes, nose, and mouth		
		Preventing Contamination by Hands		
6	IN OUT N/O	Hands clean and properly washed		
7	IN OUT N/A N/O	No bare hand contact with ready-to-eat food or approved alternate method properly followed		
8	IN OUT	Adequate handwashing facilities supplied & accessible		
		Approved Source		
9	IN OUT	Food obtained from approved source		
10	IN OUT N/A N/O	Food received at proper temperature		
11	IN OUT	Food in good condition, safe, and unadulterated		
12	IN OUT N/A N/O	Required records available: shellstock tags, parasite destruction		
		Protection from Contamination		
13	IN OUT N/A	Food separated and protected		
14	IN OUT N/A	Food-contact surfaces: cleaned & sanitized		
15	IN OUT	Proper disposition of returned, previously served, reconditioned, and unsafe food		

	Compliance Status		COS	R
		Potentially Hazardous Food (TCS food)		
16	IN OUT N/A N/O	Proper cooking time and temperatures		
17	IN OUT N/A N/O	Proper reheating procedures for hot holding		
18	IN OUT N/A N/O	Proper cooling time and temperatures		
19	IN OUT N/A N/O	Proper hot holding temperatures		
20	IN OUT N/A	Proper cold holding temperatures		
21	IN OUT N/A N/O	Proper date marking and disposition		
22	IN OUT N/A N/O	Time as a public health control: procedures & records		
		Consumer Advisory		
23	IN OUT N/A	Consumer advisory provided for raw or undercooked foods		
		Highly Susceptible Populations		
24	IN OUT N/A	Pasteurized foods used; prohibited foods not offered		
		Chemical		
25	IN OUT N/A	Food additives: approved and properly used		
26	IN OUT	Toxic substances properly identified, stored, used		
		Conformance with Approved Procedures		
27	IN OUT N/A	Compliance with variance, specialized process, and HACCP plan		

Risk factors are food preparation practices and employees behaviors most commonly reported to the Centers for Disease Control and Prevention as contributing factors in foodborne illness outbreaks. **Public health interventions** are control measures to prevent foodborne illness or injury.

GOOD RETAIL PRACTICES

Good Retail Practices are preventative measures to control the introduction of pathogens, chemicals, and physical objects into food.
Mark "X" in box if numbered item is **not** in compliance Mark "X" in appropriate box for COS and/or R COS=corrected on-site during inspection R=repeat violation

			COS	R
		Safe Food and Water		
28		Pasteurized eggs used where required		
29		Water and ice from approved source		
30		Variance obtained for specialized processing methods		
		Food Temperature Control		
31		Proper cooling methods used; adequate equipment for temperature control		
32		Plant food properly cooked for hot holding		
33		Approved thawing methods used		
34		Thermometers provided and accurate		
		Food Identification		
35		Food properly labeled; original container		
		Prevention of Food Contamination		
36		Insects, rodents, and animals not present		
37		Contamination prevented during food preparation, storage & display		
38		Personal cleanliness		
39		Wiping cloths: properly used and stored		
40		Washing fruits and vegetables		

			COS	R
		Proper Use of Utensils		
41		In-use utensils: properly stored		
42		Utensils, equipment and linens: properly stored, dried, handled		
43		Single-use/single-service articles: properly stored, used		
44		Gloves used properly		
		Utensils, Equipment and Vending		
45		Food and nonfood-contact surfaces cleanable, properly designed, constructed, and used		
46		Warewashing facilities: installed, maintained, used; test strips		
47		Nonfood-contact surfaces clean		
		Physical Facilities		
48		Hot and cold water available; adequate pressure		
49		Plumbing installed; proper backflow devices		
50		Sewage and waste water properly disposed		
51		Toilet facilities: properly constructed, supplied, cleaned		
52		Garbage/refuse properly disposed; facilities maintained		
53		Physical facilities installed, maintained, and clean		
54		Adequate ventilation and lighting; designated areas use		

Person in Charge (Signature) Date:

Follow-up: YES NO (Circle one) Follow-up Date:

Inspector (Signature)

Courtesy of the Food and Drug Administration

HACCP-Based Inspections

Some health departments use HACCP-based inspections to evaluate establishments. This type of inspection focuses on the control of hazards throughout the flow of food rather than on the sanitary appearance of the facility. Inspectors observe the way an establishment receives, stores, prepares, cooks, holds, cools, reheats, and serves food. They also assess whether critical control points are actually being controlled.

Scoring is done differently for a HACCP-based inspection. Instead of using a point system with demerits, these inspections determine critical violations. Many of these violations are similar to the four- and five-point items identified in the traditional inspection system.

Since a HACCP-based inspection could be viewed as complex and time-consuming, it might only be performed under special circumstances. For example, a health department might perform a HACCP-based inspection to:

- Trace the source of contamination following a report of foodborne illness
- Evaluate an establishment with significant hazards
- Assist an establishment in developing or implementing a HACCP system
- Decide how often certain establishments should be inspected
- Determine how inspections should be conducted

Inspection Frequency

Some health departments are required to conduct inspections at least every six months. However, the frequency will vary depending on the area, type of establishment, or food served. Many health departments use a risk-based approach to inspection frequency. Determining factors can include:

- **Size and complexity of the operation.** Larger operations offering a considerable number of potentially hazardous food items might be inspected more frequently.
- **An establishment's inspection history.** Establishments with a history of low sanitation scores or consecutive violations might be inspected more frequently.
- **Clientele's susceptibility to foodborne illness.** Nursing homes, schools, daycare centers, and hospitals might receive more frequent inspections.
- **Workload of the local health department and the number of inspectors available.**

The Inspection

In most cases, an inspector will arrive without prior warning and ask for the manager or the person in charge of the operation. In your absence, make sure your employees know who is in charge. Some companies have specific policies on how to handle an inspection. Make sure you are aware of those policies.

Do not refuse entry to the inspector. In some jurisdictions, inspectors may have the authority to gain access or to revoke the establishment's permit for refusal to allow an inspection.

Accompany the health inspector during the inspection so you can answer any questions.

It is best to accompany the health inspector during his or her inspection of the establishment. This allows you to answer any questions and, possibly, correct deficiencies immediately. Accompanying the inspector will also give you the opportunity to learn from the inspector's comments and suggestions and to get valuable food safety advice.

Steps in the Inspection Process

The following suggestions will enable managers and operators to get the most out of food safety inspections:

1. **Ask for identification.** Many inspectors will volunteer their credentials. Do not let anyone enter the back of the facility without proper identification. Clarify the purpose of the visit. Make sure you know whether it is a routine inspection, the result of a customer complaint, or for some other purpose.

2. **Cooperate.** Most inspectors have learned to expect some defensiveness and resentment from restaurant and foodservice operators. This can be interpreted as having something to hide. Answer all of the inspector's questions to the best of your ability. Instruct employees to do the same. Explain to the inspector that you wish to accompany him or her during the inspection. This will encourage open communication and a good working relationship. If a deficiency can be corrected quickly, do so, or tell the inspector when it can be corrected.

3. **Take notes.** As you accompany the inspector, make a note of any problem pointed out. (See *Exhibit 14c.*) Taking your own notes will help you remember exactly what was said. Make it clear you are willing to correct problems. If you believe the inspector is incorrect about something, note what was mentioned. Then ask the inspector's supervisor for a second opinion.

4. **Keep the relationship professional.**

5. **Be prepared to provide records requested by the inspector.** You might ask the inspector why these records are needed. If a request appears inappropriate, you can check with the inspector's supervisor or with your lawyer about limits on confidential information. Remember, any records you provide to the inspector will become part of the public record. Inspectors might ask for purchase records to verify food has been received from an approved source, records of pest control treatments, or a list of all chemicals used in the facility. HACCP records could be requested in some cases. In fact, HACCP records very well could be an important part of an inspection because they document the establishment's efforts to ensure food safety.

Exhibit 14c

Foodservice Inspection

As you accompany the inspector, make a note of any problem pointed out.

6 **Discuss violations and time frames for correction with the inspector.** After the inspection, the inspector will discuss the results and the score (if a score is given) and arrange for any follow-up if necessary. The inspection report should be studied closely. Deficiencies and comments should be discussed in detail with the inspector. To make complete, permanent corrections, you will need to know the exact nature of the violation, how it impacts food safety, how to correct it, and whether or not the inspector will follow up. The inspector may offer expert advice on how to correct deficiencies.

You will be asked to sign the inspection report to acknowledge you have received it. Follow your company's policy regarding this. A copy of the report is then given to you or the person in charge at the time of the inspection. Copies of all reports should be kept on file and referred to when planning improvements and assessing facility goals. Copies are also kept by the health department. These are public documents and may be available to the public upon request.

Exhibit 14d

Inspection Follow-Up

Determine why a violation occurred by evaluating procedures, the master cleaning schedule, and employee foodhandling practices. Establish new procedures or revise existing ones.

7 **Follow up.** You must act on all deficiencies noted on the inspection report. Keep the report with you as you correct the problems. Critical deficiencies should be corrected immediately or within forty-eight hours or in the time frame outline by the inspector. All other deficiencies should be corrected as soon as possible. Determine why they occurred by evaluating procedures, the master cleaning schedule, employee training, and foodhandling practices. (See *Exhibit 14d.*) Establish new procedures or revise existing ones to correct the problem permanently. Inform employees and retrain them if necessary.

Closure

In some states, if the inspector determines a facility poses an immediate and substantial health hazard to the public, he or she may ask for a voluntary closure or issue an immediate suspension of the permit to operate. Examples of hazards calling for closure include:

- Significant lack of refrigeration
- Backup of sewage into the establishment or its water supply
- Emergency, such as a building fire or flood
- Significant infestation of insects or rodents
- Long interruption of electrical or water service
- Clear evidence of a foodborne-illness outbreak related to the establishment

A suspension requires the approval of the local health department. If an establishment receives a suspension, it must cease operations immediately. However, the owner can request a hearing if he or she believes the suspension was unjustified. There is often a time limit to request such a hearing (usually within five to ten days of the inspection). Check local regulations to determine these limits.

The suspension order may be posted at a public entrance to the establishment; however, this is not required if the establishment closes voluntarily. Regulations vary from one jurisdiction to another, but to reinstate a permit to operate, the establishment must eliminate the hazards causing the suspension and then pass a reinspection.

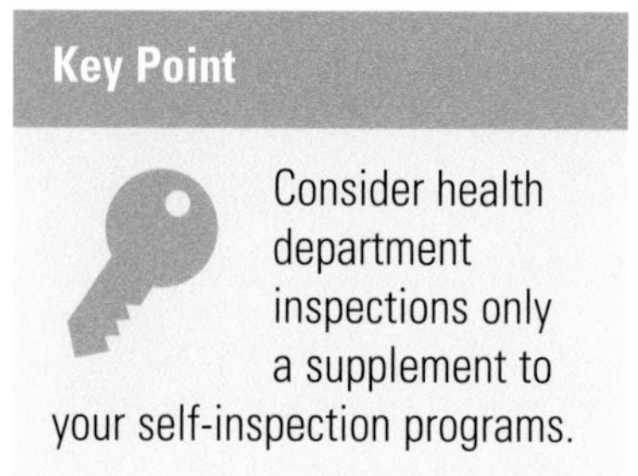

SELF-INSPECTIONS

In addition to the regular inspections performed by the health department, well-managed establishments will perform continuous self-inspections to keep food safe. These establishments consider health department inspections only a supplement to their self-inspection programs. A good self-inspection program has many benefits, including:

- Safer food
- Improved food quality

- A cleaner environment for employees and customers
- Higher inspection scores

Strive to exceed the expectations of the health department. The higher you set your standards, the more likely you are to do well on health department inspections. Additionally, your customers will notice your commitment to providing them with a safe dining experience.

Something to Think About...

Cool It!

The owner of a small, traditional Italian restaurant in the Midwest had built her business around weekend diners. The specialty of the house was a very thick meat sauce that was served over nearly all of the dishes offered.

To have sufficient quantities of sauce available for the weekend rush, the staff had developed a pattern of preparing a large batch of meat sauce on Sunday night and storing it in five-gallon buckets for the following weekend. Unfortunately, no plan had been established for proper cooling before placing the buckets into the restaurant's small refrigerator. As a result, temperatures at the center of the buckets of sauce were found to be as warm as 70°F (21°C) when the buckets were opened the following weekend—five days after being placed in the refrigerator.

The local health inspector was quick to spot the time-temperature abuse. Since purchasing a blast chiller was not an option, another solution was developed using the existing equipment. Working together, the inspector and restaurant owner established a system that involved adding ice to the cooked sauce to help it cool and filling storage containers only half full. They also determined that the sauce should be cooked in smaller batches—cooking it twice per week rather than in a single batch. As a result, the sauce was cooled properly, preventing a potentially hazardous situation.

FEDERAL REGULATORY AGENCIES

Several federal government agencies work to protect the sanitary quality of food in an establishment. Earlier in this chapter, the roles of the FDA and the USDA were discussed. A few other agencies concerned with food safety should also be mentioned.

The **Centers for Disease Control and Prevention (CDC)**, located in Atlanta, Georgia, are agencies of the U.S. Department of Health and Human Services. The Centers provide these services:

- Investigate outbreaks of foodborne illness
- Study the causes and control of disease
- Publish statistical data and case studies in the *Morbidity and Mortality Weekly Report* (*MMWR*)
- Provide educational services in the field of sanitation
- Conduct the Vessel Sanitation Program—an inspection program for cruise ships

The **Environmental Protection Agency (EPA)** sets air- and water-quality standards and regulates the use of pesticides (including sanitizers) and the handling of wastes.

The **National Marine Fisheries Service (NMFS)**, of the U.S. Department of Commerce, implements a voluntary inspection program that includes product standards and sanitary requirements for fish-processing operations.

VOLUNTARY CONTROLS WITHIN THE INDUSTRY

Few industries have devoted as much effort to regulating themselves as the restaurant and foodservice and food-processing industries. Scientific and trade associations, manufacturing firms, and foodservice corporations have vigorously pursued programs to raise the standards of the industry through research, education, and cooperation with government. Although participation in these programs is voluntary, these organizations have actively promoted professional standards, recommended legislative policy, sponsored uniform enforcement procedures, and provided educational opportunities.

The overall results in food safety have included:

- Increased understanding of foodborne illness and its prevention
- Improvements in the sanitary design of equipment and facilities
- Industry-wide efforts to maintain the sanitary quality of food during processing, shipment, storage, and service
- Efforts to make foodservice laws more practical, uniform, and science based

While many organizations have contributed to these endeavors, those having the most relevance for the foodservice operator can be found in the resources section at the end of each chapter.

SUMMARY

Government control regarding food safety in the United States is exercised at three levels: federal, state, and local. Recommendations for restaurant and foodservice regulations are issued at the federal level by the FDA in the form of the *FDA Food Code.* Regulations are written at the state level, and enforcement is usually carried out at the state and local levels. Some agencies at the federal level, such as the FDA and the USDA, are directly involved in the inspection process.

All establishments must follow practices critical to the safety of the food served. The inspection process lets the establishment know how well it is following these practices.

During the inspection, cooperate with the health inspector and keep the relationship professional. Accompany him or her during the inspection. When the inspector points out a problem, take notes. If a deficiency can be corrected immediately, do so. Be prepared to provide records that are requested. Discuss violations and time frames, and then follow up.

Well-managed establishments will perform continuous self-inspections to protect food safety, in addition to the regular inspections performed by the health department. Establishments with high standards for sanitation and food safety consider health department inspections only a supplement to their own self-inspection programs.

Apply Your Knowledge

1. Did Jerry handle the inspection correctly?
2. What does he need to do following this inspection?

A Case in Point

Carolyn, the inspector from the city's public-health department, was standing at the door of Jerry's Diner. Jerry greeted her, and they both walked to the kitchen. Carolyn pulled out a HACCP worksheet. While Jerry explained that the cook was preparing stir-fried chicken and vegetables, Carolyn noted the ingredients and their sources. Then she observed the preparation procedures and noted times and temperatures, which she plotted on a time-temperature graph. She also filled in a product flowchart and indicated the critical control points for the stir-fried chicken. Jerry told her what corrective actions his employees were trained to carry out if critical control points were not met.

Next, Carolyn checked the concentration of the sanitizing solution in the three-compartment sink the dish washer used to clean, rinse, and sanitize equipment. She also checked the handwashing station.

Jerry and Carolyn went to Jerry's office, where they discussed the report. Jerry compared his flowchart for the stir-fried chicken and vegetables with the one Carolyn had done, determining where changes could be made to improve monitoring.

For answers, please turn to the Answer Key.

Apply Your Knowledge

Use these questions to review the concepts presented in this chapter.

Discussion Questions

1. What is the role of the FDA regarding food protection?
2. What are the roles of federal, state, and local agencies regarding the regulation of food safety in establishments?
3. What should a manager do during and after an inspection?
4. What are some factors that determine the frequency of health inspections in an establishment?

For answers, please turn to the Answer Key.

Apply Your Knowledge

Use these questions to test your knowledge of the concepts presented in this chapter.

Multiple-Choice Study Questions

1. An establishment can be closed for all of these reasons *except*
   A. significant lack of refrigeration.
   B. backup of sewage.
   C. significant infestation of insects or rodents.
   D. minor violations not corrected within forty-eight hours or less.
2. Which is a goal of the food safety inspection program?
   A. To evaluate whether an establishment is meeting minimum food safety standards
   B. To protect the public's health
   C. To convey new food safety information to establishments
   D. All of the above
3. Which operation would most likely be subject to a food safety inspection by a federal agency?
   A. Hospital
   B. Passenger train traveling from New York to Chicago
   C. Local ice cream store with a history of food safety violations
   D. Food kitchen run by church volunteers
4. A person shows up at a restaurant claiming to be a health inspector. What should the manager do?
   A. Ask to see identification.
   B. Ask to see an inspection warrant.
   C. Ask for a hearing to determine if the inspection is necessary.
   D. Ask for a one-day postponement to prepare for the inspection.
5. Which agency enforces food safety in a restaurant?
   A. FDA
   B. CDC
   C. State or local health department
   D. USDA

Apply Your Knowledge

Multiple-Choice Study Questions

6. Violations noted on the health-inspection report should be
   A. discussed in detail with the inspector.
   B. corrected within forty-eight hours or when indicated by the inspector if they are critical.
   C. explored to determine why they occurred.
   D. All of the above

7. The responsibility for keeping food safe in an establishment rests with the
   A. FDA.
   B. manager/operator.
   C. health inspector.
   D. state health department.

8. Food regulations developed by state agencies are
   A. minimum standards necessary to ensure food safety.
   B. maximum standards necessary to ensure food safety.
   C. voluntary guidelines for establishments to follow.
   D. inspection practices for grading meats and meat products.

For answers, please turn to the Answer Key.

ADDITIONAL RESOURCES

Articles and Texts

Bryan, Frank L. 2000. Conducting Effective Foodborne-Illness Investigations. *Journal of Environmental Health.* 63 (1): 9.

Jacobs, Don. 2003. Food Safety I.Q. The Health Inspection Process is No Longer What it Once Was. *Restaurant Hospitality.* 87 (11): 76.

Jacobs, Don. 2004. Food Safety—Establishing Good Communication and a Solid Relationship with Your Local Health Authority Is the Smart Thing to Do for the Future of Your Business. *Restaurant Hospitality.* 88 (1): 78.

Web Sites

Association of Food and Drug Officials

www.afdo.org

The Association of Food and Drug Officials (AFDO) fosters uniformity in the adoption and enforcement of food, drug, medical devices, cosmetics and product safety laws, rules, and regulations. AFDO also provides the mechanism and the forum in which regional, national, and international issues are deliberated and resolved uniformly to provide the best public health and consumer protection. Visit this Web site for information on the association's annual conference, its position statements, and food safety related information.

Conference for Food Protection

www.foodprotect.org

The Conference for Food Protection (CFP) is a nonprofit organization that provides a forum for regulators, industry professionals, academia, professional organizations, and consumers to identify problems, formulate recommendations, and develop and implement practices that ensure food safety. Though the conference has no formal regulatory authority, it is an organization that influences model laws and regulations among all government agencies. Visit this Web site for information regarding previously recommended changes to the *FDA Food Code,* standards for permanent outdoor cooking facilities, and other guidance documents, along with how to participate in the conference.

Environmental Protection Agency

www.epa.gov

Since 1970, the U.S. Environmental Protection Agency (EPA) has been instrumental in creating a cleaner environment and protecting the health of all Americans. EPA staff research and set national standards for a variety of environmental programs. Visit this Web site to find out the latest news and information relating to food safety and the environment.

Food and Drug Administration

www.fda.gov

The Food and Drug Administration (FDA) is responsible for promoting and protecting public health by helping safe and effective products reach consumers, monitoring products for continued safety, and helping the public obtain accurate, science-based information needed to improve health. Visit this Web site for information about all programs regulated by the FDA, including food-product labeling, bottled water, and food products except meat and poultry.

FDA Center for Food Safety and Applied Nutrition

www.cfsan.fda.gov/list.html

As the center within the Food and Drug Administration (FDA) responsible for food safety, the Center for Food Safety and Applied Nutrition (CFSAN) promotes and protects the public health by researching and implementing guidelines, policies, and standards to ensure that food is safe, nutritious, wholesome, and properly labeled. This Web site provides information relevant to all aspects of food safety and security, including corresponding guidelines, policies, and standards.

Food Marketing Institute

www.fmi.org

The Food Marketing Institute (FMI) represents food retailers and wholesalers in large multistore chains, regional firms, and independent supermarkets. FMI provides its members with a forum to work effectively with the government, suppliers, employees, customers, and their communities. This Web site provides publications and resources that support FMI's efforts in research, education, public information, government relations, and industry relations.

Food Products Association

www.fpa-food.org

The Food Products Association (FPA) represents the food and beverage industry in the United States and worldwide. FPA's laboratory centers, scientists, and professional staff provide technical and regulatory assistance to member companies and represent the food industry on scientific and public policy issues involving food safety, food security, nutrition, consumer affairs, and international trade. This Web site posts recent news releases and provides the latest information and resources on food safety and security issues.

Food Safety and Inspection Service

www.fsis.usda.gov

The Food Safety and Inspection Service (FSIS) is the public health agency within the U.S. Department of Agriculture responsible for ensuring that the nation's commercial supply of meat, poultry, and egg products is safe, wholesome, and correctly labeled and packaged. Visit this Web site for food safety information and regulations related to meat, poultry, and eggs.

Gateway to Government Food Safety Information

www.foodsafety.gov

This Web site provides links to selected government food safety-related information.

Grocery Manufacturers of America

www.gmabrands.com

The Grocery Manufacturers of America (GMA) is an association that represents the food, beverage, and consumer products industry. Visit this Web site for articles and public policy on food safety issues, and to sign up for the GMA smart brief (receive free daily e-briefings on the food, beverage, and consumer products industry).

Institute of Food Technologists

www.ift.org/cms

The Institute of Food Technologists (IFT) is a nonprofit international scientific society with 22,000 members working in food science, technology and related professions in industry, academia, and government. As the society for food science and technology, IFT brings sound science to the public discussion of food issues. Informational scientific reports and summaries can be found on this Web site, as well as other useful publications.

International Association for Food Protection

www.foodprotection.org/main/default.asp

The International Association for Food Protection (IAFP) is a nonprofit association of food safety professionals. Through the association, members are able to keep informed of the latest scientific, technical, and practical developments in food safety and sanitation. The International Food Safety Icons, which are simple pictorial representations of food safety tasks understood regardless of a person's native language, can be accessed from this Web site.

International Food Information Center

www.ific.org

Working with an extensive roster of scientific experts who serve to translate research into understandable and useful information, the International Food Information Center (IFIC) collects and disseminates scientific information on food safety, nutrition, and health. This Web site provides information on topics in food safety and health.

National Environmental Health Association

www.neha.org

National Environmental Health Association's (NEHA) mission is to advance the environmental health and protection of professionals for the purpose of providing a healthful environment for all. In pursuit of its mission, NEHA offers a variety of programs, credentials, and educational opportunities. This Web site includes many useful resources for the restaurant and foodservice industry.

National Restaurant Association Educational Foundation

www.nraef.org

The National Restaurant Association Educational Foundation (NRAEF) is a nonprofit organization dedicated to fulfilling the educational mission of the National Restaurant Association. Focusing on three key strategies of risk management, recruitment, and retention, the NRAEF is the premier provider of educational resources, materials and programs that address attracting, developing and retaining the industry's workforce. The purchase of any NRAEF product helps fund outreach and scholarship activities on a national and state level.

Occupational Safety and Health Administration

www.osha.gov

The Occupational Safety and Health Administration's (OSHA) role is to assure the safety and health of American workers by setting and enforcing standards; providing training, outreach, and education; and encouraging continual improvement in workplace safety and health. This Web site provides links to regulation and compliance documents and valuable information on the control of occupational hazards.

U.S. Department of Health and Human Services

www.hhs.gov

The Department of Health and Human Services (HHS) is the federal government's principal agency for protecting the health of all Americans. The department oversees more than three hundred programs executed by such agencies as the Food and Drug Administration (FDA) and the Centers for Disease Control (CDC). Visit this Web site to access food safety information provided by the various agencies and departments overseen by HHS.

Documents and Other Resources

Code of Federal Regulations

www.access.gpo.gov/nara/cfr/cfr-table-search.html

The Code of Federal Regulations (CFR) is the codification of the general and permanent rules published in the Federal Register by the executive departments and agencies of the federal government. It is divided into fifty titles that represent broad areas subject to federal regulation. Visit this Web site to access this government information. Those titles relevant to the restaurant and foodservice industry include 7, 21, 40, and 42.

FDA Enforcement Report Index

www.fda.gov/opacom/Enforce.html

The Food and Drug Administration's Center for Food Safety and Applied Nutrition (FDA CFSAN) provides the *FDA Enforcement Report.* Published weekly on this Web site, the report contains information on actions such as recalls, injunctions, and seizures taken against food and drug products that have not met FDA regulatory requirements.

2005 *FDA Food Code*

www.cfsan.fda.gov/~dms/fc05-toc.html

The Food and Drug Administration (FDA) publishes the *FDA Food Code,* a scientifically sound technical and legal document that serves as a model for regulating the retail and foodservice industry at the federal, state, and local level. Updates to the code are issued every other year on the odd-numbered years. The *FDA Food Code* provides a system of safeguards designed to minimize foodborne illness and ensure employee health, food protection manager knowledge, safe food, nontoxic and cleanable equipment, and appropriate sanitation of the food establishment. It is used as the basis for information in this textbook.

Hazard Analysis and Critical Control Point Principles (HACCP) and Application Guidelines

www.fsis.usda.gov/OPHS/NACMCF/past/JFP0998.pdf

Adopted by the National Advisory committee on Microbiological Criteria for Foods, these guidelines found on this Web site are intended to facilitate the development and implementation of an effective HACCP plan.

Food Establishment Plan Review Guide

www.cfsan.fda.gov/~dms/prev-toc.html

The *Food Establishment Plan Review Guide,* provided by the Food and Drug Administration (FDA) and the Conference for Food Protection (CFP), is available at this Web site. This guide has been developed to provide guidance and assistance in complying with nationally recognized food safety standards. It includes design, installation, and construction recommendations regarding food equipment and facilities.

Managing Food Safety: A Manual for the Voluntary Use of HACCP Principles for Operators of Food Service and Retail Establishments

www.cfsan.fda.gov/~dms/hret2toc.html

The Food and Drug Administration's Center for Food Safety and Applied Nutrition (FDA CFSAN) provides this manual at this Web site. The manual, which was developed by the FDA and reviewed and endorsed by the Conference for Food Protection (CFP), provides the restaurant and foodservice industry with a roadmap for developing and voluntarily implementing a food safety management system based on HACCP principles.

Notes

Employee Food Safety Training

Inside this chapter:

- Initial and Ongoing Employee Training
- Delivering Training
- Training Follow-Up
- Food Safety Certification

After completing this chapter, you should be able to:

- Recognize a manager's responsibility to provide food safety training to employees.
- Identify the need to maintain food safety training records.
- Identify appropriate training tools for teaching food safety.
- Recognize that foodhandlers require initial and ongoing food safety training.

Key Terms

- Training need
- Training objective
- Evaluation
- Training plan
- Training delivery methods

Apply Your Knowledge

Check to see how much you know about the concepts in this chapter. Use the page references provided with each question to explore the topic.

Test Your Food Safety Knowledge

1. **True or False:** A major advantage of Web-based food safety training is that the content is delivered the same way every time. *(See page 15-14.)*

2. **True or False:** An employee who receives food safety training upon being hired does not require further training. *(See page 15-5.)*

3. **True or False:** Training videos will be less effective if the trainer stops them at different points to discuss the concepts presented. *(See page 15-14.)*

4. **True or False:** It is important for legal reasons to keep records of food safety training conducted at the establishment. *(See page 15-5.)*

5. **True or False:** It is the manager's responsibility to provide employees with food safety training. *(See page 15-3.)*

For answers, please turn to the Answer Key.

INTRODUCTION

Training in the establishment is difficult at best. Time, tools, and resources are often lacking. In addition, managers are faced with turnover and staffing problems that frequently take precedence. Training can also be costly. It often takes staff away from regular tasks. Sometimes it requires the use of Web-based programs or the services of professional trainers. Typically, it involves the use of training tools to reinforce knowledge, such as videos, slides, books, and CD-ROMs. However, food safety training will have a positive return on investment in the long run. The benefits can include:

- **Avoiding the costs associated with a foodborne-illness outbreak.** These costs can be significant and may include legal fees and medical bills.
- **Preventing the loss of revenue and reputation resulting from a foodborne-illness outbreak.** These losses can have a devastating impact on an establishment.

Exhibit 15a

Identifying Training Needs

Training needs can be identified by observing employee job performance.

Exhibit 15b

Food Safety Training

All employees require general food safety knowledge, while some knowledge will be specific to the job position.

- **Improving employee morale and reducing turnover.** Most employees want to do the job right and expect to receive training.
- **Increasing customer satisfaction.** When customers see an establishment is committed to serving safe food, satisfaction will be higher.

INITIAL AND ONGOING EMPLOYEE TRAINING

As a manager, it is your responsibility to ensure that employees have the knowledge and skills needed to handle food safely in your establishment. You must also keep them informed of changes in the science of food safety and best practices in the industry.

Your first task is to assess the training needs in your establishment. A **training need** is a gap between what employees are required to know to perform their jobs and what they actually know. For new hires, the need might be apparent. For employees already on staff, the need is not always obvious.

Identifying food safety training needs may require work. There are several ways to accomplish this. They can include:

- Testing employees' food safety knowledge
- Observing employees' performance on the job (see *Exhibit 15a*)
- Surveying employees to identify areas of weakness

All employees require general food safety knowledge. Other knowledge will be specific to the tasks performed on the job. For example, all employees need to know the proper way to wash their hands, but only dishwashers need to know how to load a dishrack to ensure items are properly cleaned and sanitized. (See *Exhibit 15b.*)

It is dangerous to assume new employees will understand your establishment's food safety procedures without proper training. From their first day on the job, they should understand the importance of food safety and receive training in several critical areas. (See *Exhibit 15c* on the next page.)

Exhibit 15c

Critical Food Safety Knowledge for Employees

Proper personal hygiene

- Maintaining health
- Personal cleanliness
- Proper work attire
- Hygienic practices (including handwashing)

Safe food preparation

- Time-temperature control
- Preventing cross-contamination
- Handling food safely during:
  - Preparation and cooking
  - Holding and cooling
  - Reheating and service

Proper cleaning and sanitizing

- Procedures for cleaning and sanitizing food-contact surfaces

Safe chemical handling

- Procedures for safely handling chemicals used in the establishment

Pest identification and prevention

Regardless of their position, employees need to be retrained periodically on these food safety practices. This can be accomplished by scheduling short retraining sessions, holding meetings to update staff on new procedures, or conducting motivational sessions that reinforce food safety practices.

Keep records of all food safety training conducted at your establishment. For legal reasons, it is important to document that employees have completed this training.

To be successful, a food safety training program requires these essential elements:

- **Clearly defined and measurable objectives.** Training objectives need to be identified before any training takes place. **Training objectives** are statements that describe what employees should be able to do after training has been completed. They should be clearly defined and measurable. For example, if you want employees to be able to wash their hands according to the five steps outlined in *ServSafe Essentials,* then a clearly defined training objective might state: **Employees will be able to wash their hands following the five steps for proper handwashing.**
- **Training that supports the objectives.** Training objectives serve as a guide for delivering content. The content should follow the objectives closely. Content that goes beyond the scope of your training objectives burdens employees with nonessential information. It also wastes time and money. Content that fails to meet objectives results in employees who are not prepared to handle food safely.
- **Evaluation to ensure the objectives have been achieved.** **Evaluation** tells you if training has provided employees with the knowledge and skills needed to handle food safely. To evaluate training, you must carefully compare employee performance to the training objectives. There are several ways in which employee performance can be measured. It is often measured through written or oral tests. (See *Exhibit 15d.*) It can also be measured by evaluating the employee while he or she performs a task required by the objective. In the handwashing example, the best way to measure whether employees can wash their hands properly might be to

Exhibit 15d

Evaluation

Written tests can be used to evaluate whether employees have the knowledge to handle food safely.

evaluate them while they wash their hands. During the evaluation, you would watch to see if they accurately followed the five steps for proper handwashing.

- **A work climate that reinforces training.** The work climate will have a dramatic impact on whether employees will do what they have been trained to do. If there are penalties for doing things the right way, employees may be less likely to follow through on the job. For example, employees may not change disposable gloves when necessary if they are continually disciplined for "wasting" gloves. They also will be less likely to follow the proper practices if they are not reinforced on the job. For example, a foodhandler may begin to wear bracelets and rings to work after she begins to see that coworkers are not penalized for doing so. Employees may also have a difficult time doing what they have been trained to do when the proper tools are not available. For example, if thermometers are broken or not available, an employee cannot spot-check the temperature of food during the receiving process.
- **Management support.** When it comes to training, never underestimate the importance of leading by example. Employees must see that the establishment's commitment to food safety comes from the top down. If managers show that commitment through behavior and attitude, employees are likely to follow. (See *Exhibit 15e.*)

Exhibit 15e

If managers show a commitment to food safety through behavior and attitude, employees are likely to follow.

DELIVERING TRAINING

Before any training actually occurs, you should start with the necessary planning. This includes choosing appropriate training delivery methods (covered later in this chapter); selecting training materials; and determining the best method for evaluating employee knowledge after the session has ended. Planning also includes making administrative decisions, such as where to hold the training and how to schedule it.

Once these decisions have been made, the next step is to create an agenda or training plan. A **training plan** is a very specific list of events that will take place during the training session. When developing a training plan, you should list:

- **Specific learning objective(s).**

- **Training tools needed for the session.** This may include books, audiovisual materials, and food safety equipment or props.
- **Specific training points that should be covered.** The training plan should also identify the amount of time that should be spent on each point.

A sample training plan for calibrating a thermometer has been included in *Exhibit 15f.*

Exhibit 15f

Sample Training Plan for Calibrating a Thermometer

TRAINING PLAN: Calibrating a Thermometer

Task
Calibrate a bimetallic stemmed thermometer.

Learning Objective
Given a bimetallic stemmed thermometer, an adjustable wrench, and ice water, the employee will be able to accurately calibrate the thermometer using the ice-point method.

Training Tools
Bimetallic stemmed thermometers, large containers, adjustable wrenches, ice, and water

Training Points	Time Frames
1 Tell the employees that thermometers must be calibrated before each shift. 2 Explain to the employees the three steps for calibrating a bimetallic stemmed thermometer. **Steps for calibrating a bimetallic stemmed thermometer:** **Step 1:** Fill a large container with crushed ice. Add tap water. **Step 2:** Put the thermometer stem into the ice water so the sensing area is completely submerged. **Step 3:** Hold the adjusting nut with a wrench and rotate the head of the thermometer until it reads 32°F (0°C).	5 minutes
3 Demonstrate the three steps for calibrating a bimetallic stemmed thermometer. 4 Ask the employees to explain the three steps for calibrating a bimetallic stemmed thermometer.	5 minutes
5 Have the trainees practice calibrating a bimetallic stemmed thermometer.	10 minutes

A training plan should include learning objectives, training tools, and a list of specific training points that will be covered.

Guidelines for Effective Training

There are several things you can do when delivering training to make it easier for your employees to learn:

- **Prepare for the presentation.** Make sure you are knowledgeable in all areas of food safety. Practice the presentation using your training plan and rehearse until it feels natural.
- **Help employees see "what's in it for me."** To get employees' attention, you must show them how the training will help them do their jobs faster, easier, or better.
- **Let employees practice and apply information.** As a general rule, one-third of the time spent training should be devoted to presenting content. The remaining two-thirds should be devoted to activities that allow employees to apply what they have learned and receive feedback. (See *Exhibit 15g.*) Application or practice without feedback will be ineffective. Feedback must be specific, immediate, and worded positively. It should also be given for both correct and incorrect performance.
- **Follow a logical sequence when presenting information.** Information should be presented step by step from start to finish.
- **Tie new knowledge to past knowledge when presenting information.** Build on employees' past knowledge and experiences.
- **Present information using the "zoom" principle.** Think in terms of a zoom lens on a camera. Begin by zooming out—explaining the big picture to employees. This is a high-level overview of the information you are presenting. Next, "zoom" in on some detail and then back out to the big picture before covering the next detail.
- **Present information in "chunks."** Break lessons into smaller, more manageable pieces of information. Allow employees to practice each chunk until they have mastered it. Do not teach the next chunk until they have mastered the previous one. Each piece of information should cover between five to nine major points.

Exhibit 15g

Practice

Two-thirds of the time spent training should be devoted to activities that allow employees to apply what they have learned and receive feedback.

- **Keep training sessions short.** The ideal length is probably twenty to thirty minutes. Longer sessions can be effective if they are well planned and incorporate a variety of learning activities.
- **Wrap up the training session properly.** Give an overview of what was covered during the session, highlighting important information. Employees should be able to answer questions about the material. They should also be able to properly demonstrate tasks they practiced during the training session. Remind employees why the information they learned is important.

Training Delivery Methods

There are a variety of **training delivery methods**—ways that training can be delivered. Some establishments opt for a more traditional approach, using lecture, demonstration, or role-play in either a one-on-one or group format. Others use technology to deliver training. No single method of delivery is best for all employees because each learns differently. Using several methods will result in more effective learning. The delivery methods you choose should allow employees to:

- Reflect on content and determine how it applies to their job
- Talk to each other about content
- See tasks being performed and perform the tasks themselves
- Hear instructions
- Read materials and take notes
- Reason through real-life situations

Exhibit 15h

One-on-One Training

One advantage of one-on-one training is that it takes into consideration the needs of individual employees.

One-on-One Training

Many establishments deliver food safety training to employees on a one-on-one basis. This method has many advantages and is often preferred because it:

- Takes into consideration the needs of individual employees (see *Exhibit 15h*)
- Can take place on the job, thus eliminating the need for a separate training location
- Enables the manager to monitor employee progress
- Allows for immediate feedback
- Offers the opportunity to apply information that has been learned

One-on-one training does have some disadvantages, however. Most important, its effectiveness depends on the ability of the person delivering the training. Therefore, the trainer must be selected very carefully. Many establishments certify trainers or validate that they have the appropriate skills before allowing them to train.

Group Training

If several employees require food safety training, group sessions might be a more practical way to deliver it. When training is delivered this way:

- It is more cost effective.
- Training is more uniform.
- You know precisely what employees have been taught.

As with one-on-one training, the effectiveness of group training depends upon the skill and ability of the trainer. It can also be a challenge to deliver training this way since employees may bring a mix of learning styles to the session. Additionally, group training often does not take into account the needs of the individual learner. Slower learners, less-skilled employees, or those with limited proficiency in English may not understand all the material presented.

Trainers must remember to keep employees involved during the training session. If involved, they will have a greater chance of retaining the information. Studies on training effectiveness show that trainees retain:

- Ten percent of what they read
- Twenty percent of what they hear
- Thirty percent of what they see
- Fifty percent of what they hear and see
- Seventy percent of what they say
- Ninety percent of what they say and do

Lecture

A lecture is a prepared oral presentation used to deliver content to a group of participants. Lectures are most effective when mixed with other presentation methods and media and are better received and accepted when the following techniques are used:

- Start with an interesting statement, observation, quotation, or question.
- Use relevant humor where appropriate.
- Use interesting and relevant examples, anecdotes, analogies, and statistics.
- Ask frequent questions to solicit audience participation.
- Use frequent small-group discussions and activities.
- Build in a review.

Demonstration

Many times you may be required to teach specific food safety tasks by demonstrating them to a person or group. Demonstrations will be most effective if you follow the "Tell/Show/Tell/Show" model. (See *Exhibit 15i* on the next page.)

Exhibit 15i

Demonstrating a Task Using the Tell/Show/Tell/Show Model

1 Tell them how to do it.

Explain the overall steps using written procedures, diagrams, forms, etc.

2 Show them how to do it.

Demonstrate the task slowly, so employees can see what is happening. Then repeat the demonstration at normal speed.

Emphasize key points as you demonstrate the task. Explain how each step fits into the task sequence.

3 Have them tell you how to do it.

Ask employees to explain each of the steps in sequence.

4 Have them show you how to do it.

Ask employees to demonstrate the task. Provide appropriate feedback, correcting errors as they occur.

Role-Play

In role-plays, trainees act out a situation to try out new skills or apply what has been learned. They usually are set up so one employee is confronted by another employee and must answer questions, handle problems, provide satisfaction, solve a complaint, etc. Different types of role-plays are suitable for different types of learning situations. For example, a role-play can be used to teach a manager how to work with a health inspector during an inspection. When using a role-play, you should:

- Keep the role-play simple.
- Provide employees with detailed instructions.
- Explain and model the situation before employees begin.

Exhibit 15j

Job Aids

Job aids can be used to train employees and provide a reference while back on the job.

Job Aids

Job aids can also be used to train employees. They include worksheets, checklists, flowcharts, written procedures, glossaries, diagrams, decision tables, etc. Posters that illustrate the steps for proper handwashing or washing dishes in a three-compartment sink are good examples. (See *Exhibit 15j*.) Job aids have the added benefit of serving as a reference for employees while on the job. They are particularly useful when:

- Consequences of making a mistake are severe.
- Safety is a concern.
- Tasks are performed infrequently.
- Tasks are complex.
- Sequence is critical when performing the task.

Training Videos and DVDs

Training videos and DVDs can be used to introduce information, reinforce information during the session, or review information at the end of the training session. They can be great tools if used properly. However, they will be much less effective if they are simply played for employees without your direct involvement. To use videos and DVDs effectively, you should:

- Familiarize yourself with the content.
- Prepare employees by explaining what they will learn and why it is important.
- Select stopping points to discuss concepts or practices that require emphasis.
- Ask questions afterward to reinforce content.

Exhibit 15k

Technology-Based Training

Web-based training and interactive CD-ROMs provide consistent training delivery and feedback.

Technology-Based Training

Web-based training, interactive CD-ROMs, and other technology-based training programs offer yet another way to deliver training. The benefits of technology-based training include:

- **Consistent delivery and feedback.** (See *Exhibit 15k.*) The training and feedback are delivered the same way every time.
- **Learner control.** Employees may have control over their learning path.
- **Interactive instruction.** Employees are able to interact, review, and explore the content.
- **Increased practice.** Employees are allowed to practice a skill until they are proficient.

- **Self-paced training.** Employees can progress at their own pace.
- **Training records are easily created and stored.** Technology-based training can automatically track an individual's training progress and exam scores. Accessing this information and generating the necessary reports is often very easy.
- **Training can be delivered anytime, anywhere.** Web-based training is available anytime and anywhere there is an Internet connection.
- **Reduced cost.** Technology-based training can reduce training costs by freeing up trainers and eliminating travel costs.
- **Training that engages and supports different learning styles.** Technology-based training can be very engaging since it often includes a variety of media, such as video, sound, animation, photos, etc. For this reason, it often works well for the visual and auditory learner.
- **Multilingual training.** The training can be delivered in more than one language.

Exhibit 15l

Games

Games can be used to practice principles learned in the training session and create excitement.

Games

Games can be used to create excitement and get employees' attention when the material is difficult or mundane. Games can also be used to practice previously learned principles. (See *Exhibit 15l.*) Games should be short, simple, and provide some type of reward. To use games effectively, you should:

- Explain how they relate to the information being presented.
- Explain the rules carefully.
- Have a practice round before actually playing the game.
- Make sure employees do not lose sight of the game's purpose while playing.
- Discuss the game after it has finished.

Case Studies

Case studies are a good way to let employees apply what they have learned. By this method, employees are presented with a fictional or real-world story that poses specific problems. They are then asked to apply what they have learned to solve them.

The trainer provides feedback on how employees dealt with the problems and discusses other ways they might have been handled. When conducting case studies:

- Provide clear instructions.
- Make sure employees see the case study's relevance.
- Make sure employees identify realistic solution(s).
- Carefully facilitate the discussion.

Selecting Training Materials

Training materials help participants learn and retain information, add interest, save time, and make the trainer's job easier. To be useful, they must be:

- **Accurate.** Materials must be factual, current, and complete. To ensure they are accurate, food safety training materials should be based on the latest in the science of food safety. Sources for information should include recognized authorities from associations, universities, or federal, state, and local health departments. To ensure that materials are relevant, they should also be based on best practices in the industry.

  Regulatory requirements or guidance documents can be used to deliver food safety training. Materials can also be developed internally, or purchased from professional or educational organizations, industry suppliers, or other authorities in foodservice training.

- **Appropriate.** Written materials must be matched to the reading-comprehension levels of employees. For employees with limited English proficiency, training materials in other languages should be used. (See *Exhibit 15m.*) In addition, materials should suit the abilities of the trainer. Limitations imposed by the training location will influence your choice of training materials. For example, in order to use technology-based tools, the proper equipment must be available.
- **Attractive.** When teaching subjects that generally do not have a wide appeal, it is critical that you hold employees' attention. To do this, create eye-catching training materials. Design the information to be memorable, and make sure the materials involve the employees by providing meaningful activities.

Exhibit 15m

Using Appropriate Training Materials

If employees are not proficient in English, use training materials translated into their native language.

TRAINING FOLLOW-UP

It is critical to follow up with employees after the training session has ended. By doing so, you can determine whether or not they can still do what they were trained to do. Proper follow-up can also help ensure training will have a lasting effect. You can follow up with employees by:

- Giving periodic quizzes to test their knowledge of the learned material
- Periodically asking them to demonstrate the new tasks they have learned
- Comparing on-the-job performance to company policy

Follow-up should occur within thirty days of the training session. Research suggests that after thirty days, there is less of a chance that employees will retain the information learned.

Exhibit 15n

Food Safety Certification

Food safety certification demonstrates that a person comprehends basic food safety principles.

FOOD SAFETY CERTIFICATION

The National Restaurant Association and federal and state regulatory officials recommend food safety training and certification, particularly for managers and supervisors. Certification demonstrates that a person comprehends basic food safety principles and recommended food safety practices that prevent foodborne illness.

Manager certification is already a requirement in some states. In other states, some cities and counties require certification, although the state as a whole does not require it. (See *Exhibit 15n.*) The National Restaurant Association Educational Foundation Web site provides a jurisdictional summary of training and certification requirements for the entire country.

Regardless of state regulations, many proactive establishments and managers have made a strong commitment to food safety training. You should make the same commitment to training, as well as certification even if your city or state does not require it. As a conscientious restaurant or foodservice manager, you should train your employees, monitor their practices, and make food safety a part of everyone's job description. In this way, you will be able to ensure that the food you serve is safe.

SUMMARY

As a manager, it is your responsibility to ensure that employees have the knowledge and skills needed to handle food safely in your establishment. Some of the benefits can include preventing a foodborne-illness outbreak, avoiding the loss of revenue and reputation following an outbreak, improving employee morale, and increasing customer satisfaction.

Your first task is to assess the training needs in your establishment. A training need is a gap between what employees are required to know to perform their jobs and what they actually know. To identify food safety training needs, you can test your employees' knowledge, observe their performance, or survey them to identify areas of weakness.

All employees require general food safety knowledge. Other knowledge will be specific to the tasks performed on the job. From their first day, new employees should receive training on the importance of food safety, proper personal hygiene, and safe food preparation practices. They should also receive training on proper cleaning and sanitizing practices, safe chemical handling, and pest prevention.

Regardless of their position, employees need to be retrained periodically. This can be accomplished by scheduling short retraining sessions, holding meetings to update staff on new procedures, or conducting motivational sessions that reinforce food safety practices. Keep records of all training conducted at your establishment. For legal reasons, it is important to document that employees have completed food safety training.

A successful food safety training program will be led by clearly defined and measurable objectives. Training must support these objectives, and there must be methods for evaluating the training to ensure objectives have been met. True success cannot be achieved unless the work climate supports the training, and it has management support.

Before any training actually occurs, you should begin with the necessary planning. This includes choosing appropriate training delivery methods, selecting training materials, and determining the best method for evaluating employee knowledge after the session. Planning also includes making administrative decisions such as where to hold the training, and how to schedule it. Once these decisions have been made, the next step is to create an agenda or training plan.

Training can be delivered using a variety of methods. Some establishments use technology to deliver it. Others opt for a more traditional approach, including lecture, demonstration, or role-play methods—either one-on-one or in a group training format. No single method of delivery is best for all employees because each learns differently. Using several methods will result in more effective learning.

Following up with employees within thirty days of the training session is critical in determining if employees can still do what they were trained to do. Follow up by giving periodic quizzes, asking employees to demonstrate the new tasks they learned, and by comparing on-the-job performance to company policy.

The National Restaurant Association and federal and state regulatory officials recommend food safety training and certification, particularly for managers. Certification shows that a person comprehends basic food safety principles and recommended food safety practices that prevent foodborne illnesses.

Apply Your Knowledge

Use these questions to review the concepts presented in this chapter.

Discussion Questions

1. What is the "zoom" principle?
2. What is the "Tell/Show/Tell/Show" method, and when might it be used?
3. How can an establishment determine its food safety training needs?
4. What is an objective? Give an example of a good objective.
5. What are some methods that can be used to deliver training?

For answers, please turn to the Answer Key.

Apply Your Knowledge

Use these questions to test your knowledge of the concepts presented in this chapter.

Multiple-Choice Study Questions

1. Which is true regarding food safety training?
   A. The ideal length for a training session is one to two hours.
   B. Training records should be used to document training.
   C. Employees only require food safety knowledge that is specific to their job tasks.
   D. Further training is unnecessary if employees received training upon being hired.
2. New employees should receive training on
   A. HACCP.
   B. crisis management.
   C. active managerial control.
   D. pest identification and prevention.
3. Which is true about group training?
   A. It ensures that training is more uniform.
   B. It allows employees to learn at their own pace.
   C. It allows employees to choose their own learning path.
   D. It is better able to meet the needs of the individual learner.
4. When demonstrating a task, you should do all of these *except*
   A. explain task steps before demonstrating them.
   B. demonstrate the task and then explain what was demonstrated.
   C. have the employee explain the steps before demonstrating them.
   D. demonstrate the task slowly the first time and then again at normal speed.
5. Which is true when delivering food safety training?
   A. The lecture method should always be used to deliver it.
   B. A Web-based method should always be used to deliver it.
   C. To prevent confusion, do not use more than one method of delivery.
   D. Use several delivery methods because each employee learns differently.

For answers, please turn to the Answer Key.

ADDITIONAL RESOURCES

Articles and Texts

2004. Food Safety—Getting Food Safety Knowledge Out to Your Front Line Employees. *Restaurant Hospitality.* 88 (6): 110.

Jenkins-McLean, Terri, Chris Skilton, and Clarence Sellars. 2004. Engaging Food Service Workers in Behavioral-Change Partnerships. *Journal of Environmental Health.* 66 (9): 15.

Web Sites

The American Society for Training and Development

www.astd.org/astd

The American Society for Training and Development (ASTD) is dedicated to workplace learning and performance. ASTD is the voice of the training profession and seeks to continually advance the knowledge and techniques to convert learning and capability into performance and practice. This Web site lists resources, conferences, and educational-development opportunities for training professionals.

Conference for Food Protection

www.foodprotect.org

The Conference for Food Protection (CFP) is a nonprofit organization that provides a forum for regulators, industry professionals, academia, professional organizations, and consumers to identify problems, formulate recommendations, and develop and implement practices that ensure food safety. Though the conference has no formal regulatory authority, it is an organization that influences model laws and regulations among all government agencies. Visit this Web site for information regarding previously recommended changes to the *FDA Food Code,* standards for permanent outdoor cooking facilities, and other guidance documents, along with how to participate in the conference.

The Council of Hotel and Restaurant Trainers

www.chart.org

The Council of Hotel and Restaurant Trainers (CHART) is an organization dedicated to training in the hospitality industry. CHART's objective is to provide hospitality trainers access to the industry's training professionals and solutions. Visit this Web site for upcoming conferences and resources on training and professional development.

The International Council on Hotel, Restaurant, and Institutional Education

www.chrie.org

The International Council on Hotel, Restaurant, and Institutional Education (I-CHRIE) is the global advocate of hospitality and tourism education for schools, colleges, and universities offering programs in hotel and restaurant management, foodservice management, and culinary arts. This Web site lists publications available that promote the hotel and restaurant industry.

International Food Safety Council

www.nraef.org/ifsc/ifsc_about.asp

In 1993, the National Restaurant Association Educational Foundation recognized the need for food safety awareness and created the International Food Safety Council as its food safety awareness initiative. The council's mission is to heighten the awareness of the importance of food safety education throughout the restaurant and foodservice industry through its educational programs, publications, and awareness campaigns, such as National Food Safety Education Month held each year in September.

National Restaurant Association Educational Foundation

www.nraef.org

The National Restaurant Association Educational Foundation (NRAEF) is a nonprofit organization dedicated to fulfilling the educational mission of the National Restaurant Association. Focusing on three key strategies of risk management, recruitment, and retention, the NRAEF is the premier provider of educational resources, materials, and programs that address attracting, developing, and retaining the industry's workforce. The purchase of any NRAEF product helps fund outreach and scholarship activities on a national and state level.

Documents and Other Resources

2005 *FDA Food Code*

www.cfsan.fda.gov/~dms/fc05-toc.html

The Food and Drug Administration (FDA) publishes the *FDA Food Code,* a scientifically sound technical and legal document that serves as a model for regulating the retail and foodservice industry at the federal, state, and local level. Updates to the code are issued on the odd-numbered years. The *FDA Food Code* provides a system of safeguards designed to minimize foodborne illness; ensure employee health, food protection manager knowledge, safe food, nontoxic and cleanable equipment, and appropriate sanitation of the food establishment. It is used as the basis for information in this textbook.

Notes

Answer Key

Chapter 1
Providing Safe Food

Page	Activity
1-2	**Test Your Food Safety Knowledge**

1. True 2. True 3. False 4. True 5. False

Page	Activity
1-14	**Discussion Questions**

1. Potentially hazardous food typically:
   - Contains moisture
   - Contains protein
   - Has a neutral or slightly acidic pH
   - Requires time-temperature control to prevent the growth of microorganisms and the production of toxins
2. The potential costs of a foodborne-illness outbreak include the following:
   - Loss of customers and sales
   - Loss of prestige and reputation
   - Embarrassment
   - Lawsuits resulting in legal fees
   - Increased insurance premiums
   - Lowered employee morale
   - Employee absenteeism
   - Need for retraining employees
   - Food costs that result from discarding food supplies that may or may not be contaminated
   - Fees for testing food supplies and employees

3. The elderly are at higher risk for contracting a foodborne illness because their immune systems and resistance to illness may have weakened with age. Also, as people age, their senses of smell and taste are diminished, so they may be less likely to detect "off" odors or tastes, which indicate that food may be spoiled.
4. The three major types of hazards to food safety are biological hazards, chemical hazards, and physical hazards.

1-15 Multiple-Choice Study Questions

1. B 2. B 3. B 4. B 5. D

Chapter 2

The Microworld

Page Activity

2-2 Test Your Food Safety Knowledge

1. True 2. False 3. True 4. True 5. True

2-40 A Case in Point

1. The illness was caused by bacteria.
2. *Bacillus cereus* was the microorganism responsible for the outbreak.
3. *Bacillus cereus* is an intoxication, which is why the children became ill so quickly.

2-41 Discussion Questions

1. If potentially hazardous food is left in the temperature danger zone for four hours or more, bacteria can grow to levels high enough to make someone ill.
2. A foodborne infection results when a person eats food containing pathogens, which then grow in the intestine and cause illness. Typically, symptoms of a foodborne infection do not appear immediately. A foodborne intoxication results when a person eats food containing toxins that cause illness. The toxin may have been produced by pathogens found on the food or may be the result of a chemical contamination. The toxin might also be a natural part of a plant or animal consumed. Typically, symptoms of foodborne intoxication appear quickly, within a few hours. A foodborne toxin-mediated infection results when a person eats food containing pathogens, which then produce illness-causing toxins in the intestine.
3. An outbreak of salmonellosis can be prevented by the following:
   - Cooking raw beef, poultry, and eggs to their required minimum internal temperature
4. The following are basic characteristics of viruses:
   - Some may survive freezing and cooking.
   - They can be transmitted from person to person, from people to food, and from people to food-contact surfaces.

- They usually contaminate food through a foodhandler's improper personal hygiene.
- They can contaminate both food and water supplies.

5. The two FAT TOM requirements for growth that are easiest for an establishment to control are time and temperature.

   To control time:

   - Limit the amount of time food spends in the temperature danger zone.

   To control temperature:

   - Cook food to the proper temperature.
   - Freeze food properly.
   - Refrigerate food at 41°F (5°C) or lower.

2-42 Multiple-Choice Study Questions

1. C	3. C	5. C	7. D	9. C
2. D	4. C	6. C	8. C	10. A

Chapter 3

Contamination, Food Allergens, and Foodborne Illness

Page	Activity
3-2	**Test Your Food Safety Knowledge**

1. False 2. False 3. True 4. False 5. True

3-20 A Case in Point

1. Two factors contributed to the outbreak.
   - When the mahi-mahi (a scombroid species of fish) was time-temperature abused, the bacteria associated with the fish produced the toxin histamine.
   - Since cooking does not destroy this toxin, the consumption of the fish resulted in a scombroid poisoning (intoxication).

3-21 Discussion Questions

1. To guard against a seafood-specific foodborne illness, establishments should do the following:
   - Purchase seafood from approved, reputable suppliers.
   - Refuse fish that show evidence of thawing and refreezing.
   - Refuse fresh fish that have not been received at 41°F (5°C) or lower.
   - Prevent time-temperature abuse during storage and preparation.

2. Toxic-metal poisoning can occur when food is stored in or prepared with nonfood-grade vessels and utensils, such as pewter pitchers, copper saucepans, or galvanized buckets. Improperly installed carbonated-beverage dispensers can cause contamination if carbonated water is allowed to flow back into copper supply lines.
3. To keep chemicals from contaminating food:
   - Follow directions supplied by the manufacturer when using chemicals.
   - Exercise caution when using chemicals during operating hours.
   - Store chemicals in a separate storage area in their original containers, away from food, utensils, and equipment used for food.
   - If chemicals are transferred to smaller containers or spray bottles, label each container appropriately.
   - Never apply pesticides yourself; always use a licensed pest control operator (PCO).
4. To keep customers with food allergies safe:
   - You and your employees must be aware of the most common food allergens.
   - You and your employees should be able to inform customers of menu items that contain potential allergens. You should also be able to describe each menu item.
   - If you or your employees do not know if a food item is allergen free, say so.
   - When preparing food for a customer with allergies, ensure that the food does not come in contact with the ingredient to which the customer is allergic. All cookware, utensils, and tableware must be allergen free to prevent contamination.

3-22 Multiple-Choice Study Questions

1. B	3. C	5. A	7. D
2. D	4. C	6. D	8. B

Chapter 4
The Safe Foodhandler

Page Activity

4-2 Test Your Food Safety Knowledge

1. False 2. True 3. True 4. True 5. False

4-16 A Case in Point 1

1. Yes, this situation represents a threat to food safety.

2. Chris did several things correctly.
   - She properly washed her hands and put on a new pair of single-use gloves at the beginning of her shift.
   - She stepped away from the food-preparation area to sneeze.
3. Chris did several things wrong.
   - After using the tissue, Chris should have taken off her gloves, washed her hands, and put on a new pair of gloves, since she may carry pathogens such as *Staphylococcus aureus* in her nose and mouth.
   - She should have informed her manager of her condition.
   - While some jurisdictions allow foodhandlers to drink from a covered container with a straw, Chris was drinking from an open container in a food-preparation area.
   - Chris should have placed her medication inside a covered, leak-proof container that was clearly labeled before placing it inside the walk-in refrigerator.

4-17 A Case in Point 2

1. The employee should have notified her supervisor immediately when she became ill with diarrhea. Her supervisor should not have allowed her to work while ill.

4-18 Discussion Questions

1. Employees must meet the following work attire requirements:
   - Wear a clean hat or other hair restraint.
   - Wear clean clothing daily.
   - Remove aprons when leaving food-preparation areas.
   - Wear clean, closed-toe shoes with a sensible, nonslip sole.
   - Remove jewelry from hands and arms prior to preparing or serving food or while around food-preparation areas.
2. The following personal behaviors can contaminate food:
   - Nose picking
   - Rubbing an ear
   - Scratching the scalp
   - Touching a pimple or an open sore
   - Running fingers through the hair
3. All cuts and infected wounds should be covered with a clean, dry bandage. Disposable gloves or finger cots should be worn over bandaged cuts on hands.
4. Foodhandlers must follow these procedures when using gloves:
   - Gloves must never be used in place of handwashing.
   - Hands must be washed before putting on gloves and when changing to a new pair.

- Gloves used to handle food are for single use only and should never be washed and reused.
- Gloves should be removed by grasping them at the cuff and peeling them off inside out over the fingers while avoiding contact with the palm and fingers.
- Gloves should be changed:
  - As soon as they become soiled or torn
  - Before beginning a different task
  - At least every four hours during continual use, and more often when necessary
  - After handling raw meat and before handling cooked or ready-to-eat food

5. Managers must exclude from the establishment foodhandlers who have been diagnosed with a foodborne illness caused by one of the following pathogens:
   - *Salmonella* Typhi
   - Shiga toxin-producing *E. coli*
   - *Shigella* spp.
   - Hepatitis A virus
   - Norovirus

   Managers must exclude foodhandlers from the establishment if they are vomiting, have diarrhea, or are jaundiced.

   Managers must restrict foodhandlers from working with or around food if they have a sore throat with fever.

4-19 Multiple-Choice Study Questions

1. D	4. D	7. B	10. A	13. D
2. D	5. B	8. A	11. B	14. D
3. C	6. C	9. A	12. D	15. C

Chapter 5

The Flow of Food: An Introduction

Page Activity

5-2 Test Your Food Safety Knowledge

1. True 2. False 3. False 4. False 5. False

5-12 Discussion Questions

1. Food can be time-temperature abused when it is not:
   - Cooked to its required minimum internal temperature
   - Cooled properly
   - Reheated properly
   - Held at the proper temperature

2. Cross-contamination can be prevented in an establishment by:
   - Assigning specific equipment to each type of food product prepared
   - Cleaning and sanitizing all work surfaces, equipment, and utensils after each task
   - Preparing raw meat, fish, and poultry and ready-to-eat food at different times
   - Purchasing ingredients that require minimal preparation
3. The steps for calibrating a thermometer using the ice-point method are:
   1. Fill a large container with crushed ice. Add clean tap water until the container is full. Stir the mixture well.
   2. Put the thermometer stem or probe into the ice water so the sensing area is completely submerged. Wait thirty seconds, or until the indicator stops moving. Do not let the stem or probe touch the sides or bottom of the container. Keep the stem or probe in the ice water.
   3. Hold the calibration nut securely with a wrench or other tool and rotate the head of the thermometer until it reads 32°F (0°C). On some thermocouples or thermistors, it may be possible to press a reset button to adjust the readout.

5-13 Multiple-Choice Study Questions

1. A	3. C	5. D	7. D
2. C	4. B	6. C	

Chapter 6

The Flow of Food: Purchasing and Receiving

Page	Activity

6-2 Test Your Food Safety Knowledge

1. True 2. False 3. True 4. True 5. True

6-22 A Case in Point 1

1. Betty should have asked the delivery driver to come back later. Products must be delivered when employees have adequate time to inspect them. Deliveries must be inspected immediately. By putting the deliveries away without inspecting them, Sunnydale missed an important opportunity to identify food that:
   - Was not delivered at the proper temperature
   - Was damaged or mishandled
   - Had been thawed and refrozen
   - Had expired code dates
   - Showed signs of an insect infestation

6-23 A Case in Point 2

1. John had good intentions and did most things correctly. However, he did make mistakes that could result in a foodborne illness. John should have done the following:
   - Notified the kitchen manager that the shipment had arrived.
   - Made sure that the bimetallic stemmed thermometer he took from the kitchen had been properly calibrated, as well as cleaned and sanitized before using it.
   - Cleaned and sanitized the thermometer after checking the temperature of each product. He should not have wiped the thermometer on his apron.
   - Inserted the thermometer stem into the middle of the bucket of live oysters between the shellfish for an ambient reading instead of trying to judge how cold they were with his hand. The temperature of the oysters should have been 45°F (7°C) or lower.
   - Marked the delivery date on the shellstock identification tags attached to the shipment of oysters and mussels.

6-24 Discussion Questions

1. General guidelines for receiving food safely include the following:
   - Training employees to inspect deliveries properly
   - Planning ahead for shipments
   - Scheduling deliveries for off-peak hours
   - Planning a back-up menu in case you have to return some food items
   - Receiving only one delivery at a time
   - Inspecting deliveries immediately
   - Correcting mistakes immediately
   - Putting products away as quickly as possible
   - Keeping the receiving area clean and well lit to discourage pests
2. When checking the temperature of:
   - Poultry: Insert the thermometer stem or probe into the thickest part of the product. The temperature should be 41°F (5°C) or lower.
   - Bulk milk: Fold the bag or pouch around the thermometer stem or probe, being careful not to puncture the bag. The temperature should be 41°F (5°C) or lower unless otherwise specified.
3. Fresh poultry should be rejected for all of the following conditions:
   - Purple or green discoloration around the neck
   - Dark wing tips (red wing tips are acceptable)
   - Stickiness under wings or around joints
   - Abnormal unpleasant odor

4. Cans should be rejected if they have any of the following damage:
   - Swollen ends
   - Leaks and flawed seals
   - Rust
   - Dents

6-25 **Multiple-Choice Study Questions**

1. D	4. C	7. A	10. C
2. C	5. B	8. D	11. B
3. D	6. B	9. B	12. B

Chapter 7
The Flow of Food: Storage

Page	Activity

7-2 **Test Your Food Safety Knowledge**

1. False 2. False 3. True 4. True 5. False

7-15 **A Case in Point 1**

1. Pete and Angie made several errors.
   - Pete should not have placed the hot stockpot of soup into the refrigerator to cool. Not only is this an unsafe way to cool hot, potentially hazardous food, but the hot pot could warm up the interior of the refrigerator enough to put other stored food in the temperature danger zone. Pete should have cooled the soup properly (see Chapter 8) before storing it in the refrigerator.
   - Angie should not have placed the uncovered pan of raw chicken on the top shelf in the refrigerator. Nor should she have stored the carrot cake below the raw chicken breasts, since juices could have dripped onto the cake, contaminating it. Cooked or ready-to-eat food must be stored above raw meat, poultry, and fish if these items are stored in the same unit. Ideally, raw product like fresh meat, fish, and poultry should be stored in a separate unit from cooked and ready-to-eat food. All stored food should also be wrapped properly to prevent cross-contamination.
2. Unfortunately, their mistakes could affect all of the food stored in the refrigerator.

7-16 **A Case in Point 2**

1. The following storage errors occurred:
   - Alyce placed the case of sour cream into an already overloaded refrigerator.
   - A hot stockpot of soup was stored in the walk-in refrigerator. Hot food should never be placed in a refrigerator.
   - Mary was lining the shelving with aluminum foil. This can restrict airflow in the unit.

- The temperature in the dry-storage room was 85°F (29°C), which is too warm. Dry-storage areas should be between 50°F and 70°F (10°C and 21°C).

7-17 Discussion Questions

1. The recommended top-to-bottom order for storing the items in the same refrigerator is:
   - Raw trout
   - Uncooked beef roast
   - Raw ground beef
   - Raw chicken
2. If food is removed from its original packaging, it should be placed in a clean and sanitized container and covered. The new container must be labeled with the name of the food being stored and its original use-by or expiration date.
3. Frozen dairy products such as ice cream and frozen yogurt can be stored at 6°F to 10°F (–14°C to –12°C).
4. The first-in, first out (FIFO) method is commonly used to ensure that refrigerated, frozen, and dry products are properly rotated during storage. By this method, a product's use-by or expiration date is first identified. The products are then stored to ensure that the oldest are used first. One way to do this is to train employees to store products with the earliest use-by or expiration dates in front of products with later dates. Once shelved, those stored in front are used first.

7-18 Multiple-Choice Study Questions

1. D	3. A	5. D	7. C	9. C
2. C	4. B	6. A	8. C	10. B

Chapter 8

The Flow of Food: Preparation

Page Activity

8-2 Test Your Food Safety Knowledge

1. False 2. False 3. False 4. False 5. False

8-21 A Case in Point 1

1. Here is what John did wrong:
   - He failed to wash his hands before starting work.
   - He failed to thaw the shrimp properly. When food is thawed under running water, the temperature of the water should be 70°F (21°C) or lower.
   - He took out more whole fish from the walk-in refrigerator than he could prepare in a short period of time, unnecessarily subjecting the fish to time-temperature abuse.

- He failed to clean and sanitize the boning knife, cutting board, and worktable properly after cleaning and filleting the fish. Microorganisms that may have been present on the fish could have been transferred to the shrimp that John prepared with the contaminated knife and cutting board.

8-22 A Case in Point 2

1. Here is what Angie did wrong:
   - She cooled the leftover chicken breasts improperly. Food should never be left out to cool at room temperature. Angie could have divided the chicken into smaller portions and refrigerated them or used a blast chiller.
   - She unnecessarily subjected the chicken-salad ingredients to time-temperature abuse by leaving them out on the counter while she performed other duties. Angie should have left the ingredients in the refrigerator until she was ready to prepare the salad.
   - Angie undercooked the shell eggs she was using for scrambled eggs. If shell eggs are used in operations that serve a high-risk population, such as a nursing home, the eggs must be cooked all the way through.
   - She failed to handle the large number of pooled eggs properly. She left a bowl of eggs near a warm stove in the temperature danger zone. She should have pooled a smaller number of eggs and kept them in an ice bath away from the stove. Also, she failed to check the temperature of the eggs prior to placing them on the steam table. Eggs that will be cooked and held for later service must be cooked to at least 155°F (68°C) for fifteen seconds.
   - She failed to wash her hands before starting to prepare food and did not wash her hands between foodhandling tasks. If there were any microorganisms on her hands, she would have transferred them to the food and food-contact surfaces she handled. Also, Angie wiped her hands on her apron. Wiping hands is not an adequate substitute for proper handwashing.

8-23 Discussion Questions

1. The required minimum internal cooking temperatures are:
   - Poultry: 165°F (74°C) for fifteen seconds
   - Fish: 145°F (63°C) for fifteen seconds
   - Pork: 145°F (63°C) for fifteen seconds (roasts for four minutes)
   - Ground beef: 155°F (68°C) for fifteen seconds
2. The four proper methods for thawing food are:
   - Thaw it in the refrigerator at 41°F (5°C) or lower.
   - Submerge it under running, potable water at a temperature of 70°F (21°C) or lower.
   - Thaw it in a microwave oven if it will be cooked immediately afterward.

- Thaw it as part of the cooking process as long as the product reaches the required minimum internal cooking temperature.

3. There are a number of methods that can be used to cool food, including:
   - Using ice-water baths
   - Stirring food with an ice paddle
   - Using a blast chiller or tumble chiller
   - Adding ice or cold water as an ingredient
   - Using properly equipped, steam-jacketed kettles by running cold water through the jacket
4. The steps for properly cooking food in a microwave include:
   1. Cover food to prevent the surface from drying out.
   2. Rotate or stir food halfway through the cooking process to distribute heat more evenly.
   3. Let food stand for at least two minutes after cooking to let product temperature equalize.

8-24 Multiple-Choice Study Questions

1. B	3. D	5. D	7. A	9. D
2. C	4. D	6. C	8. C	10. B

Chapter 9

The Flow of Food: Service

Page Activity

9-2 Test Your Food Safety Knowledge

1. True 2. False 3. False 4. True 5. True

9-15 A Case in Point 1

1. Here is what Jill did wrong:
   - She packed the deliveries in cardboard boxes instead of rigid, insulated carriers.
   - She used the wrong utensil to fill the soup baine.
   - She failed to make sure the internal temperature of the food on the steam table was checked at least every four hours. This would have alerted her to the fact that the steam table was not maintaining the proper temperature and that the casserole was in the temperature danger zone.
2. Jill should have done the following:
   - She should have kept the delivery meals in a hot-holding cabinet or left the food in a steam table until suitable containers were found or the driver arrived.

- She should have used a long-handled ladle, which would have kept her hands away from the soup, preventing possible contamination as she ladled it out.
- She should have discarded the casserole and any other food that was not at the right temperature, since she did not know how long the food was in the temperature danger zone.
- She should have made sure that an employee was assigned to monitor the food bar to ensure that customers, such as the children, followed proper etiquette.

9-16 A Case in Point 2

1. Megan made the following errors:
   - She wore a watch and rings while serving food.
   - She tasted the food on the customer's plate.
   - She failed to wash her hands after clearing the dirty dishes from the table.
   - She failed to clean the table properly after bussing it. Megan should not have wiped the table with the cloth she kept in her apron.
   - She improperly scooped ice into glassware. Megan should not have used the glass itself to retrieve ice from the bin. Using a glass this way could cause it to chip or break in the ice.
   - She re-served bread and butter that had been previously served to a customer. Uneaten bread or rolls should never be re-served to other customers.
   - She failed to wash her hands after scratching a sore. By scratching it and not washing her hands afterward, she could easily have contaminated everything else she touched.
2. Here is what Megan should have done before beginning her shift:
   - She should have removed her jewelry before starting work.
   - She should have washed her hands after clearing the table and before she touched the water glasses.
   - When cleaning tables between guest seatings, Megan should wipe up spills with a disposable, dry cloth. The table should then be cleaned with a clean cloth stored in a sanitizer solution (see Chapter 11).
   - She should have used tongs or an ice scoop to get ice.
   - She should have served a fresh basket of bread.
   - She should have washed her hands immediately after scratching the sore.

9-17 Discussion Questions

1. The following practices can minimize contamination in self-service areas:
   - Protect food on display with sneeze guards or food shields.

- Assign an employee to replenish food-bar items and to hand out fresh plates and silverware for return visits.
- Identify all food items on display.
- Keep raw meat, fish, and poultry separate from cooked and ready-to-eat food.

2. When transporting food, protect it from contamination and time-temperature abuse during transport by doing the following:
   - Use rigid, insulated storage containers capable of maintaining food temperatures of 135°F (57°C) or higher or 41°F (5°C) or lower.
   - Clean and sanitize the inside of delivery vehicles.
   - Check internal food temperatures regularly.
   - Make sure employees practice good personal hygiene.
   - Keep raw and ready-to-eat products separate during delivery and storage.
3. Ready-to-eat, potentially hazardous food can be displayed or held for consumption without temperature control (up to six hours for cold food and up to four hours for hot food) under the following conditions:

   Cold Food
   - It was held at 41°F (5°C) or lower prior to removing it from refrigeration.
   - It does not exceed 70°F (21°C) during the six hours.
   - It has a label that specifies both the time it was removed from refrigeration and the time it must be thrown out.
   - It is sold, served, or discarded within six hours.

   Hot Food
   - It was held at 135°F (57°C) or higher prior to removing it from temperature control.
   - It has a label that specifies when the item must be thrown out.
   - It is sold, served, or discarded within four hours.
4. When serving food off site, you must protect it from contamination and time-temperature abuse. In addition:
   - Make sure employees practice good personal hygiene.
   - Ensure there is safe drinking water for cooking, dishwashing, and handwashing.
   - Check internal food temperatures regularly.
   - Label food with storage, shelf-life, and reheating instructions for employees at off-site locations.
   - Provide food safety guidelines for consumers.
   - Ensure there is adequate power for holding and cooking equipment.
   - Provide adequate garbage storage and disposal.

Page	Activity				
9-18	**Multiple-Choice Study Questions**				
	1. A	3. C	5. D	7. A	
	2. B	4. B	6. A		

Chapter 10

Food Safety Management Systems

Page	Activity				
10-2	**Test Your Food Safety Knowledge**				
	1. True	2. True	3. True	4. True	5. True

10-25 Discussion Questions

1. In order for a food safety management system to be effective, the following programs must be in place:
   - Personal hygiene program
   - Supplier selection and specification programs
   - Sanitation and pest control programs
   - Facility design and equipment maintenance programs
   - Food safety training programs
2. The five foodborne illness risk factors identified by the Centers for Disease Control and Prevention (CDC) are:
   - Purchasing food from unsafe sources
   - Failing to cook food adequately
   - Holding food at improper temperatures
   - Using contaminated equipment
   - Poor personal hygiene
3. The four steps that should be taken when using active managerial control are:
   - Consider the five risk factors as they apply throughout the flow of food, and identify any issues that could impact food safety.
   - Develop policies and procedures that address the issues that were identified.
   - Regularly monitor the policies and procedures that have been developed.
   - Verify that the policies and procedures you have established are actually controlling the risk factors.
4. The seven HACCP principles are:
   - Principle One: Conduct a hazard analysis.
   - Principle Two: Determine critical control points (CCPs).
   - Principle Three: Establish critical limits.
   - Principle Four: Establish monitoring procedures.
   - Principle Five: Identify corrective actions.

- Principle Six: Verify that the system works.
- Principle Seven: Establish procedures for record keeping and documentation.

5. An establishment is required to have a HACCP plan in place if they perform the following activities:
   - Smoke or cure food as a method of food preservation
   - Use food additives as a method of food preservation
   - Package food using a reduced-oxygen packaging method
   - Offer live, molluscan shellfish from a display tank
   - Custom-process animals for personal use
   - Package unpasteurized juice for sale to the consumer without a warning label
   - Sprout beans or seeds

10-26 Multiple-Choice Study Questions

1. B 3. B 5. A 7. C
2. B 4. C 6. B 8. D

Chapter 11

Sanitary Facilities and Equipment

Page Activity

11-2 Test Your Food Safety Knowledge

1. True 2. False 3. True 4. False 5. True

11-31 A Case in Point

1. The people who had iced drinks at the bar became ill from the chemical drain cleaner used to clean the glasswasher drain. It is likely that the grease trap on the kitchen sink was blocked with grease. This caused wastewater (and drain cleaner) to back up into the glasswasher drain and the drain for the icemaker since this equipment shared the same piping system. Carlos had failed to install an air gap between the icemaker and the floor drain, which resulted in wastewater backing up into the icemaker, contaminating the ice. Carlos should not have tried to install and maintain the plumbing in the establishment himself. Establishments must rely on professionals to do this.
2. Aside from handling a potential lawsuit, Carlos will require the services of a professional to fix the problem and will have to clean and sanitize the icemaker storage bin thoroughly before it can be used again.

11-32 Discussion Questions

1. One of the most important factors to consider when selecting flooring for food-preparation areas is the material's porosity, or the extent to which it can become saturated with liquids. The *FDA Food Code* recommends the use of nonporous (nonabsorbent) flooring in food-preparation areas.
2. A backup of raw sewage in an establishment is cause for immediate closure, correction of the problem, and thorough cleaning.
3. To prevent backflow in an establishment:
   - Install vacuum breakers or other approved backflow-prevention devices on threaded faucets and connections between two piping systems.
   - Install air gaps wherever practical and possible. *This is the only completely reliable method.* An air gap is a space used to separate a water-supply outlet from any potentially contaminated source.
4. Sources of potable water include:
   - Approved public water mains
   - Private water sources regularly maintained and tested
   - Bottled drinking water
   - Closed portable water containers filled with potable water
   - On-site water storage tanks
   - Properly maintained water transport vehicles
   - If an establishment uses a private water supply such as a well rather than an approved public source, it should check with the local regulatory agency for information on inspections, testing, and other requirements. Generally, nonpublic water systems should be tested at least annually, and the report kept on file in the establishment.
5. The requirements of a handwashing station include:
   - Hot and cold running water supplied through a mixing valve or combination faucet at a temperature of at least 100°F (38°C).
   - Soap in liquid, bar, or powder form.
   - A means to dry hands is required. Most local codes require establishments to supply disposable paper towels in handwashing stations.
   - A waste container is required if disposable paper towels are provided.
   - Signage must indicate employees are required to wash their hands before returning to work.
   - Handwashing stations are required in food-preparation areas, service areas, dishwashing areas, and restrooms.

6. When installing stationary equipment:
   - It must be mounted on legs, at least six inches (fifteen centimeters) off the floor, or it must be sealed to a masonry base.

11-33 Multiple-Choice Study Questions

1. A	4. B	7. C	10. B
2. D	5. C	8. D	11. C
3. D	6. B	9. A	12. A

Chapter 12
Cleaning and Sanitizing

Page Activity

12-2 Test Your Food Safety Knowledge

1. False 2. False 3. False 4. False 5. False

12-29 A Case in Point 1

1. Schedules do not clean dining rooms, people do. Tim had made his schedule too rigid and failed to monitor it. He should have made the necessary adjustments to take late-night banquets into account.
2. Tim should enlist the cooperation of Norman, the night shift manager, to make sure the cleaning program is followed. Norman should bring problems to Tim's attention. Shift supervisors, employees, and managers must communicate effectively with one another.
3. Shifts should be scheduled to include all cleaning duties. If the banquet room closes at 1:00 a.m., the shift should extend beyond the closing time to take cleaning into account. Tim should also encourage his employees to follow a clean-as-you-go approach. In this way, the soiled tableware no longer being used by the banquet attendees could have been brought to the dishwasher before midnight.

12-30 A Case in Point 2

1. The tableware can be washed, rinsed, and sanitized in a three-compartment sink until the machine has been repaired. Before cleaning and sanitizing the items, clean and sanitize each compartment and all work surfaces. Then follow these steps:
   1. Rinse, scrape, or soak the items.
   2. Wash the items in the first sink in a detergent solution at least 110°F (43°C).
   3. Immerse or spray-rinse the items in the second sink.
   4. Immerse the items in the third sink in hot water or a chemical sanitizing solution. If hot-water immersion is used, the water must be at least 171°F (77°C). If chemical sanitizing is used, the sanitizer must be mixed at the proper concentration and tested with a sanitizer test kit. The sanitizing solution must also be at the proper temperature.
   5. Air-dry the items.

12-31 Discussion Questions

1. Food-contact surfaces must be cleaned and sanitized:
   - After each use
   - Anytime you begin working with another type of food
   - Anytime you are interrupted during a task and the tools or items you have been working with may have been contaminated
   - At four-hour intervals, if the items are in constant use
2. Cleaning is the process of removing food and other types of soil from a surface, while sanitizing is the process of reducing the number of microorganisms on that surface to safe levels.
3. When cleaning and sanitizing items in a three-compartment sink, follow these steps:
   1. Rinse, scrape, or soak the items.
   2. Wash the items in the first sink.
   3. Immerse or spray-rinse the items in the second sink.
   4. Immerse the items in the third sink in hot-water or a chemical sanitizing solution. If hot water immersion is used, the water must be at least 171°F (77°C). If chemical sanitizing is used, the sanitizer must be mixed at the proper concentration and tested with a sanitizer test kit. The sanitizing solution must also be at the proper temperature.
   5. Air-dry the items.
4. To store clean and sanitized tableware, utensils, and equipment:
   - Store tableware and utensils at least six inches off the floor. Keep them covered or otherwise protected from dirt and condensation.
   - Clean and sanitize drawers and shelves before clean items are stored.
   - Clean and sanitize trays and carts used to carry clean tableware and utensils. Do this daily or as often as necessary.
   - Store glasses and cups upside down. Store flatware and utensils with handles up so employees can pick them up without touching food-contact surfaces.
   - Keep the food-contact surfaces of clean-in-place equipment covered until ready for use.
5. Factors that affect the efficiency of sanitizers include the following:
   - **Contact time.** For a sanitizer to kill microorganisms, it must make contact with the object for a specific amount of time.
   - **Temperature.** To be effective, the sanitizing solution must be at the proper temperature.
   - **Concentration.** Concentrations below those recommended could fail to sanitize objects, while concentrations higher than recommended can be unsafe, and might corrode metals.

6. Cleaning tools should be cleaned before being stored in a designated area away from food and food-preparation sites. When storing cleaning materials:
   - Air-dry wiping cloths overnight.
   - Hang mops, brooms, and brushes on hooks to air-dry.
   - Clean, rinse, and sanitize buckets.

12-32 Multiple-Choice Study Questions

1. D	3. B	5. A	7. B	9. C
2. C	4. C	6. D	8. C	10. D

Chapter 13:
Integrated Pest Management

Page	Activity

13-2 Test Your Food Safety Knowledge

1. True 2. False 3. False 4. False 5. False

13-21 A Case in Point

1. Fred should have been working with a PCO and had an integrated pest management (IPM) program in place prior to his discovery of the roach infestation. While it sounds as if Fred and his staff are doing a good job keeping the establishment clean, some other measures he could have taken to prevent the infestation include:
   - Screening windows and vents
   - Installing self-closing doors and door sweeps
   - Keeping exterior openings closed tightly
   - Filling holes around pipes
   - Sealing cracks in floors and walls
   - Disposing of garbage quickly and correctly
   - Making sure that shipments are inspected for signs of pest infestation
2. Fred needs the help of a licensed PCO to help eliminate the roach infestation. The PCO may use repellents, sprays, bait, and/or traps to eliminate the roaches.

13-22 Discussion Questions

1. The purpose of an IPM program is to do the following:
   - Prevent pests from entering the establishment.
   - Deny pests food, water, and a hiding or nesting place.
   - Work with a licensed PCO to eliminate any pests that do infest it.
2. To prevent pests from entering an establishment:
   - Screen all windows and vents with at least sixteen mesh per square inch screening.

- Install self-closing devices or door sweeps on all doors.
- Install air curtains above or alongside doors.
- Keep drive-through windows closed when not in use.
- Keep all exterior openings closed tightly.
- Use concrete to fill holes or sheet metal to cover openings around pipes.
- Install screens over ventilation pipes and ducts on the roof.
- Cover floor drains with hinged grates.
- Seal all cracks in floors and walls.
- Properly seal spaces or cracks where stationary equipment is fitted to the floor.

3. Signs of a cockroach infestation include:
   - A strong oily odor
   - Droppings (feces), which look like grains of black pepper
   - Capsule-shaped egg cases that are brown, dark red, or black and may appear leathery, smooth, or shiny

   Signs of a rodent infestation include:
   - Signs of gnawing
   - Droppings that are shiny and black (fresh) or gray (older)
   - Tracks
   - Nesting materials, such as scraps of paper, cloth, hair, and other soft materials
   - Holes in quiet places, near food and water, and next to buildings
4. Your PCO should store and dispose of all pesticides used in the facility. If they are stored on the premises, follow these guidelines:
   - Keep them in their original container.
   - Store them in locked cabinets away from areas where food is stored and prepared.
   - Check local regulations before disposing of pesticides.
5. To minimize the hazard to people, have your PCO use pesticides only when you are closed for business, and employees are not on site. When pesticides will be applied, prepare the area to be sprayed by removing all food and food-contact surfaces. Cover equipment and food-contact surfaces that cannot be moved. Wash, rinse, and sanitize food-contact surfaces after the area has been sprayed. Anytime pesticides are used or stored on the premises, you should have a corresponding Material Safety Data Sheets (MSDS) since they are hazardous materials.

13-23 Multiple-Choice Study Questions

1. B	3. B	5. D	7. C	9. B
2. C	4. D	6. D	8. A	

Chapter 14

Food Safety Regulation and Standards

Page	Activity
14-2	**Test Your Food Safety Knowledge**

1. False 2. False 3. True 4. False 5. True

14-18 A Case in Point

1. Jerry did a pretty good job handling the inspection. He was cooperative and professional, answering the inspector's questions to the best of his ability. However, he should have asked the inspector for identification and inquired about the purpose of the visit. He also should have taken notes during the inspection. This would have helped him remember later exactly what was said.
2. Jerry did discuss violations with the inspector and now needs to follow up on the findings. He should walk through his facility and determine why each problem occurred. Then he should establish new procedures or revise existing ones to permanently correct problems.

14-19 Discussion Questions

1. The Food and Drug Administration (FDA) issues the *FDA Food Code.* In addition, it inspects foodservice operations that cross state borders (interstate establishments such as those on planes and trains, as well as food manufacturers and processors) because they overlap the jurisdictions of two or more states. The FDA shares responsibility with the U.S. Department of Agriculture (USDA) for inspecting food-processing plants to ensure standards of purity, wholesomeness, and compliance with labeling requirements.
2. Recommendations for restaurant and foodservice regulations are issued at the federal level, regulations are written at the state level, and enforcement is carried out at the state and local level.
3. During an inspection, a manager should do the following:
   - Ask for identification.
   - Cooperate.
   - Take notes.
   - Keep the relationship professional.
   - Be prepared to provide records requested by the inspector.

   After the inspection, the manager should do the following:
   - Discuss violations and time frames for correction with the inspector.
   - Follow up by determining why each problem occurred, and then establish new procedures or revise existing ones. It may also be necessary to retrain employees.

4. Factors that determine the frequency of a health inspection include:
   - Size and complexity of the operation. Larger operations offering a large number of potentially hazardous food items might be inspected more frequently.
   - Inspection history of the establishment. Establishments with a history of low sanitation scores or consecutive violations might be inspected more frequently.
   - Clientele's susceptibility to foodborne illness. Nursing homes, schools, daycare centers, and hospitals might receive more frequent inspections.
   - Workload of the local health department and the number of inspectors available.

14-20 Multiple-Choice Study Questions

1. D 3. B 5. C 7. B
2. D 4. A 6. D 8. A

Chapter 15
Employee Food Safety Training

Page	Activity

15-2 Test Your Food Safety Knowledge

1. True 2. False 3. False 4. True 5. True

15-20 Discussion Questions

1. The "zoom" principle is a training technique modeled after the zoom lens on a camera. First, the instructor "zooms" out, explaining the big picture. Next, the instructor "zooms" in to explain a detail. Finally, the instructor "zooms" back out, showing where the detail fits in to the big picture.
2. The Tell/Show/Tell/Show method is an instructional technique used to demonstrate specific tasks. It involves first telling the employee how to perform a task, then showing the employee how to do it. Next, it is the employee's turn to tell the instructor how to perform the task, and finally the employee shows the instructor how the task is performed.
3. An establishment can determine its food safety training needs by doing the following:
   - Testing employees' food safety knowledge
   - Observing employee job performance
   - Questioning or surveying employees to identify areas of weakness
4. An objective states what the learner will be able to do after instruction is completed. Objectives need to be stated clearly and in measurable terms. Examples of clearly stated objectives include:
   - Demonstrate the proper procedure for calibrating a bimetallic stemmed thermometer using the ice-point method.

- Identify the minimum internal cooking temperatures for meat, seafood, and poultry.

5. Methods that can be used to deliver training include:
   - Demonstration
   - Lecture
   - Role-play
   - Job aids
   - One-on-one training
   - Group training
   - Technology-based training
   - Training videos and DVDs
   - Games
   - Case studies

15-21 Multiple-Choice Study Questions

1. B 2. D 3. A 4. B 5. D

Notes

Glossary

Note: The number(s) in bold at the end of each entry refers to the chapter in which the term is discussed in detail.

A

Abrasive cleaners. Cleaners containing a scouring agent used to scrub off hard-to-remove soils. They may scratch some surfaces. **12**

Acid cleaners. Used on mineral deposits and other soils that alkaline cleaners cannot remove, such as scale, rust, and tarnish. **12**

Acidity. Level of acid in a food. An acidic substance has a pH below 7.0. Foodborne microorganisms typically do not grow in highly acidic food, while they grow best in food with a neutral to slightly acidic pH. **2**

Active managerial control. Food safety management system designed to prevent foodborne illness by addressing the five most common risk factors identified by the Centers for Disease Control and Prevention (CDC). **10**

Air curtains. Devices installed above or alongside doors that blow a steady stream of air across an entryway, creating an air shield around open doors. Insects avoid them. Also called air doors or fly fans. **13**

Air gap. Air space used to separate a water-supply outlet from any potentially contaminated source. The air space between the floor drain and the drainpipe of a sink is an example. An air gap is the only completely reliable method for preventing backflow. **11**

Alkalinity. Level of alkali in food. An alkaline substance has a pH above 7.0. Most food is not alkaline. **2**

Americans with Disabilities Act (ADA). Federal law requiring reasonable accommodation for access to a facility by patrons and employees with disabilities. **11**

Application. Applying what was learned, with feedback from the instructor/trainer. **15**

Aseptically packaged food. Food that has been sealed under sterile conditions, usually after UHT-pasteurization. UHT stands for ultra-high temperature. **6**

B

Backflow. Unwanted reverse flow of contaminants through a cross-connection into a potable water system. It occurs when the pressure in the potable water supply drops below the pressure of the contaminated supply. **11**

Bacteria. Single-celled, living microorganisms that can spoil food and cause foodborne illness. Bacteria present in food can quickly multiply to dangerous levels when food is improperly cooked, held, and reheated. Some form spores that can survive freezing and very high temperatures. **2**

Bacterial growth. Reproduction of bacteria by splitting in two. When conditions are favorable, bacterial growth can be rapid—doubling the population as often as every twenty minutes. Their growth can be broken down into four phases: lag phase, log phase, stationary phase, and death phase. **2**

Bimetallic stemmed thermometer. The most common and versatile type of thermometer, measuring temperature through a metal probe with a sensor in the end. Most can measure temperatures from 0°F to 220°F (–18°C to 104°C) and are accurate to within ±2°F (±1°C). They are easily calibrated. **5**

Biological hazard. Pathogenic microorganisms that can contaminate food, such as certain bacteria, viruses, parasites, and fungi, as well as toxins found in certain plants, mushrooms, and fish. **1**

Biological toxins. Poisons produced by pathogens, plants, or animals. They can also occur in animals as a result of their diet. **3**

Blast chiller. Equipment designed to cool food quickly. Many are able to cool food from 135°F to 37°F (57°C to 3°C) within ninety minutes. **11**

Boiling-point method. Method of calibrating a thermometer based on the boiling point of water. **5**

Booster heater. Water heater attached to hot-water lines leading to dishwashing machines or sinks. Raises water to temperature required for heat sanitizing of tableware and utensils. **11**

C

Calibration. Process of ensuring that a thermometer gives accurate readings by adjusting it to a known standard, such as the freezing point or boiling point of water. **5**

Cantilever-mounted equipment. Equipment that is attached to a mount or a wall with a bracket, allowing for easier cleaning behind and underneath. **11**

Carriers. People who carry pathogens and infect others yet never become ill themselves. **4**

Centers for Disease Control and Prevention (CDC). Agencies of the U.S. Public Health Service that investigate foodborne-illness outbreaks, study the causes and control of disease, publish statistical data, and conduct the Vessel Sanitation Program. **14**

Chemical hazard. Chemical substances that can contaminate food, such as pesticides, food additives, preservatives, cleaning supplies, and toxic metals that leach from cookware and equipment. **1**

Chemical sanitizing. Using a chemical solution to reduce the number of microorganisms on a clean surface to safe levels. Items can be sanitized by immersing in a specific concentration of sanitizing solution for a required period of time or by rinsing, swabbing, or spraying the items with a specific concentration of sanitizing solution. **12**

Chemical toxins. Poisons found in some cleaning agents and pesticides, as well as the by-products of toxic-metal reactions. **3**

Chlorine. Commonly used chemical sanitizer due to its low cost and effectiveness. It kills a wide range of microorganisms. **12**

Ciguatera poisoning. Illness that occurs when a person eats fish that has consumed the ciguatera toxin. This toxin occurs in certain predatory tropical reef fish, such as amberjack, barracuda, grouper, and snapper. **3**

Clean. Free of visible soil. It refers only to the appearance of a surface. **1, 12**

Cleaning. Process of removing food and other types of soil from a surface, such as a countertop or plate. **12**

Cleaning agents. Chemical compounds that remove food, soil, rust stains, minerals, or other deposits from surfaces. **12**

Cold-holding equipment. Equipment specifically designed to hold cold food at an internal temperature of 41°F (5°C) or lower. **9**

Contact spray. Spray used to kill insects on contact. Usually used on groups of insects, such as clusters of roaches and nests of ants. **13**

Contamination. Presence of harmful substances in food. Some food safety hazards occur naturally, while others are introduced by humans or the environment. **1**

Corrective action. Predetermined step taken when food does not meet a critical limit. **10**

Coving. Curved, sealed edge placed between the floor and wall to eliminate sharp corners or gaps that would be impossible to clean. Coving also eliminates hiding places for pests and prevents moisture from deteriorating walls. **11**

Critical control point (CCP). In a HACCP system, the points in the process where you can intervene to prevent, eliminate, or reduce identified hazards to safe levels. **10**

Critical limit. In a HACCP system, the minimum or maximum limit a critical control point (CCP) must meet in order to prevent, eliminate, or reduce a hazard to an acceptable level. **10**

Cross-connection. Physical link through which contaminants from drains, sewers, or other waste-water sources can enter a potable water supply. A hose connected to a faucet and submerged in a mop bucket is an example. **11**

Cross-contamination. Occurs when microorganisms are transferred from one food or surface to another. **5**

D

Death phase. The phase in bacterial growth in which the number of bacteria dying exceeds the number growing, resulting in a population decline. **2**

Demonstration. Process of illustrating a skill or task in front of another person or a group. **15**

Detergent. Cleaning agent designed to penetrate and soften soil to help remove it from a surface. **12**

Dry storage. Storage used to hold dry and canned food at temperatures between 50°F and 70°F (10°C and 21°C) and at a relative humidity of 50 to 60 percent. **7**

E

Electronic insect eliminator ("zapper"). Mechanical device that uses light to attract flying insects to an electrically charged grid that kills them. **13**

Environmental Protection Agency (EPA). Federal agency that sets standards for environmental quality, including air and water quality, and regulates pesticide use and waste handling. **14**

Evaluation. Method used to determine if employees have the knowledge and skills needed to meet the objectives of the training program. Written, oral, and performance-based tests are often used. **15**

Exclusion. Prohibiting foodhandlers from working in the establishment due to specific medical conditions. **4**

F

FAT TOM. Acronym for the conditions needed by most foodborne microorganisms to grow: food, acidity, temperature, time, oxygen, moisture. **2**

FDA Food Code. Science-based reference for retail food establishments on how to prevent foodborne illness. These recommendations are issued by the FDA to assist state health departments in developing regulations for a foodservice inspection program. **1, 14**

Feedback. Evaluation given to employees about their performance, including constructive criticism given to correct a mistake or praise to reinforce proper performance of a skill or procedure. **15**

Finger cot. Protective covering used to cover a properly bandaged cut or wound on the finger. **4**

First in, first out (FIFO). Method of stock rotation in which products are shelved based on their use-by or expiration dates, so oldest products are used first. **7**

Flood rim. Spill-over point of a sink. **11**

Flow of food. Path food takes through an establishment, from purchasing and receiving through storing, preparing, cooking, holding, cooling, reheating, and serving. **5**

Food allergy. The body's negative reaction to a particular food protein. **3**

Food and Drug Administration (FDA). Federal agency that issues the *FDA Food Code* working jointly with the U.S. Department of Agriculture (USDA) and the Centers for Disease Control and Prevention (CDC). The FDA also inspects foodservice operations that cross state borders—interstate establishments such as food manufacturers and processors, and planes and trains—because they overlap the jurisdictions of two or more states. **14**

Food bar. Self-service buffet at which patrons can choose what they want to eat as they serve themselves. **9**

Food-contact surface. Surface that comes into direct contact with food, such as a cutting board. **1**

Food-grade sealant. Nontoxic sealant used to seal equipment to a countertop or a masonry base. **11**

Food irradiation. Process of exposing food to an electron beam or gamma rays to reduce pathogenic and spoilage microorganisms. Also known as cold pasteurization. **2**

Food Safety and Inspection Service (FSIS). Agency of the U.S. Department of Agriculture (USDA) that inspects and grades meat, meat products, poultry, dairy products, eggs and egg products, and fruit and vegetables shipped across state boundaries. **14**

Food safety management system. Group of programs, procedures, and measures designed to prevent foodborne illness by actively controlling risks and hazards throughout the flow of food. **10**

Food security. The prevention or elimination of the deliberate contamination of food. **3**

Foodborne illness. Illness carried or transmitted to people by food. **1**

Foodborne-illness outbreak. According to the Centers for Disease Control and Prevention (CDC), an incident in which two or more people experience the same illness after eating the same food. **1**

Foodborne infection. Result of a person eating food containing pathogens, which then grow in the intestines and cause illness. Typically, symptoms of a foodborne infection do not appear immediately. **2**

Foodborne intoxication. Result of a person eating food containing toxins (poisons) that cause an illness. The toxins may have been produced by pathogens found on the food or may be the result of a chemical contamination. The toxins might also be a natural part of the plant or animal consumed. Typically, symptoms of foodborne intoxication appear quickly, within a few hours. **2**

Foodborne toxin-mediated infection. Result of a person eating food containing pathogens, which then produce illness-causing toxins in the intestines. **2**

Foot-candle. Unit of lighting equal to the illumination one foot from a uniform light source. **11**

Frozen storage. Storage typically designed to hold food at temperatures that will keep it frozen. **7**

Fungi. Ranging in size from microscopic, single-celled organisms to very large, multicellular organisms. Fungi most often cause food to spoil. Molds, yeasts, and mushrooms are examples. **2**

Gastrointestinal illness. Illness relating to the stomach or intestine. **4**

Glue board. Pest-control device in which mice are trapped by glue and then die from exhaustion or lack of water or air. They are also used to identify the type of cockroaches that might be present. **13**

H

HACCP plan. Written document based on HACCP principles describing procedures a particular establishment will follow to ensure the safety of food served. **10**

Hair restraint. Device used to keep a foodhandler's hair away from food and to keep the individual from touching it. **4**

Hand antiseptic. Liquid or gel used to lower the number of microorganisms on the skin's surface. Hand antiseptics should only be used after proper handwashing, not in place of it. Only those hand antiseptics that are compliant with the Food and Drug Administration (FDA) should be used. **4**

Handwashing station. Sink designated for handwashing only. Handwashing stations must be conveniently located in restrooms, food-preparation areas, service areas, and dishwashing areas. **11**

Hard water. Water containing minerals such as calcium and iron in concentrations higher than 120 parts per million (ppm). **12**

Hazard analysis. Process of identifying and evaluating potential hazards associated with food in order to determine what must be addressed in the HACCP plan. **10**

Hazard analysis critical control point (HACCP). Food safety management system based on the idea that if significant biological, chemical, or physical hazards are identified at specific points within a product's flow through the operation, they can be prevented, eliminated, or reduced to safe levels. **10**

Hazard Communication Standard (HCS). OSHA standard, also known as Right-to-Know or HAZCOM, requiring employers to tell their employees about potential chemical hazards at the establishment. It also requires employers to train employees in how to use chemicals safely. **12**

Health inspector. City, county, or state employee who conducts foodservice inspections. Health inspectors are also known as sanitarians, health officials, and environmental health specialists. They are generally trained in food safety, sanitation, and public health principles. **14**

Heat sanitizing. Using heat to reduce the number of microorganisms on a clean surface to safe levels. One common way to heat sanitize tableware, utensils, or equipment is to submerge them in or spray them with hot water. **12**

Heat-treated. Food that has been cooked, partially cooked, or warmed. **1**

Hepatitis A. Disease-causing inflammation of the liver. It is transmitted to food by poor personal hygiene or contact with contaminated water. **4**

High-risk population. People susceptible to foodborne illness due to the effects of age or health on their immune systems, including infants and preschool-age children, pregnant women, older people, people taking certain medications, and those with certain diseases or weakened immune systems. **1**

Histamine. Biological toxin associated with temperature-abused scombroid fish (and other affected species), which cause scombroid poisoning. **3**

Host. Person, animal, or plant on which another organism lives and takes nourishment. **2**

Hot-holding equipment. Equipment such as chafing dishes, steam tables, and heated cabinets specifically designed to hold food at an internal temperature of 135°F (57°C) or higher. **9**

Hygrometer. Instrument used to measure relative humidity in storage areas. **7**

Ice-point method. Method of calibrating thermometers based on the freezing point of water. **5**

Ice-water bath. Method of cooling food in which a container holding hot food is placed into a sink or larger container of ice water. The ice water surrounding the hot food container disperses the heat quickly. **8**

Ice paddle. Plastic paddle filled with ice or water and then frozen. Used to stir hot food to cool it quickly. **8**

Immune system. The body's defense system against illness. People with compromised immune systems are more susceptible to foodborne illness. **1**

Infected lesion. Wound or injury contaminated with a pathogen. **4**

Infestation. Situation that exists when pests overrun or inhabit an establishment in large numbers. **13**

Integrated pest management (IPM). Program using prevention measures to keep pests from entering an establishment and control measures to eliminate any pests that do get inside. **13**

Iodine. Sanitizer effective at low concentrations and not as quickly inactivated by soil as chlorine. It might stain surfaces and is less effective than chlorine. **12**

J

Jaundice. Yellowing of the skin and eyes that could indicate a person is ill with hepatitis A. **4**

Job aids. Materials or visual reminders used to deliver training content to employees. **15**

L

Lag phase. Phase in bacterial growth in which bacteria are first introduced to a new environment. In this phase, bacteria go through an adjustment period in which their numbers are stable as they prepare to grow. To control the growth of bacteria, prolong the lag phase as long as possible. **2**

Lecture. Prepared oral presentation used to deliver content to a group. **15**

Log phase. Phase in bacterial growth in which conditions are favorable for bacteria to multiply very rapidly. Food quickly becomes unsafe during this phase. **2**

M

Master cleaning schedule. Detailed schedule listing all cleaning tasks in an establishment, when and how they are to be performed, and who will perform them. **12**

Material Safety Data Sheets (MSDS). Sheets supplied by the chemical manufacturer listing the chemical and its common names, its potential physical and health hazards, information about using and handling it safely, and other important information. OSHA requires employers to store these sheets so they are accessible to employees. **12**

Microorganisms. Small, living organisms that can be seen only with the aid of a microscope. There are four types of microorganisms that can contaminate food and cause foodborne illness: bacteria, viruses, parasites, and fungi. **2**

Minimum internal cooking temperature. The required minimum temperature the internal portion of food must reach to sufficiently reduce the number of microorganisms that might be present. This temperature is specific to the type of food being cooked. Food must reach and hold its required internal temperature for a specified amount of time. **8**

Mobile unit. Portable foodservice facilities, ranging from concession vans to full field kitchens capable of preparing and cooking elaborate meals. **9**

Modified atmosphere packaging (MAP). Packaging method by which the air inside of a package is altered using gases, such as carbon dioxide and nitrogen. Many fresh-cut produce items are packaged this way. **6**

Mold. Type of fungus that causes food spoilage. Some molds produce toxins that can cause foodborne illness. **2**

Monitoring. In a HACCP system, the process of analyzing whether critical limits are being met and things are being done right. **10**

N

National Marine Fisheries Service (NMFS). Agency of the U.S. Department of Commerce that provides a voluntary inspection program that includes product standards and sanitary requirements for fish processing operations. **14**

NSF International. Organization that develops and publishes standards for sanitary equipment design. They also assess and certify that equipment has met these standards. Restaurant and foodservice managers should look for an NSF International mark (or UL EPH product mark) on commercial foodservice equipment. UL stands for Underwriters Laboratories. **11**

O

Objective. Statement of what a trainee will be able to do after training or instruction is completed. **15**

Occupational Safety and Health Administration (OSHA). Federal agency that regulates and monitors workplace safety. **12**

Off-site service. Service of food to someplace other than where it is prepared or cooked, including catering and vending. **9**

P

Parasite. Organism that needs to live in a host organism to survive. Parasites can be found in water and inside many animals, such as cows, chickens, pigs, and fish. Proper cooking and freezing will kill parasites. Avoiding cross-contamination and practicing proper handwashing can also prevent illness. **2**

Pathogens. Illness causing microorganisms. **2**

Personal hygiene. Habits that include keeping the hands, hair, and body clean and wearing clean and appropriate uniforms. Avoiding unsanitary actions and reporting illness and injury are also features of good personal hygiene. **4**

Pest control operator (PCO). Licensed professional who uses safe, current methods to prevent and control pests. **13**

Pesticide. Chemical used to control pests, usually insects. **13**

pH. Measure of a food's acidity or alkalinity. The pH scale ranges from 0 to 14.0. A pH between 7.1 and 14 is alkaline, while a pH between 0.0 and 6.9 is acidic. A pH of 7.0 is neutral. Foodborne microorganisms grow well in food that has a neutral to slightly acidic pH (7.5 to 4.6). **2**

Physical hazard. Foreign objects that can accidentally get into food and contaminate it, such as hair, dirt, metal staples, and broken glass, as well as naturally occurring objects, such as bones in fillets. **1**

Plant toxins. Poisons found naturally in some plants. **3**

Pooled eggs. Eggs that have been cracked open and combined in a common container. **8**

Porosity. Extent to which water and other liquids are absorbed by a substance. Term usually used in relation to flooring material. **11**

Potable water. Water that is safe to drink. **11**

Potentially hazardous food. Food that contains moisture and protein and has a neutral or slightly acidic pH. Such food requires time-temperature control to prevent the growth of microorganisms and the production of toxins. **1**

Presentation. Delivery of content to the learner, which can be accomplished through a variety of methods. **15**

Pulper. Device used to grind food and other waste into small parts that are flushed with water, which is then removed. The processed, solid wastes weigh less and are more compact for easier disposal. **11**

Q

Quaternary ammonium compounds (quats). Group of sanitizers all having the same basic chemical structure. They work in most temperature and pH ranges, are noncorrosive, and remain active for short periods of time after they have dried. However, quats may not kill certain types of microorganisms, and they leave a film on surfaces. **12**

R

Ready-to-eat food. Any food that is edible without further washing or cooking. It includes washed, whole or cut fruit; vegetables; deli meats; and bakery items. Sugars, spices, seasonings, and properly cooked food items are also considered ready-to-eat. **1**

Reasonable care defense. Defense against a food-related lawsuit stating that an establishment did everything that could be reasonably expected to ensure that the food served was safe. **1**

Record keeping. In a HACCP system, the process of collecting documents that allow you to show you are continuously preparing and serving safe food. **10**

Refrigerated storage. Storage used to hold potentially hazardous food at an internal temperature of 41°F (5°C) or lower. **7**

Regulations. Laws determining standards of behavior. Restaurant and foodservice regulations are typically written at the state level and based on the *FDA Food Code.* **14**

Residual spray. Type of pesticide spray that leaves behind a film that insects absorb as they crawl across it. Used in cracks and crevices like those along baseboards, these sprays can be liquid or a dust, such as boric acid. **13**

Resiliency. Ability of a surface to react to a shock without breaking or cracking, usually used in relation to a flooring material. **11**

Restriction. Prohibiting foodhandlers from working with or around food, food equipment, and utensils. **4**

Role-play. Training method in which trainees enact a situation to try out new skills or apply new knowledge. **15**

ROP. Stands for "reduced oxygen packaging." Packaging method that reduces the amount of oxygen available in order to slow microbial growth. ROP methods include *sous vide,* MAP, and vacuum packaging. **6**

S

Sanitary. State that exists when the number of pathogens on a clean surface has been reduced to safe levels. **1, 12**

Sanitizer. Chemical used to sanitize. Chlorine, iodine, and quats are the three most common types of chemical sanitizer in the restaurant and foodservice industry. **12**

Sanitizing. Process of reducing the number of microorganisms on a clean surface to safe levels. **12**

Scombroid poisoning. Illness that occurs when a person eats a scombroid fish (or certain other species) that has been time-temperature abused. Scombroid fish include tuna, mackerel, bluefish, skipjack, and bonito. **3**

Service sink. Sink used exclusively for cleaning mops and disposing of wastewater. At least one service sink or one curbed drain area is required in an establishment. **11**

Shelf life. Recommended period of time during which food can be stored and remain suitable for use. **7**

Shellstock identification tags. Each container of live, molluscan shellfish received must have an ID tag that must remain attached to the container until all the shellfish have been used. Tags are to be kept on file for ninety days from the harvest date of the shellfish. **6**

Single-use gloves. Disposable gloves designed for one-time use. They provide a barrier between hands and the food they touch. Gloves should never be used in place of handwashing. Foodhandlers should wash hands before putting on gloves and when changing to a new pair. **4**

Single-use item. Disposable tableware designed to be used only once. It includes plastic flatware, paper or plastic cups, plates, and bowls. **9**

Single-use paper towel. Paper towel designed to be used once, then discarded. **4**

Slacking. Process of gradually thawing frozen food in preparation for deep-frying. **8**

Sneeze guard. Food shield placed over self-service displays and food bars that extends seven inches beyond the food and fourteen inches above the food counter. **9**

Solvent cleaners. Alkaline detergents, often called degreasers, that contain a grease-dissolving agent. **12**

***Sous vide* food.** Packaging method by which cooked or partially cooked food is vacuum packed in individual pouches and then chilled. This food is heated for service in the establishment. Frozen, precooked meals are often packaged this way. **6**

Spoilage microorganism. Foodborne microorganism that causes food to spoil but typically does not cause foodborne illness. **2**

Spore. Form that some bacteria can take to protect themselves when nutrients are not available. Spores are commonly found in soil and can contaminate food grown there. A spore can resist heat, allowing it to survive cooking temperatures. Spores can also revert back to a form capable of growth. This can occur when food is not held at the proper temperature or cooled or reheated properly. **2**

Stationary phase. Phase of bacterial growth in which just as many bacteria are growing as are dying. Follows the log phase of bacterial growth. **2**

T

Technology-based training. Training programs delivered via a computer or other technology. **15**

Temperature danger zone. The temperature range between 41°F and 135°F (5°C to 57°C) within which most foodborne microorganisms rapidly grow. **2**

Temporary unit. Establishment operating in one location for no more than fourteen consecutive days in conjunction with a special event or celebration. Usually serves prepackaged food or food requiring only limited preparation. **9**

Thermometer. Device for accurately measuring the internal temperature of food, the air temperature inside a freezer or cooler, or the temperature of equipment. Bimetallic stemmed thermometers, thermocouples, and thermistors are common types of thermometers used in the restaurant and foodservice industry. **5**

Time-temperature abuse. Food has been time-temperature abused any time it has been allowed to remain too long at a temperature favorable to the growth of foodborne microorganisms. **1**

Time-temperature indicator (TTI). Time and temperature monitoring device attached to a food shipment to determine if the product's temperature has exceeded safe limits during shipment or subsequent storage. **5**

Toxic-metal poisoning. Illness caused when toxic metals are leached from utensils or equipment containing them. **3**

Toxins. Poisons produced by pathogens, plants, or animals. Most occur naturally and are not caused by the presence of microorganisms. Some occur in animals as a result of their diet. Many chemicals are also toxic. **2, 3**

Training delivery methods. Approaches for providing training to employees. This can include more traditional methods such as lectures, demonstrations, or role-play or more technology-based approaches such as Web-based training and interactive CD-ROMs. Regardless of the approach, it is important to use more than one method of delivery since employees learn differently. **15**

Training need. Gap between what employees are required to know to do their jobs and what they actually know. There are several ways to identify food safety training needs, including observing job performance, testing food safety knowledge, and surveying employees to identify areas of weakness. **15**

Training objective. Statement that describes what employees should be able to do after training has been completed. They serve as the guide when delivering content, and, therefore, must be clearly defined and measureable. **15**

Training plan. Specific list of events that will take place during the training session. Training plans should include the specific learning objectives, a list of training tools needed during the training session, and specific talking points that should be covered. **15**

Training program. Structured sequence of events that leads to learning. **15**

Tumble chiller. Equipment designed to cool food quickly. Prepackaged hot food is placed into a drum rotating inside a reservoir of chilled water. The tumbling action increases the effectiveness of the chilled water in cooling the food. **11**

Two-stage cooling. Criteria by which cooked food is cooled from 135°F to 70°F (57°C to 21°C) within two hours and from 70°F to 41°F (21°C to 5°C) or lower within the next four hours, for a total cooling time of six hours. **8**

U

Ultra-high temperature (UHT) pasteurized food. Food that is heat treated at very high temperatures (pasteurized) to kill microorganisms. This food is often also aseptically packaged—sealed under sterile conditions to keep it from being contaminated. **6**

Underwriters Laboratories (UL). Provides sanitation classification listings for equipment found in compliance with NSF International standards. Also lists products complying with their own published environmental and public health standards. **11**

U.S. Department of Agriculture (USDA). Federal agency responsible for the inspection and quality grading of meat, meat products, poultry, dairy products, eggs and egg products, and fruit and vegetables shipped across state lines. **14**

V

Vacuum breaker. Device preventing the backflow of contaminants into a potable water system. **11**

Vacuum-packed food. Food processed by removing air from around it while sealed in a package. This process increases the product's shelf life. **6**

Variance. Document issued by a regulatory agency that allows a requirement to be waived or modified. **10**

Vegetative microorganisms. Bacteria in the process of reproducing (growing) by splitting in two. **2**

Vending machine. Machines that dispense hot and cold food, beverages, and snacks. **9**

Verification. In a HACCP system, the process of confirming that critical control points and critical limits are appropriate, that monitoring is alerting you to hazards, that corrective actions are adequate to prevent foodborne illness from occurring, and that employees are following established procedures. **10**

Virus. Smallest of the microbial food contaminants. Viruses rely on a living host to reproduce. They usually contaminate food through a foodhandler's improper personal hygiene. Some survive freezing and cooking temperatures. **2**

Warranty of sale. Rules stating how food must be handled in an establishment. **1**

Water activity (a_w). Amount of moisture available in food for microorganisms to grow. It is measured in a scale from 0.0 to 1.0, with water having a water activity (a_w) of 1.0. Potentially hazardous food typically has a water-activity value of 0.85 or higher. **2**

Yeast. Type of fungus that causes food spoilage. **2**

Notes

Index

A

B

C

G

H

V

W

Y